T0310244

NANOSCALE MULTIFUNCTIONAL MATERIALS

NANOSCALE MULTIFUNCTIONAL MATERIALS

Science and Applications

Edited by

SHARMILA M. MUKHOPADHYAY

A JOHN WILEY & SONS, INC., PUBLICATION

For general information on our other products and services or for technical support, please contact our Customer Care Department within the United States at (800) 762-2974, outside the United States at (317) 572-3993 or fax (317) 572-4002.

Wiley also publishes its books in a variety of electronic formats. Some content that appears in print may not be available in electronic formats. For more information about Wiley products, visit our web site at www.wiley.com.

Library of Congress Cataloging-in-Publication Data:

Nanoscale multifunctional materials: science & applications / edited by Sharmila Mukhopadhyay
    p. cm.
    Includes bibliographical references and index.
    ISBN 978-0-470-50891-6 (hardback)
    1. Nanostructured materials.   I. Mukhopadhyay, Sharmila M.
    TA418.9.N35N34554 2011
    620.1′15—dc23

                                                                        2011021421

Printed in the United States of America

oBook ISBN: 9781118114063
ePDF ISBN: 9781118114032
ePub ISBN: 9781118114056
MOBI ISBN: 9781118114049

10 9 8 7 6 5 4 3 2 1

CONTENTS

CONTRIBUTORS

Abinash Agrawal, Department of Earth and Environmental Sciences, Wright State University, Dayton, Ohio

Ian T. Barney, Center for Nanoscale Multifunctional Materials, Wright State University, Dayton, Ohio

Liming Dai, Case Western Reserve University, Cleveland, Ohio

Sabyasachi Ganguli, University of Dayton Research Institute, Dayton, Ohio

Hong Huang, College of Engineering and Computer Science, Wright State University, Dayton, Ohio

Allen G. Jackson, Center for Nanoscale Multifunctional Materials, Wright State University, Dayton, Ohio

Sadhan C. Jana, Department of Polymer Engineering, University of Akron, Akron, Ohio

Bor Z. Jang, College of Engineering and Computer Science, Wright State University, Dayton, Ohio

Guillermo A. Jimenez, Department of Polymer Engineering, University of Akron, Akron, Ohio; currently at the School of Chemistry, National University of Costa Rica, Heredia, Costa Rica

Sushil R. Kanel, Department of Systems and Engineering Management, Air Force Institute of Technology, Wright-Patterson Air Force Base, Dayton, Ohio

Dong-Shik Kim, Department of Chemical and Environmental Engineering, University of Toledo, Toledo, Ohio

Byoung J. Lee, Department of Polymer Engineering, University of Akron, Akron, Ohio; currently at Goodyear Tire and Rubber Company, Akron, Ohio

Sharmila M. Mukhopadhyay, Center for Nanoscale Multifunctional Materials, Wright State University, Dayton, Ohio

Paul T. Murray, University of Dayton Research Institute, Dayton, Ohio

Byung-Wook Park, Department of Chemical and Environmental Engineering, University of Toledo, Toledo, Ohio

Upendra Patel, Dr. Jivraj Mehta Institute of Technology, Gujarat, India

Soumya S. Patnaik, Air Force Research Laboratory, Wright-Patterson Air Force Base, Dayton, Ohio

Liangti Qu, Beijing Institute of Technology, Beijing, China

Ajit K. Roy, Air Force Research Laboratory, Wright-Patterson Air Force Base, Dayton, Ohio

Jayesh P. Ruparelia, Institute of Technology, Nirma University, Ahmedabad, India

Sangwook Sihn, University of Dayton Research Institute, Dayton, Ohio

Chunming Su, U.S. Environmental Protection Agency, Ada, Oklahoma

Mesfin Tsige, Department of Polymer Science, University of Akron, Akron, Ohio

Venkata K. K. Upadhyayula, Oak Ridge Institute of Science and Education, Oak Ridge, Tennessee

PREFACE

The possible impact of nanomaterials on future products and services is enormous. This has led to a tremendous expansion in research and development efforts related to nanoscale materials and devices. It has also raised new questions as to how these products will influence the environment, human health, business, education, and other areas of society. As with any emerging field, a vast amount of disconnected information is emerging, often spread across multiple disciplines. The influence of these materials, however, is truly multidisciplinary, and progress will depend upon cross-communication between fundamental science and diverse applications. This work is an attempt at consolidating several diverse areas of nanomaterials in order to provide the reader with a broader perspective. It will also help in envisioning future products and possibilities that may not exist today.

The book has been divided into three sections: Section I provides a panoramic overview of nanomaterials, beginning with a chapter highlighting the scientific phenomena that make nanomaterials different from conventional solids. This is followed by a chapter detailing the impact made by these materials on various areas of society. Section II provides articles related to the processing and analysis of nanoscale materials. These involve multiple experimental approaches in fabrication and characterization, as well as theoretical modeling and simulations. Section III provides discussions of technological areas where nanomaterials can be used. Examples are included dealing with the advanced energy, thermal management, environmental and biomedical areas. Selected figures are reproduced in color free of charge at ftp://ftp.wiley.com/public/sci_tech_med/nanoscale_multifunctional.

The book is aimed at the following readers:

- Advanced students and instructors in the fields of science and engineering.
- Professional scientists and engineers, who may be trained in more traditional disciplines but who need to learn about this emerging area.
- Policymakers and management experts looking for an understanding of scientific challenges, prospective uses, and emerging markets for nanomaterials.

The overall goal is to capture the multidisciplinary and multifunctional flavor of nanomaterials while providing in-depth expert discussion of select areas.

Acknowledgments

Several people have made this book possible. I express my sincere gratitude to the contributors for their effort and dedication in preparing the chapters, to Wright State University, and to my students for the teaching experience that motivated me to create this book. Special thanks to Anil Karumuri, Ph.D. student, for his continuous assistance and hard work in putting the manuscript together. I also thank Anita Lekhwani of Wiley, for giving me the idea of starting a book, and her publication, team for their continuous help and support. Finally, I am indebted to my husband, Bhaskar Mukhopadhyay, for his constant encouragement and support, and to my daughter, Amrita, for her endless enthusiasm and help with graphic design.

Dayton, Ohio   Sharmila M. Mukhopadhyay
November 2010

SECTION I

OVERVIEW

1

KEY ATTRIBUTES OF NANOSCALE MATERIALS AND SPECIAL FUNCTIONALITIES EMERGING FROM THEM

SHARMILA M. MUKHOPADHYAY

Center for Nanoscale Multifunctional Materials, Wright State University, Dayton, Ohio

Nanoscale Multifunctional Materials: Science and Applications, First Edition.
Edited by Sharmila M. Mukhopadhyay.
© 2012 John Wiley & Sons, Inc. Published 2012 by John Wiley & Sons, Inc.

1 BACKGROUND

Nanoscale materials, which can be either stand-alone solids or subcomponents in other materials, are less than 100 nm in one or more dimensions. To put this dimension in perspective, a nanometer (nm) is one billionth of a meter and one millionth of a millimeter. In terms of familiar objects, the diameter of human hair ranges between 50,000 and 100,000 nm, and of what we call "a speck of dust" ranges between 1000 and 100,000 nm. This implies that a nanomaterial is 500 to 1000 times thinner than human hair in one or more relevant dimension(s)!

Many such materials have always existed in nature, both in the living and nonliving world. In fact, most biological phenomena occur at these scales. The most sophisticated biological machines, such as protein assembly and photosynthesis, involve nanoscale structural units. Geological solids such as clays and minerals also occur as nanoscale entities. Even some historically engineered products such as ceramic and glass artwork from earlier centuries had incorporated pigments that today we would label as nanomaterials.

Despite their abundance in nature and a few historical products, the size-related aspects of this family of solids had not been focused on explicitly by the scientific community. Some scientists and visionaries, such as Richard Feynman [1], had predicted that "there is plenty of room at the bottom," implying that to create new materials, one can build them from the bottom up (i.e., atom by atom). Such comments made academic sense and grabbed headlines, but such technology could evolve only after new tools capable of monitoring and manipulating materials at the nanoscale became available. Finally, in the 1990s it was pointed out that something as mundane as carbon can be created in multiple nanoscale structures that are not only elegant but also capable of unprecedented properties [2–4]. This discovery energized everyone's interest in nanoscale and opened the floodgates of scientific curiosity into carbon and all other materials that can be created at these scales. Since then, it has become more and more obvious every day that a large number of unprecedented game-changing applications are possible using nanoscale materials and structures [5–10].

It is important to note that even well-known conventional materials may exhibit completely altered properties when broken into minute sizes. In addition, a large number of new structures that do not have larger counterparts are possible at nanometer scales. Based on these, some attributes and properties that change significantly in nanomaterials have been summarized in the following sections. The discussions have been classified into two groups: nanoscale particles, which can be regarded as fragmented pieces of conventional materials; and uniquely assembled solids, which do not have regular-sized counterparts in larger conventional materials. The first category refers to property changes related directly to size (i.e., changes that occur when the size of the solid is reduced to nanoscale). The second category deals with uniquely structured solids that have been assembled atom by atom to create a completely different material, not just a smaller fragment of a larger solid. Examples in the first category are nanoparticles, or nanoclusters of conventional materials, and are sometimes classified as *zero-dimensional materials*. Common examples from the second category are nanotubes, nanocages, and superlattices. For example, carbon nanostructures such as nanotubes and buckyballs do not have much scaling relationship with bulk graphite. Similarly, peptide nanotubes, self-assembled monolayers, and semiconducting superlattices do not have corresponding bulk counterparts. In both of these categories of nanosolids, unique scale-related properties provide the community with a new inventory of materials for future devices.

2   NANOSCALE PARTICLES AND FRAGMENTS: INFLUENCE OF SIZE

As mentioned earlier, in this section we point out mainly properties that can change significantly simply by reducing the size of material. It is assumed that the core material is the same as that of the larger conventional material in terms of interatomic bonding and chemical composition. Properties that change significantly with size have been classified into the following categories:

1. Basic physical parameters: size, shape, and surface area
2. Thermodynamic quantities
3. Kinetic properties: diffusion
4. Chemical properties: reactivity and catalysis
5. Electronic and optical properties
6. Magnetic properties

Many of the effects may be interrelated at a deeper level. The attempt here is to point out explicitly some properties that change due to nanoscale dimensions, and potential applications resulting from them.

2.1   Basic Physical Parameters

One of the obvious effects of reducing the size of any solid is that the specific surface area (exposed area per unit bulk volume) will be increased. For common geometrical shapes, values of specific surface area (surface area/volume of given solid) are shown in Table 1. Surface area per unit mass can be obtained by dividing these by the

TABLE 1   Specific Surface Area of Some Solids as a Function of Size

Shape of Solid	Characteristic Length (units of length)	Surface Area/Volume (units of 1/length)
Sphere (spherical nanoparticle)	Diameter (d)	$6/d$
Cube (cubic nanoparticle)	Edge (a)	$6/a$
Ultrathin sheet (nanofilm/nanolayer)	Thickness (t), width (w) length (l); $t \ll w, l$	$2/t$ (assuming that $t \ll w, l$)
Utrathin cylinder (nanotube/nanowire)	Length (l), radius (r) $r \ll l$	$2/r$ (assuming that $r \ll l$)

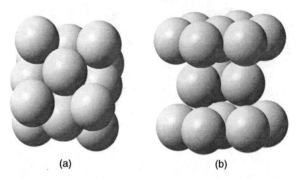

(a)   (b)

FIGURE 1 Closest packing of spherical atoms possible: (a) face-entered cubic and (b) hexagonal close-packed structures.

density of the compact solid. It can be seen that the specific surface area is inversely proportional to its smallest dimension.

In addition to solid geometry, atomic-scale features become more pronounced at nanoscale dimensions. Certain sizes and shapes of atomic clusters can be seen to be more stable than others, and therefore more frequent. Conventional materials can be regarded as infinite arrays: atoms or molecular units occupying three-dimensional space. That concept is not applicable here, since the solid is terminated after a finite number of repeat units. So there is a need to rethink the entire phenomenon of solid assembly in terms of energy minimization of nanoclusters [9]. This is an area overlapping colloid chemistry and solid-state physics. The concepts a of "magic numbers," "critical sizes," and so on, evolve from this type of treatment, details of which will be different for each specific nanomaterial in a given environment.

A simple model using elemental metals can demonstrate the basic concept. If the metallic atoms are considered to be hard spheres with nondirectional packing, close-packed clusters will occur. The overall geometry of packing within the cluster will be either face-centered cubic (fcc) or hexagonal close packed (hcp) in arrangement, as seen in Figure 1. These can be created by stacking two-dimensional layers of close-packed atoms over one another. Figure 2 shows the individual layers, with possible

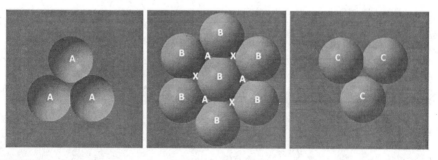

FIGURE 2 Three layers of close-packed spherical atoms. Atoms of layer C will be on location X in fcc structure, and location A in hcp.

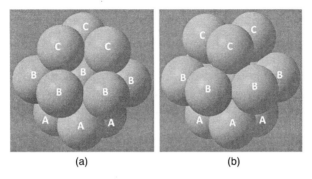

(a)   (b)

FIGURE 3   Stable cluster of 13 atoms shown in the three layers of Figure 2: (a) fcc packing; (b) hcp packing.

locations of how they can be stacked. Isolating a single disk or layer of atoms will involve breaking too many bonds and is not practical. It can be seen that the smallest chunk of atoms, or nanocluster, that can be isolated with some stability consists of a three-layer cluster: a triangle of three atoms in layer 1, a hexagonal layer consisting of seven atoms (six at corners and one in the middle) over that, and the capping triangular layer of three atoms on the top. The combined cluster, in two alternative stacking styles, can be seen in Figure 3. The number of atoms in this smallest stable cluster, 13, becomes the first *magic number* for this type of metal. Note that of the 13 atoms in this cluster, 12 are on the surface and are therefore expected to have incomplete bonding and higher reactivity than that of conventional metal. Therefore, 12 of 13 or 92% of atoms are now available for bonding with atomic and molecular species in the environment (e.g., air, vacuum, plasma, solution, fluids). If this concept of building the most stable cluster of atoms having progressively larger sizes is extended, one can predict comparatively compact and stable clusters having five layers (55 or 57 atoms, depending on fcc or hcp arrangement), then seven atomic layers (147 or 153 atoms for fcc and hcp, respectively), and so on. The fraction of atoms on the surface will decrease with increasing number of layers. Figure 4 plots the percentage of surface atoms as a function of particle size, and the trend is clear.

It must be noted here that as one moves on from ideal pure metals and builds more complex solid clusters (involving covalent or ionic bonding of more than one atomic species), the exact model will have to be modified accordingly. However, depending on atoms and the interatomic bonds and bond angles involved, a sequence of magic numbers or stable cluster sizes can be predicted, and the surface area will always increase rapidly with decreasing cluster size. This concept has been developed in more detail in several recent books [9,11–13] and may provide the building blocks for common nanomaterials in the future.

In summary, it can be generalized that nanoscale solids can be considered as finite clusters of atoms. Based on their interatomic bonds, certain sizes and shapes are expected to be more stable than others. More important, as the cluster size gets smaller, a larger percentage of atoms are on the surface. These surface atoms have fewer primary bonds than do interior atoms, resulting in higher energy, lower stability,

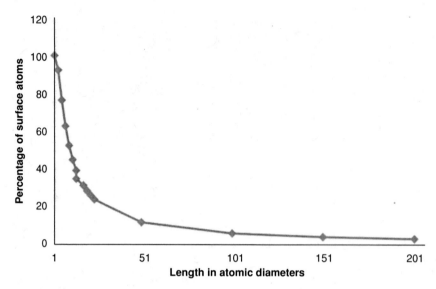

FIGURE 4   Percentage of surface/interface atoms that are available for bonding with environmental species, as a function of cluster size. (Adapted from [9].)

and a greater tendency to bond and interact with whatever is available (i.e., each other, environmental gases, contaminants, etc.).

It is worth pointing out that for nanoscale phases incorporated in a larger material, the high surface area is replaced by an *interfacial area* that links the nanoparticle with the larger matrix. Interfacial atoms are different from those in the interior of a solid phase, resulting in interface-controlled properties similar to the surface-controlled properties of isolated particles. Many of the engineering properties of nanomaterials discussed in the next sections can be attributed to this surface/interface effect.

2.2   Thermodynamic Quantities

The thermodynamic properties of solids can change as the cluster size is reduced to nanometer scale. A fundamental quantity is the cohesive energy, which is the energy involved in holding atoms together to form a solid or cluster of any size [9]. This is the energy difference between the sum of individual isolated atoms (without any bonding) and the solid cluster after atoms have bonded. Therefore, each interatomic bond contributes to reduction in the total energy of the system. The stronger the bond between atoms, the greater the reduction in the total energy of the system per bond, implying "higher" overall cohesive energy.

This type of energy is often estimated analytically and/or computationally, and can be related directly and indirectly to experimental thermodynamic quantities such as latent heat and melting temperature. However, in ionic compounds or complex compositions containing multiple atoms, factors such as variations in stoichiometry and defect- or strain-related effects may complicate this correlation. The lattice parameter is also sometimes related to cohesive energy, but the relationship can be very material specific. The most general observations are that smaller particle sizes result in less overall cohesive energy; this implies less latent heat and a lower melting temperature for nanoscale materials.

Energy calculations beginning at the atomistic level are not new to the fields of thermodynamics, statistical mechanics, and quantum mechanics [14,15], and earlier results have paved the path for theoretical modeling of nanoscale solid clusters [16–20]. Property predictions by computational methods such as the cellular method, density functional theory (DFT), linear combination of atomic orbitals (LCAO), and several others have been investigated for decades, but the age of nanomaterials has caused significant expansion of this field.

It must be emphasized that the success of theoretical predictions will depend upon the understanding and utilization of realistic forms of interatomic potential (energy) between individual atoms. For purely ionic solids, the predominant energy is from Coulomb interactions and can be summed up in terms of the Madelung constant and the equilibrium distance between nearest neighbors. In covalent bonding, the wavefunctions of the outer electrons overlap, and cohesive energy per monomer or single component is often calculated from molecular orbital theory. In metals the picture is a little different, and needs to be modeled as positive ions in a collective negative field or "sea of electrons.' Here the cohesive energy becomes a function of interatomic distance and the Fermi energy. It must be noted that all these seemingly diverse systems yield similar qualitative outcomes, in terms of trends at nanoscale. These are summarized below.

- Cohesive energy is independent of the cluster size for larger solids, but when cluster size is reduced to a few nanometers, cohesive energy decreases rapidly with decreasing size.
- Latent heat of fusion and melting temperature follow a similar pattern. As clusters reach nanometer dimensions, the latent heat of fusion and melting temperature are both lowered significantly.
- The important factor to note in this discussion is the dimensions at which detectable size-related changes start occurring. For simple polymers such as polyacetylene and for pure alkali metals such as Li, the changes noted above can occur around 2 to 3 nm. On the other hand, for group IV elements such as tin and ionic solids such as NaCl and CeO_2, these changes occur at much larger dimensions, perhaps 10 to 20 nm [9,21,22]. Figure 5 shows a representative plot adapted from data published in the literature.

2.3 Kinetic Properties

In traditional materials, *diffusion* almost always implies the transport of atoms or ions through the solid. In the context of nanomaterials, another aspect of matter transport that may be important is transport of the material itself through another host phase (i.e., air, water, fluid, tissue, etc.). Since this section is about extrapolating properties of larger solids to the nanoscale, the former phenomenon (diffusion of atomic species *in* nanomaterials) is discussed. Transport (or migration) *of* nanomaterials themselves through another phase is relevant for biological and environmental concerns but is not discussed here.

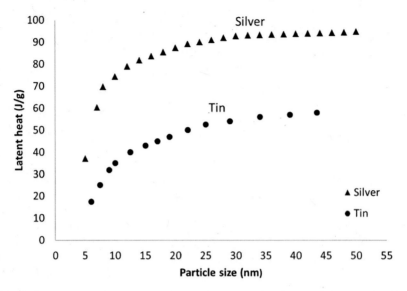

FIGURE 5   Variation of latent heat of two different metals (silver and tin) as a function of cluster radius. (Adapted from [21] and [22].)

In terms of the diffusion coefficient of a given species through a solid material, several paths are possible. They include transport through the interior of the solid (lattice diffusion, D_l), through the outer surface region (surface diffusion, D_s), or through grain boundaries and interfaces (boundary diffusion, D_b). Each of these diffusion coefficients (D_i) is strongly temperature dependent and can be generalized to have the following relationship:

$$D_i = D_{0i} \exp\left(\frac{-Q_i}{RT}\right)$$

where D_{0i} is a constant for that particular path, T the absolute temperature, and Q_i the activation energy, which reflects how easy it is for the diffusing atom to move through the path specified.

Among the three diffusion paths listed, D_s is obviously the highest because the surface region is most distorted and least densely packed. D_b is lower than D_s but higher than D_l since the atomic packing density at boundaries tends to be intermediate between those of the lattice and the outer surface.

When the size of the solid becomes very small (nanoscale), several factors need to be considered:

- Nanomaterials have lower cohesive energies and larger lattice parameters, implying lower packing densities of atoms. This means that for diffusion of atoms through a nanosize lattice, a lower activation energy (Q_l) is expected. This implies higher lattice diffusion coefficients.

- However, it must be pointed out that these changes are seen only for really small dimensions (<5 nm for metals and covalent molecules, 10 to 15 nm for ionic compounds) and are not important above that. Moreover, lattice diffusion as a whole may be less important for nanoscale materials, for the reason given below.
- Nanoscale particles have a higher surface/volume ratio than that of conventional materials; therefore, a greater contribution comes from surface diffusion. Conventional scaling considerations provided in most textbooks [23] indicate that the overall diffusion coefficient observed (D_{obs}) in a solid having lattice and surface diffusion paths will be given as follows:

$$D_{obs} = D_l + \frac{\delta}{d}D_s$$

Here d is the dimension of the lattice or solid particle and δ is the thickness of the surface region, where atoms are distorted and the diffusion coefficient corresponds to the faster value, D_s. δ is typically 3 to 5 nm in pure metals, and perhaps 5 to 10 nm in ceramic compounds. When particle size (d) is several micrometers or higher (perhaps in the millimeter range), as in conventional-size materials, the second term is negligible even if D_s is higher than D_l. However, as d approaches δ, the second term becomes prominent, and when they become comparable, the second term can shadow the lattice diffusion contribution completely. (Note that the same argument applies to grain boundary diffusion for a solid having nanometer-scale grain sizes.)

In summary, it is safe to assume that in most cases, overall diffusion observed through nanoscale material is significantly faster than that through regular-size material. This implies that all thermally driven processes in nanomaterials will be faster and will require less time and/or lower temperatures. These effects have significant technological implications for the materials-processing community, as shown in the following examples.

- Homogenization of mixed phases is significantly easier in nanomaterials. This arises from two factors: the small length scales needed for complete inter diffusion, and higher overall diffusion coefficients. Therefore, obtaining homogeneous multication oxide phases (for magnetic, superconducting, or electrochemical devices) is easier if nanoscale-component powders are available.
- Sintering and solidification rates for any powder agglomerate will be higher if the starting raw material happens to be smaller. The smaller starting particles have a higher surface area and can be packed more densely in the "green compact." Moreover, surface energies of smaller particles themselves are higher. Together, these increase the overall driving force for sintering and densification processes, resulting in the lower sintering temperatures and/or shorter heat treatment times needed for a desired microstructure.

These phenomena have created a large market demand for nanoscale powders as raw materials, especially for advanced ceramics.

2.4   Chemical Properties

The chemical reactivity of a solid is relevant to many phenomena, such as, but not limited to, phase transformations, corrosion, electrochemistry, and catalysis. It is widely recognized to be a surface-driven process. It is also known that as the size of the solid gets smaller, its specific surface area increases inversely proportional to its size. Therefore, the general rule of thumb is to assume that chemical reactivity increases as the particle size is decreased. However, it must be noted that this inversely proportional relationship works only up to a certain point.

As the particle size gets really small (on the order of 1 to 2 nm), the size dependence becomes erratic. Several factors other than overall size can become important. For example, the equilibrium shape can change, as discussed earlier. The crystallographic orientations, curvatures, and surface chemical activities of exposed planes can determine its overall chemical property. Quantum phenomena can also become apparent. For ideal elemental solids, some nanoscale chemical effects are explained by the "jellium" model [9], which treats a cluster of a few atoms as a large atom with its own set of electronic energy levels. This model often predicts that certain specific "magic" cluster sizes will have more unfilled orbitals than others, resulting in peaks of chemical activity at certain sizes.

In addition to inherent size-related effects, the role of coatings can become more pronounced. The surface of a material is rarely a termination of the bulk composition and is expected to have an outer coating of adsorbed atoms based on the environment. These layers themselves can be size and shape dependent, making prediction complicated. A typical example is aluminum, where nanoparticles can be very energetic and useful as explosives [24]. All aluminum particles are expected to have an outer atomic layer of aluminum oxide that passivates the surface. The reactivity of the particle does not depend too much on its overall size but on what factors can cause cracks in the oxide layer. This, in turn, is more dependent on particle shape than on size. Another example can be seen in gold nanoparticles, which are widely investigated due to their extensive applications in medicine [25,26]. It is reported that the reactivity of gold with different organic species and its interaction with live cells depend not only on overall size but also on the aspect ratio of the particle. Therefore, gold nanorods and nanospheres can have completely different responses in the body, and the shape and size effects of gold–protein interaction is becoming an emerging field in nanomedicine.

A large industry depending on the surface activity of nanomaterials is that of catalysis. Investigation of catalysis has always focused on surface effects [27]. Manufacturers of traditional catalysts, sensors, and electrochemical charge storage devices have always tried to incorporate very high porosity on surfaces to increase their specific surface area. With the advent of nanomaterials, an entirely new generation of very powerful catalytic materials and structures may become possible since the specific surface area of these is significantly higher. For example, if a precious metal useful for catalysis can now be deposited as nanoclusters on a highly porous

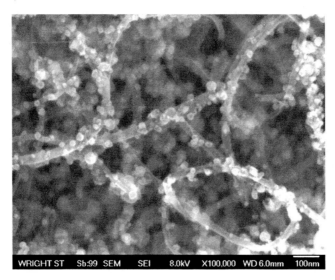

WRIGHT ST   Sb:99 SEM   SEI   8.0kV   X100,000 WD 6.0mm   100nm

FIGURE 6   Pd nanoparticles attached to carbon nanotubes grafted on porous carbon foam. The small amount of Pd can provide an unusually high catalytic surface area in compact space.

hierarchical substrate, the surface area obtainable from the same total mass of the precious metal can be increased manyfold. Figure 6 shows palladium nanoparticles attached to carbon nanotubes grafted on porous carbon foam [28]. By using this type of assembly, a very small amount of Pd can provide an unusually high catalytic surface area in a compact space. This can result in reduced cost and weight, with a simultaneous gain in catalytic efficiency.

In this field, too, the complicated relationships of size, shape, and surface chemistry need to be understood. As the particle size decreases, the overall shape of the nanocatalyst as well as surface layers can become significant factors, and there may be optimum sizes. For example, the catalytic dissociation rate of CO gas on Rh-particle catalysts is seen to increase with decreasing particle size until the size reaches about that of a 100-atom cluster. After that, in the range of 100 to 10 atomic clusters, decrease in size causes a reduced catalytic rate [9]. An additional important factor is the surface electronic state of the catalyst particle, which often depends on the electronic property of the underlying support. Therefore, for catalysis-related applications, decrease in particle size can often be generalized to provide an increase in catalytic activity only up to a certain critical size. Below that, additional factors mentioned earlier need to be investigated for specific catalyst-support systems. For example, it is known that the active phase of Pt and other transition metal catalysts may depend on suboxides formed on the surface. It is also known that the nature of suboxides is not only size-sensitive but also strongly dependent on the supporting substrate. Therefore, nanocatalysts may sometimes show very different and counterintuitive behavior for specific reactions compared to bulk catalysts. These behaviors are often specific to the reaction being catalyzed and require extensive investigation.

2.5   Electronic and Optical Properties

Electronic properties of solids are often explained in terms of band theory. In isolated atoms, electrons are known to occupy discrete energy levels [29]. As atoms form bonds during solidification, the outer electronic levels overlap, giving rise to electronic energy bands [30–32]. The outermost filled energy band (the valence band) and the next band (the conduction band) are separated by a gap in energy that constitutes the very important concept of bandgap energy, E_g. Detailed study of the band structure predicts the various electronic energy levels and bandgaps relevant to a solid. The band structures for common solids are well established assuming an infinite array of atoms, which correspond to the bulk materials. When the cluster of atoms is too small to be considered an infinite array, the electronic band structure has to be predicted from first principles by solving Schrödinger's equation, details of which are beyond the scope of this book. The relevant outcome from that is that the band structure is expected to change as the size of the crystal is reduced to nanoscale.

Many of the effects related to the changed band structure are loosely referred to as *quantum confinement*. These become important once one or more dimension(s) of a solid are comparable to the wavelength of the electron wavefunction. There is a large body of literature investigating these quantities in nanosolids [33,34]. Such studies often include detailed modeling of the reciprocal space as a function of cluster size, and estimation of bandgaps and other quantities from them. This type of understanding is very important because the band structure will determine physical properties such as color, electrical and thermal transport, photoluminescence, fluorescence, carrier concentrations, and current–voltage response. Possible integration of nanomaterials in electronic devices [9] often depends on their band structure. The most visible and widespread application of nanoscale effects is seen in the modern arsenal of "quantum structures" that are making their way into light-emitting diodes (LEDs), displays, sensors, infrared detectors, lasers, and other electronic devices.

Electronic nanomaterials are sometimes classified based on how many dimensions are in the nanoscale. When only one dimension of the material is nanoscale (i.e., films, multilayer devices, etc.), it is called a quantum well; when it is nanoscale in two dimensions (i.e., tubes, rods, etc.), it is called a *quantum wire*; and when it is nanoscale in all three dimensions, it is a *quantum dot*. It must be noted that at these scales, the size of the cluster can drive other parameters, such as equilibrium shapes, defect densities, and local composition gradients, all of which can influence the band structure. Therefore, detailed calculation of each system of atoms can be important. Among familiar materials, gold may be the most widely discussed elemental nanocluster, due to its simplicity as well as its great potential in biological applications [25,26,35]. Even for the apparently simple elemental composition of a well-known metal, a large number of surprising observations are yet to be explained. In addition to select metals, the widely investigated nanomaterials in electronics are semiconducting compounds relevant as sensors, photovoltaics, optical displays, and energy storage devices. Despite the large variety of composition-specific issues, some simplified generalizations can be made about the band structure of nanoclusters compared with that of larger crystals. They are summarized below.

- The general rule of thumb is that bandgap energy Eg increases with diminishing crystal size. This leads to the *blue shift* or shift in any optical absorption, emission, and related peaks toward the higher-energy, lower-wavelength side. For fluorescence, photoluminescence, or related applications, this can have far-reaching consequences. These have been investigated systematically in a few compounds, such as ZnS, CdSe, and TiO_2, which have potential use in photovoltaics and sensors [36,37].

- For metals bandgap is zero anyway, but changes in band structure due to size do result in some property changes. Most obvious quantities influenced [38,39] are the Fermi level, electron affinity, and surface plasmon response. In general, as dimensions of metallic clusters approach <5 nm, electron affinity decreases with decreasing size, Fermi energy increases as crystals become smaller, and altered plasmon response results in metal-specific color changes.

2.6 Magnetic Properties at Nanoscale

Magnetic response in materials falls into one of the following categories [40]: diamagnetic, paramagnetic, ferromagnetic, ferrimagnetic, or antiferromagnetic. Magnetic materials of importance are paramagnetic, ferromagnetic, or ferrimagnetic. These have atoms or ions with unpaired electrons where spins are not canceled, resulting in a net nonzero magnetic dipole moment. In a paramagnetic state, the individual magnetic dipoles do not interact with each other and are oriented randomly, due to thermal vibrations. Therefore, in the absence of any external magnetic field, there is a net magnetization of zero. When an external magnetic field is applied, these dipoles align providing a net (nonzero) magnetization. In the ferro- or ferrimagnetic states, the adjoining magnetic dipoles have strong enough exchange interactions to counter thermal randomness. Therefore, each magnetic dipole influences its neighbors and there is spontaneous alignment of neighboring magnetic spins. The relative alignment depends on the sign of the exchange interaction potential, a purely quantum mechanical phenomenon [41,42]. If the alignment happens to be parallel, the material is ferromagnetic and magnetic moments add up, as seen in iron. If they are antiparallel and equal, the material is antiferromagnetic and magnetic moments of adjoining atoms cancel each other completely, as in manganese. If ions with unequal spins are aligned antiparallel, the material is ferromagnetic and there is net magnetic moment due to uncanceled spins, as in ceramic ferrites [43].

   In normal-sized ferromagnets or ferrimagnets, this alignment is complete within regions of the crystals called *domains*, inside which the magnetic moments are saturated. However, each domain may point in a different direction, and the total magnetization (the vector sum of all dipole moments, normalized over unit volume of the material) is less than the theoretical saturation value. It may even be zero for a completely "unpoled" specimen. As an external magnetic field is applied, the magnetic moment of the material increases by the mechanisms of domain growth and domain rotation. During growth, domains that are oriented in the direction of the field grow in size, and those oriented opposite shrink. During rotation, the magnetization vector of each domain tends to rotate toward the applied field. When the external field is removed, some remnant magnetization (M_r) remains, which has to be overcome by a reverse coercive field (H_c) if the material is to be fully demagnetized. The extent to which each mechanism can operate depends on several factors: (1) structural parameters of the magnetic material, such as grain size, shape, thermal history, strain, impurities, and inclusions; (2) strength of the external field; and (3) operating temperature, which determines diffusion rates and atomic mobility. These quantities define what is called the *hysteresis loop*, characteristic of magnetic materials.

Nanosizing can have profound effects on the magnetic properties of materials. If the size of individual magnetic particles is reduced below a certain critical length, it is energetically unfavorable to form domains, and the entire particle exists as a single domain. This critical size can range between 50 and 100 nm in magnetic ferrites such as $CoFe_2O_4$ and Fe_3O_4. These "monodomain" particles individually exhibit theoretical values of saturation magnetization and are free to align fully with external applied fields. Therefore, the hysteresis effect is reduced substantially or even eliminated, and they exhibit what is termed as a *superparamagnetic effect* [44,45] whereby ferromagnetic and ferrimagnetic particles can align fully, like paramagnets, but with extremely high magnetic moments. Figure 7 shows a very generalized schematic of how some useful properties may depend on particle size.

This effect can lead to multiple applications that involve incorporation of superparmagnetic particles in solid thin films, liquid media, or biological fluids. Careful optimization of particle size is needed in each case. On the one hand, smaller particle sizes imply fewer magnetic particles and lower saturation magnetization. On the other hand, larger particle sizes may result in domain formation that

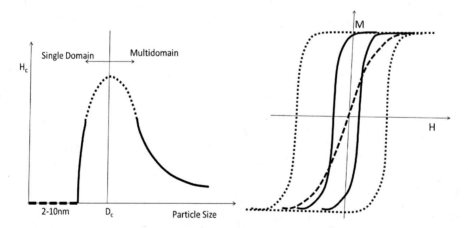

FIGURE 7 Simplified schematic of how magnetic behavior depends on particle size. Very small particle sizes (dashed lines) result in no hysteresis but high magnetization. This is the superparamagnetic effect. As particle size increases, the area inside the hysteresis loop increases. In some magnetic materials this area may show a maximum at a certain particle size (dotted lines).

prevents complete alignment. It must also be noted that smaller particles are more susceptible to thermal fluctuations. This results in lower *blocking temperatures*, defined as temperatures below which the magnetic orientation can be maintained (or magnetic information stored) without thermal losses.

The applications of nanoscale magnetic phenomena can be far reaching, affecting the areas of data storage, magnet-tunable electronics, and medical imaging as well as treatments. A few selected examples are given below.

- *High-capacity magnetic storage devices.* These are single-domain magnetic particles, especially elongated or ellipsoidal particles embedded in thin films. They can retain information based on how they are magnetized during recording. The idea is to store one *bit* of data in each nanoparticle, implying that data storage density per unit area is inversely proportional to particle size.

- *Ferrofluids.* Traditional ferrofluids of the 1950s and 1960s involved micrometer-sized magnetic particles suspended in fluids. They had magnetorheological properties that could be changed with external fields. Current ferrofluids are being investigated that are made up of magnetic nanoparticles suspended in fluids (often, transparent fluids). These can have unique optical properties that can be controlled with a magnetic field, which leads to applications such as tunable diffraction coatings and optical displays.

- *Nanobiomagnetics.* The largest area in which nanomagnetic particles are expected to have an impact is in the field of medicine. Applications range from sensors in medical diagnostic tools such as magnetic resonance imaging (MRI) [46] to directed therapeutic treatments and drug delivery [47]. For examples many investigations involve incorporation of magnetic nanoparticles functionalized with specific molecules that can be targeted to designated tissues of the body. Once they accumulate at target sites, these molecular nanomagnets can be used in MRI to improve the contrast between healthy and diseased tissue. Alternatively, these localized particles may contain therapeutic drugs to heal, or be locally heated through external fields to destroy unhealthy tissues. This concept opens up the entire field of precisely targeted treatments for cancers and other problematic diseases while minimizing side effects to other healthy tissues.

In summary, several important properties of materials change simply because of size reduction to the nanoscale. A few of them have been outlined in this section. In addition, there can be completely different types of solids that are assembled from the nanoscale in unprecedented ways. These are discussed in the following section.

3   SPECIAL NANOMATERIALS: UNIQUE ASSEMBLY AND FUNCTIONALITY

Materials that are assembled differently from conventional bulk materials can have completely different structures not seen in conventional materials. These include one-dimensional materials such as nanotubes, two-dimensional materials such as thin films and multilyered materials, and three-dimensional nanomaterials such as nanocomposites and nanocrystalline solids. Also mentioned are materials that have complex multidimensional geometries, such as hierarchical solids, nanocages, and nanosponges.

Needless to say, this field is advancing at a rapid pace, and this section provides only a small sampling of interesting structures and fast-emerging applications. Examples provided have been classified as follows:

1. Nanotubes and nanostructures
2. Thin films and multilayer materials
3. Bulk nanostructured materials
   a. Nanocrystalline solids
   b. Nanocomposites
4. Biological and biomimetic nanostructures

3.1   Nanotubes and Nanostructures

In 1985, the first nanostructure, which has 60 carbon atoms forming a soccer-ball-shaped cage (C_{60}), was formally reported [2]. Shortly after that, several other nanostructures, such as nanotubes, nanohorns, nanocones, nanotoroids, and nanohelicoids (commonly nicknamed *fullerenes*), began to be reported [3–5,48–59]. Most widespread nanostructures appear to be made of carbon, but a significant number of other materials are being synthesized in these unique forms. These include inorganic compounds such as zinc oxide, titanium dioxide, boron nitrite, and metal disulfides as well as organic molecules based on peptides and DNA. Nanomaterials had finally come to stay!

It is possible that many of these structures were always present in carbon dust, pyrolysis products, and organic matter. However, they had not been isolated and investigated as carefully as has been done in recent years, and had never been synthesized intentionally. Graphical models of some of the most intriguing and potentially useful structures are available in the published literature as well as in cyberspace, and the list is growing every day.

A vast majority of uniquely assembled nanostructures are carbon-based and are sometimes called fullerenes. Since these have the simplest composition, they can be regarded as model structures for detailed understanding. The basic background of carbon chemistry is essential for this. The carbon atom forms either sp^3-hybridized or sp^2-hybridized orbitals for bonding. So when carbon bonds with carbon, two crystallographic allotropes are possible: diamond (sp^3) or graphitic (sp^2), as shown in Figure 8. In each atom of diamond, all four sp^3 electrons are "localized" in strong and rigid covalent bonding with fixed bond angles. This rigidity gives diamond its hardness, and the strongly localized electrons give it the high bandgap energy. The high bandgap makes diamond the optically transparent and electrically nonconducting material that we know. On the other hand, the sp^2 bonding in graphite provides strong

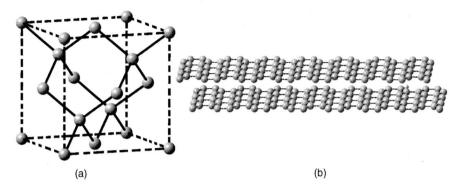

(a)   (b)

FIGURE 8   Basic structures of pure carbon: (a) diamond; (b) graphite.

covalent bonds only within the hexagonal planes. Each carbon atom is covalently bonded to three other atoms in this plane, and the fourth valence electron is the π-bonded electron, which is delocalized and is therefore free to move along the plane. Across the planes, multiple layers may be stacked with relatively weak van der Waals bonds between them, resulting in poor electron transport as well as low bond strength across the planes. This gives graphitic carbon its unique properties, such as high strength, high electrical and thermal conductivities within planes, and lower conductivity and strength normal to the planes. Lubrication is possible by sliding between planes.

Most common carbon nanomaterials are derived from sp^2-hybridized graphite. A single layer of graphite is called *graphene*. A large variety of other nanostructures can be formed from distorted, folded, or rolled-up graphene sheets, similar to origami structures made out of paper. Some commonly discussed carbon nanostructures are shown in Figure 9. The special outcome of using graphene sheets as a building block for other structures is that the overall material properties will be determined by the stacking direction of graphitic sheets in that structure. Therefore, a wide range of properties is possible if the orientation of graphitic planes can be guided during fabrication.

Among the various nanostructures, nanotubes are the most widely used. Carbon nanotubes (CNTs) can be regarded as cylindrical sheets of rolled-up graphene. A single sheet of graphene creates a single-walled nanotube, and a stack of two or more sheets gives a multiwalled nanotube. The geometrical relationship between the orientations of the hexagonal plane with respect to the cylindrical axis determines the chiral angle of the nanotube, details of which are widely available in textbooks [9]. The hexagons within the plane can occasionally be replaced by pentagons or heptagons, due to defective assembly. Some manipulation of such defect concentrations is possible by the introduction of impurity atoms. The chirality and defect concentration of nanotubes can have significant influence on their band structure and electronic behavior, giving rise to metallic and semiconducting nanotubes. Subsequent impurity levels can make the semiconducting nanotubes p-type or n-type, similar to conventional semiconductors.

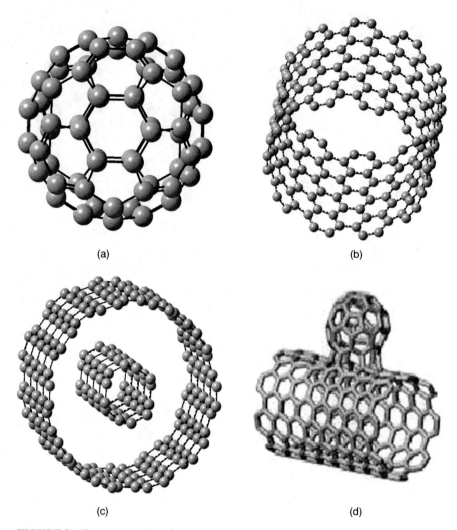

(a)   (b)

(c)   (d)

FIGURE 9   Computer models of some graphene-based nanostructures: (a) fullerene molecule (C_{60}); (b) single-walled nanotubes; (c) multiwalled nanotubes; (d) nanobuds.

CNTs can show unique mechanical, thermal, and electronic properties. For example, single-walled carbon nanotubes can be created with exceptionally high strength and with high electrical and thermal conductivities along the fibers. They can be dispersed into other matrices to form nanocomposites, or pulled into fibers and yarns to incorporate their unique properties. Multiwalled nanotubes have similar properties, sometimes to a lesser extent due to weak bonds between the concentric tubes. These are, however, available in various diameters and often are less expensive to make and purify. Therefore, depending on specifications and cost, different grades of engineering materials incorporating carbon nanotubes (CNTs) can be designed.

Their electronic properties mentioned above also make them promising substrates and components for miniature electronic devices.

Another nanostructure picking up momentum in terms of production and applications is graphene, which is a single sheet of graphite (either as flake or supported film). This has the potential of becoming a less expensive and more widely used form of nanocarbon suitable for multiple applications. Additional nanostructures that have specialized identified applications are various types of fullerene cages. These may be used to trap and transport specific atoms or clusters, therefore to provide better options for hydrogen storage, controlled drug delivery, remote sensing and imaging, and many other technologies.

The key to widespread use of any of these materials in engineering devices will lie in isolating and purifying the components, and obtaining large quantities of the structure desired. Once such processes are well established and these materials become readily available, many unprecedented properties and unique applications can be envisioned [60–63]. Following is a short list of some of the products in which use of carbon nanostructures are being investigated:

- *Sports equipment.* Traditional carbon microfiber composites have been common for several years in tennis rackets, golf equipment, fishing gear, bicycle components, helmets, and other accessories. Nanotubes and graphene plates are making some of these products stronger and lighter than ever before.

- *Clothes and fabric made of carbon nanotubes.* These can be made into waterproof, tear-resistant, combat-proof clothing. Since CNTs can be made with metallic or semiconducting properties, depending on its defect level, there is an attempt to have all the personal electronics embedded in the fabric itself. This can make clothing with built-in personal electronics, as well as communication equipment, particularly advantageous for soldiers, naturalists, or adventurers.

- *Ultrastrong yarn and ropes made of CNT fibers.* These are being investigated for possible use in space elevators. Blends of CNTs into other textile materials (e.g., polyethylene) can improve properties of other functional fabrics.

- *CNT added to traditional construction materials.* Concrete and other construction materials reinforced with CNTs are being investigated for improved toughness and for the design of built-in structural monitoring, crack healing, and so on.

- *Biomedical devices.* CNTs are suitable for certain piezoelectric devices and components, such as synthetic muscles, due to their giant elongations and contractions when a current is run through them.

- *Advanced composites for infrastructure, energy, and transportation materials.* CNT-reinforced composites can be used for bridges, decks, wind turbines, boats, automotive bodies, aircraft bodies, and other structural components, and it may replace a significant part of steel-based structures.

- *Papers made of carbon nanostructures (bucky-paper).* This is already available in the market and has the advantage of being a strong, conducting, and ultralight paper that can provide electronic chip cooling, electrostatic shielding, and a range of other applications.

- *Electronic components.* Integration of CNTs in VLSI (very large scale integrated) circuits have potential as interconnects, transistors, electrodes, substrates, and so on.

- *CNTs as substrate or template for other nanostructures.* By adding other components to CNTs (as attachments, coatings, grafts, or matrices), the next generation of lithium-ion batteries, ultracapacitors, switches, transistors, LED devices, and various other electronic components have been demonstrated.

- *Replacements for carbon black.* Conductive CNTs can replace traditional carbon black in materials for tires, motor brushes, and other products. Particularly in moving parts such as motor brushes, they improve electrical and thermal conductivity as well as tribological properties. Brushes made of CNT composites tend to be cooler-running (due to simultaneous improvement in thermal conductivity and lubrication), stronger, and less brittle.

- *Filters, sponges, and absorbers.* Coating surfaces for selective absorption of certain species can make fibers and porous CNT structures suitable for water and air filters that trap pollutants and store hydrogen and other chemicals. Fullerene cages with controlled inner dimensions are being investigated as molecular traps, biotech containers, and drug delivery agents.

It must again be reiterated that this is a very small sampling of applications, and the list keeps growing every day. Large-scale commercialization of many of the end products outlined above has been slow, due to inherent limitations in the availability of standardized materials. The majority of production techniques produce mixed nanostructures, and separating out specific fullerenes is difficult and expensive. However, it is only a matter of time when some of the materials will be readily available. It is currently seen that as soon as production issues in some structures are addressed, large numbers of papers and patents involving products and applications follow immediately. It is therefore expected that as many of these nanostructures become feasible at larger scales, the floodgate of applications will become more intense.

3.2 Thin Films and Multilayer Materials

There are many advanced materials that incorporate thin films either as coatings or as internal layers. Some materials are composed entirely of layered structures. These types of materials and applications have existed since the beginning of technology (at all dimensional scales) in the form of surface-altered layers, paints, coatings, and lamellar structures. Everyday applications of thin films include windshields, eyeglasses, electroplated metals, corrosion barriers, and sunscreens. In addition to surface films, several electronic, magnetic, and optical applications have incorporated multilayer structures such as superlattices.

Despite the earlier existence of films and multilayers at all scales, this field is getting a significant boost with our increased focus on nanomaterials. The theory and potential applications were known for some time, but few materials and systems were controllable at nanoscales. In recent years, the availability of modern thin-film deposition techniques has made it possible to precisely tailor the composition, morphology, and structure of a larger variety of materials at these scales. This has led to the explosion of truly nanoscale films and multilayers [64–73] including self-assembled monolayers, superlattices, functional coatings, hierarchical morphologies, and so on. The detailed design and physical principles behind each device are beyond the scope of this book. A common thread in all these materials is that the properties can be "tuned" or tailored by the thickness of the layers, and investigation along these lines is likely to keep increasing. A brief list of the most widespread thin-film and multilayer devices that are enhanced significantly by nanotechnology is given below.

- Semiconductor superlattices for controlled quantum well behavior, such as tunable bandgap devices and laser diodes.
- Antireflection coatings tailored for different electromagnetic wavelengths.
- Dielectric mirrors, interference filters, and tunable transmission cables.
- Substrates and buffer layers of unique electronic devices such as superconductors.
- Multilayer metals and ceramics for magnetic applications.
- Many plastics, even inexpensive consumer products, are beginning to incorporate these materials. Examples include food packages and soft-touch plastic goods such as toothbrushes and polymeric heating pipes.

3.3 Bulk Nanostructured Materials

Bulk nanostructured materials are normal-size materials with embedded nanoscale substructures. Such materials are sometimes referred to as *three-dimensional nanomaterials* in that they have finite size in all three dimensions, but their properties are dominated by the nanoscale components. When the material is a pure single-phase solid and the nanosize components are the individual crystalline grains, the material is said to be *nanocrystalline*. On the other hand, if there is more than one phase in the material, at least one of which has nanoscale dimensions, it is called a *nanocomposite*. In a broader sense, nanoscale materials may be suspended in a fluid or gel phase, which can result in a *nanocolloid* or *nanofluid*. In all of these, the shape, size, and distribution of the second nanophase is important, and several variations are possible even with the same basic components.

Nanocrystalline Solids Conventional metals and ceramics, which are naturally crystalline, contain grain sizes on the order of micrometers or higher. Reducing the size of these grains down to nanometer dimensions can have dramatic effects on many properties [74,75]. Note that atoms at grain boundaries have different types of bonding and significantly higher irregularities than those in the interior of the grain. Nanosized grains imply that a larger fraction of atoms are at boundaries (a relationship identical to that of surface fraction versus particle size, shown in Figure 4). Several thermodynamic and kinetic property changes attributed to the high surface area of nanoparticles were listed in Section 3.2. Many of these are expected to occur to a lesser degree in nanocrystalline materials, where the grain boundary area per unit volume can be very high.

Therefore, even though each nanocrystalline material has its own specific pattern of grain-size dependence, some generalizations can be made, and their technological outcomes can be significant. A few are listed below.

- In the case of metals, plastic deformation is controlled by dislocation motion. Traditionally, smaller grains (micrometer sizes) were believed to make the metal stronger, harder, and more brittle by pinning dislocations at boundaries. This phenomenon, called *grain boundary strengthening*, often represented by the well-known Hall–Petch relation [76], shows strength as being inversely proportional to the square root of the grain size. However, when the grain size is reduced to about 10 nm or lower, this rule is no longer followed [9], and many metals become softer again. This is attributed partly to grain-boundary sliding and intergranular failure, but newer models and mechanisms are under investigation.

- Fine-grained metals tend to have better electrical resistance and poorer thermal conductivity than those of larger grain sizes. This is not surprising since these physical properties are related to electron mobility in metals, which is impeded by the grain boundaries.

- In the case of ceramic materials, grain-boundary strengthening is very rare at any grain size. A common observation is that ceramics tend to become more ductile and fracture resistant as grain size is reduced. This is a very desirable property in ceramics, which are often limited by brittleness. An important phenomenon for some nanograined materials at elevated temperatures is superplasticity. This phenomenon can be very advantageous for the fabrication of complex shapes.

Nanocomposites   Nanocomposites can have a large variety of possible arrangements in terms of shape–size connectivity and structural organization of the multiple phases [77]. The simplest classification can be made in the following way:

- Nanoscale precipitates or clusters dispersed in a solid matrix
- Nanofiber or nanowhiskers reinforcing the matrix phase
- Layered nanocomposites (similar to multilayered materials discussed earlier)

The second phase (precipitate, cluster, fiber, or whisker) dispersed in the matrix may be a completely foreign filler, a product phase formed by internal reaction/phase transformation, or even just a void or pore. Many common second-phase fillers or inclusions are nanoscale particles, nanotubes, and fullerenes, discussed in earlier sections. Precipitates formed inside supersaturated solids (often controlled at nanoscale by thermal treatments) can also provide the second phase. In addition, aerogels, nanoporous foams, and colloidal solids are all special cases of nanocomposites where the second phase may be air or fluid.

Useful properties of a nanocomposite (i.e., mechanical, thermal, electrical, optical, magnetic electrochemical, catalytic, etc.) can differ completely from those of its component materials, thereby opening up unprecedented technologies. A common filler in currently available nanocomposites are carbon nanostructures that can help to enhance electrical and thermal conductivities, increase stiffness, and enhance crack deflection and toughness. Other common nanomaterials dispersed in composites are TiO_2 and ZnO for optical properties, nitrides and carbides for hardness and wear resistance, nanoclays for targeted properties, and nanometals for color.

3.4 Biological and Biomimetic Nanostructures

The biggest inspiration in design and assembly of unique nanomaterials can be drawn from nature, especially from living systems. It is noteworthy that most traditional engineering materials that shaped the earlier eras of technology (e.g., stone age, bronze age, steel and concrete age, silicon age) did not have too many functionalities. They were relatively simple materials that offered one or more select properties suitable for specific applications that managed to propel the society forward at that time. But in this age, scientists and engineers are ready to build on the earlier knowledge base to create complex multifunctional and somewhat "intelligent" devices and systems. There is therefore a demand for a higher level of sophistication in materials design incorporating multiple functionalities in the same structure.

For such complexities, it may be wise to take a closer look at nature, which produces very elegant multifunctional materials using energy-efficient, environmentally friendly techniques. A few examples of common biological nanostructures that provide good starting points for synthetic nanomaterials mimicking them have been listed below. Details of each structure may be found in the references cited.

1. *Nacre:* strong resilient shiny ceramics made from brittle components. Nacre (mother-of-pearl) is a lamellar ceramic formed in the mollusk shell. It is composed of hexagonal platelets of aragonite (calcium carbonate) that are 10 to 20 μm wide and 0.5 μm thick. These are arranged in a continuous layer, with adjoining layers separated by sheets of organic matrix composed of elastic proteins. The crystallographic c-axis of the bricks points perpendicular to the shell wall, but the direction of the other axes varies between groups. The layout of the brittle platelets in an elastic organic matrix is such that transverse crack propagation is strongly inhibited, resulting in a strong structure with a high Young's modulus along with high toughness. The biological process of growing such structures, called *encystations*, continues as long as the mollusk lives. The ceramic minerals nucleate on randomly dispersed elements within the organic matrix and grow into bricklike plates until adjacent bricks impinge on each other to form a complete layer. Once the smooth, shiny layer is complete, it provides strong, tough protection for the underlying soft delicate tissues of the oyster.

Several approaches to mimicking this structure in engineered materials are under investigation [78,79]. There can be extensive uses of such structures, especially if they can be made at large scales at lower cost.

2. *Lotus leaf:* superhydrophobic and self-cleaning surfaces. Lotus, an icon of eternal purity in ancient cultures, is also regarded as one of the most elegant examples of surface modification in nature. It grows in muddy water but is always dust-free because its surfaces are water-repellant (hydrophobic) and self-cleaning [80–85]. Detailed investigation shows that the surface of the leaf has a complex architecture with two levels of structural hierarchy combined with special chemistry, which work together to minimize adhesion. This hierarchical double structure consists of micrometer-scale bumps covered with nanoscale hairs, coated with a waxy film. This combination of unique morphology and chemistry on the surface can provide water contact angles between 160 and 170°, implying that less than 1 to 2% of the water droplet is in contact with the surface. Any water droplet will roll off the surface, carrying with it any dirt or contamination. In addition to lotus leaves, some feathers and animal structures are seen to have self-cleaning effects due to hydrophobic behavior. These are of great importance for protection of the living organism against pathogens, pollutants, and other unhealthy objects. This structure is extensively, but incompletely, mimicked by many scientists. The approach in most cases is to combine micro- and nanoscale attachments on the surface with hydrophobic chemical groups. The current thrust is to create this property in fabrics, paints, tiles, rooftops, and other surfaces. More improved versions with better controlled morphology and chemistry are also being applied to biomedical microfluidic devices such as "lab-on-a-chip" components.

3. *Bone:* hierarchical structure with tailored load-bearing properties. Bone is a very unique material [86–88] in which the structural organization is highly sophisticated, described in terms of several hierarchical levels of organization. The substructures are modified in each part of the skeletal system, tailored precisely for the type of load-bearing capacity and functionality needed at that location. At the micro and macro levels, each complete bone consists of two main regions: cortical bone and trabecular bone. Cortical is the hard outer layer composed of compact bone tissue with minimal porosity (5 to 30%), just enough to accommodate blood vessels. This tissue accounts for the smooth, white, and solid appearance and 80% of the total bone mass. On the other hand, trabecular bone is the filling inside the compact bone, which has an open cellular porous network. The network is made of rod- and platelike elements with about 30 to 90% porosity. This structure allows room for both blood vessels and bone marrow. Trabecular bone has 20% of total bone mass but nearly 10 times the surface area of compact bone. If there is any alteration in the strain in the bone, rearrangement of the trabeculae can be seen. At the nanolevel, the bone is a composite of linear collagen fibrils and mineral particles with their main orientation aligned with that of the overall trabeculae. Few hydroxyapatite/collagen-based minerals and composites have mimicked some aspects of natural bone, but the overall structure has many sophisticated features yet to be copied. An important aspect of bone is its ability to heal. Other tissues in the body may repair by producing scar tissue, but bone has the capability to fully regenerate itself in every way. Deeper

understanding of this phenomenon may offer ways of incorporating this behavior in synthetic structural materials.

4. *Wood:* multiscale cellular composite. Wood, like bone, is a cellular composite but is designed quite differently, suitable for the plant kingdom [89–91]. It is perhaps the oldest and most commonly used structural material. It happens to be a fiber-reinforced composite (cellulose fibers in a lignin matrix) that has a complex overall organization. It has tubular cells oriented parallel to the axis of the stem or branch. The cell walls are about 25-nm regions of crystalline cellulose embedded in an organic amorphous matrix of hemicellulose lignin. The cellulose fibrils are wound around the tubular wood cells at a spiral angle, the microfibril angle, which plays an important role in the mechanical property of that particular wood. Because of this offset, the axial modulus of the wood is lower than the theoretical value but the toughness is increased significantly. Wood by itself is very useful as a building material, as evidenced by its widespread use in historical as well as modern construction. In addition, engineered wood, composites made from wood-derived products, are widely used by the construction industry. Wood-derived composites can be strands, particles, or fibers of wood bonded with some type of adhesive matrix material. These products can be tailored to precise design specifications and range from traditional plywood to the modern variety of wood–plastic composites [92] available today.

It must be emphasized that the list above is a very small sampling of biological structures that can provide inspiration for the design and creation of multiscale, multifunctional materials suitable for many diverse applications. There are a large number of other examples in nature where nanoscale assembly is merged with larger structures to provide the perfect tailoring for a unique function. Scientists and engineers are only beginning to copy some of these designs with their newly acquired tools of nanotechnology.

4 SUMMARY

In this chapter, some special features of nanomaterials have been discussed. It can be seen that even familiar conventional solids can show distinctly different properties when reduced in size to nanometer scales. This nanosize effect can be attributed to higher surface area, altered shape and surface composition, changes in thermodynamic energies and electronic band structures. They are reflected in most physical properties, such as latent heat, melting temperature, diffusion coefficients, reactivity, catalysis, bandgap, optical and electronic parameters, and magnetic behavior. In addition, there are many materials that have unique architectures at the nanoscale and are very different from conventional solids. These include nanotubes, fullerenes, superlattices, nanocrystalline materials, and nanocomposites. More advanced materials may utilize hierarchical architectures where nanoscale structures are combined with larger features to provide novel designs. Many natural biomaterials utilize nanostructures in very efficient and elegant ways and can provide inspiration for multifunctional materials of the future.

REFERENCES

1. R. P. Feynman, Plenty of room at the bottom, historical talk at APS Meeting, Dec. 1959.

2. H. W. Kroto, J. R. Heath, S. C. O'Brien, R. F. Curl, and R. E. Smalley, C60: buckminster-fullerene, *Nature*, 318, 162–163, Nov. 1985.

3. R. E. Smalley and H. W. Kroto, Discovering the fullerene; symmetry, space, stars, and C_{60}, respectively, Nobel lecture, Dec. 1996.

4. M. S. Dresselhaus, G. Dresselhaus, and Ph. Avouris, *Carbon Nanotubes: Synthesis, Structure, Properties, and Applications*, Springer-Verlag, Berlin, 2001.

5. C. P. Poole, Jr., and F. J. Owens, *Introduction to Nanotechnology*, Wiley, Hoboken, NJ, 2003.

6. G. L. Hornyak, J. J. Moore, J. Dutta, and H. F. Tibbals, *Fundamentals of Nanotechnology*, CRC Press, Boca Raton, FL, 2008.

7. L. E. Foster, *Nanotechnology: Science, Innovation, and Opportunity*, Pearson Education, Upper Saddle River, NJ, 2006.

8. A. Lakhtakia, *Handbook of Nanotechnology: Nanometer Structure Theory, Modeling, and Simulation*, SPIE Press, Ballingham, WA, 2004.

9. F. J. Owens and C. P. Poole, *The Physics and Chemistry of NanoSolids*, Wiley, Hoboken, NJ, 2008.

10. P. J. Thomas and G. U. Kulkarni, From colloids to nanotechnology: investigations on magic nuclearity palladium nanocrystals, *Curr. Sci.*, 85, 12, Dec. 2003.

11. M. F. Ashby, J. S. Paulo, G. Ferreira, and D. L. Schodek, *Nanomaterials, Nanotechnologies and Design: An Introduction for Engineers*, Butterworth-Heinemann, Oxford, 2009.

12. Y. Kawazoe, T. Kondow, and K. Ohno (Eds.), *Clusters and Nanomaterials: Theory and Experiment*, Springer-Verlag, Berlin, 2002.

13. C. Bréchignac, P. Houdy, and M. Lahmani, *Nanomaterials and Nanochemistry*, Springer-Verlag, Berlin, 2007.

14. D. R. Gaskel, *Introduction to the Thermodynamics of Materials*, 5th ed., CRC Press, Boca Raton, FL, 2008.

15. J. B. Hudson, *Thermodynamics of Materials: A Classical and Statistical Synthesis*, Wiley, New York, 1996.

16. M. Hartmann, G. Mahler, and O. Hess, Fundamentals of nano-thermodynamics: on the minimal length scales, where temperature exists, *Condens. Matter Mater. Sci.*, Aug. 2004.

17. B. K. Rao, S. N. Khanna, and P. Jena, Designing new materials using atomic clusters, *J. Cluster Sci.*, 10, 477–491, Dec. 1999.

18. M. Schmidt, R. Kusche, B. von Issendorf, and H. Haberland, Irregular variations in the melting point of size-selected atomic clusters, *Nature*, 393, 238–240, 1998.

19. K. Schwab, E. A. Henriksen, J. M. Worlock, and M. L. Roukes, Measurement of the quantum of thermal conductance, *Nature*, 404, 974–976, 2000.

20. M. Berger, *Nanoscale Thermodynamics on the Back of an Envelope*, Nanowerk LLC, Honolulu, HI, 2009.

21. W. Luo, W. Hu, and S. Xiao, Size Effect on the Thermodynamic Properties of Silver Nanoparticles, *J. Phys. Chem. C*, 112, 2359–2369, 2008.

22. S. L. Lai, J. Y. Guo, V. Petrova, G. Ramanath, and L. H. Allen, Size-dependent melting properties of small tin particles: nanocalorimetric measurements, *Phys. Rev. Lett.*, 77, 1, July 1996.

23. D. A. Porter and K. E. Easterling, *Phase Transformations in Metals and Alloys*, 2nd edn., CRC Press, Boca Raton, FL, 1992.

24. A. Rai, D. Lee, K. Park, and M. R. Zachariah, Importance of phase change of aluminum in oxidation of aluminum nanoparticles, *J. Phys. Chem. B*, 108, 14793–14795, 2004.

25. N. S. Phala, and E. V. Steen, Intrinsic reactivity of gold nanoparticles: classical, semi empirical and DFT studies, *Gold Bull.*, 2007.

26. B. D. Chithrani, A. A. Ghazan, and W. C. Chan, Determining the size and shape dependence of gold nanoparticle uptake into mammalian cells, *Nano Lett.*, 6(4), 662–668, Mar. 2006.

27. W. R. Moser, *Advanced Catalysts and Nanostructured Materials: Modern Synthetic Methods*, Elsevier, New York, 1996.

28. S. M. Mukhopadhyay, *Nanoscale Multifunctional Materials: Nature Inspired Hierarchical Architectures*, Nanotechnology Thought Leaders Series, AzoNano, Mona Vale, Australia, 2009.

29. N. Bohr, Atomic structure, *Nature*, 106, 104–107, 1921.

30. C. Kittel, *Introduction to Solid State Physics*, 8th ed., Wiley, Hoboken, NJ, 2005.

31. J. Singleton, *Band Theory and Electronic Properties of Solids*, Oxford University Press, 2001.

32. Z. D. Jastrzebski, *Nature and Properties of Engineering Materials*, 3rd ed., Wiley, New York, 1987.

33. V. V. Mitin, V. A. Kochelap, and M. A. Stroscio, *Introduction to Nanoelectronics: Science, Nanotechnology, Engineering, and Applications*, Cambridge University Press, New York, 2007.

34. Z. Tang and P. Sheng, *Nanoscale Phenomena: Basic Science to Device Applications*, Springer-Verlag, New York, 2008.

35. V. R. Reddy, *Gold Nanoparticles: Synthesis and Applications*, 2006, p. 1791, and references therein.

36. D. Patidar, K. S. Rathore, N. S. Saxena, K. Sharma, and T. P. Sharma, Energy band gap studies of CdS nanomaterials, *J. Nano Res.*, 3, 97–102, Oct. 2008.

37. V. Y. Malakhov, *Structural and Optical Characterization of InN Thin Films: Novel Photovoltaic Materials for Photovoltaic and Sensor Applications*, Kluwer Academic, Norwell, MA, 2003, p. 291.

38. F. J. Owens, Effect of nanosizing on some properties of one dimensional polyacetylene chains, *Physica E*, 25, 404–408, 2005.

39. L. M. Liz-Marzán, Nanometals formation and color, *Materials Today*, Vol 7, Issue 2, pp 26–31, 2004.

40. D. Jiles, *Introduction to Magnetism and Magnetic Materials*, 2nd ed., Chapman & Hall, London, 1998.

41. N. W. Ashcroft and N. D. Mermin, *Solid State Physics*, W.B. Saunders, Philadelphia, 1976.

42. J. D. Jackson, *Classical Electrodynamics*, 3rd ed., Wiley, New York, 1999.

43. W. D. Kingery, H. K. Bowen, and D. R. Uhlmann, *Introduction to Ceramics*, 2nd ed., Wiley, New York, 1976.

44. S. R. Shindel, S. B. Ogale, J. S. Higgins, H. Zheng, A. J. Millis, V.N. Kulkarni, R. Ramesh, R. L. Greene, and T. Venkatesan, Co-occurrence of superparamagnetism and anomalous hall effect in highly reduced cobalt doped rutile TiO_2-δ films, *Phys. Rev. Lett.*, 92, 16, (2004).

45. J. P. Vejpravová, V. Sechovský, D. Niznanský, J. Plocek, A. Hutlová, and J. L. Rehspringe, Superparamagnetism of Co-ferrite nanoparticles, *WDS'05 Proceedings, of Contributed Papers*, Part III, pp. 518–523, 2005.

46. C. G. Hadjipanayis, M. J. Bonder, S. Balakrishnan, X. Wang, H. Mao, and G. C. Hadjipanayis, Metallic iron nanoparticles for mri contrast enhancement and local hyperthermia, *Small*, 4(11), 1925–1929, Nov. 2008.

47. S. Moritake, S. Taira, Y. Ichiyanagi, N. Morone, S. Y. Song, T. Hatanaka, S. Yuasa, and S. M. Mitsubishi, Functionalized nano-magnetic particles for an in vivo delivery system, *J. Nanosci. Nanotechnol.*, 7(3), 937–44, Mar. 2007.

48. S. Iijima, Helical microtubules of graphitic carbon, *Nature*, 354, 56–58, 1991.

49. C. H. Kiang, P. H. M. Van Loosdrecth, R. Beyers, J. R. Salem, D. S. Bethune, W. A. Goddard III, H. C. Dorn, P. Burbank, and S. Stevenson, Novel structures from arc-vaporized carbon and metals: single-layer nanotubes and mettalofullerenes, *Surf. Rev. Lett.*, 3, 765–769, 1996.

50. R. L. Thess, P. Nikolaev, H. Dai, P. Petit, J. Robert, C. Xu, Y. H. Lee, S. G. Kim, A. G. Rinzler, D. T. Colbert, D. E. Scuseria, D. Tománek, J. E. Fischer, and R. E. Smalley, Crystalline ropes of metallic carbon nanotubes, *Science*, 273, 483–487, 1996.

51. A. Rubio, J. L. Corkill, and M. L. Cohen, Theory of graphitic boron nitride nanotubes, *Phys. Rev. B*, 49, 5081–5084, 1994.

52. P. Ball, The perfect nanotube, *Nature*, 382, 207–208, 1996.

53. M. R. Gadiri, J. R. Granja, R. A. Milligan, D. E. McRee, and N. Khazanovich, Self-assembling organic nanotubes based on a cyclic peptide architecture, *Nature*, 366, 324–327, 1993.

54. Z. L. Wang, Zinc oxide nanostructures: growth, properties and applications, *J. Phys. Condens. Matter*, 16, 829–858, June 2004.

55. P. W. Fowler, Carbon cylinders: a new class of closed-shell clusters, *J. Chem. Soc., Faraday Trans.*, 86, 2073, 1990.

56. S. Iijima, T. Ichihashi, and Y. Ando, Pentagons, heptagons and negative curvature in graphite microtubule growth, *Nature*, 356, 776–778, 1992.

57. G. K. Mora, O. K. Varghesea, M. Paulosea, K. Shankara, and C. A. Grimes, A review on highly ordered, vertically oriented TiO_2 nanotube arrays: fabrication, material properties, and solar energy applications, *Solar Energy Mater. Solar Cells*, 90(14), 2011–2075, Sept. 2006.

58. R. Saito, M. Fujita, G. Dresselhaus, and M. S. Dresselhaus, Electronic structure of graphene tubules based on C_{60}, *Phys. Rev. B*, 46, 1804–1811, 1992.

59. N. Hamada, S. Sawada, and A. Oshiyama, New one-dimensional conductors: graphitic microtubules, *Phys. Rev. Lett.*, 68, 1579–1582, 1992.

60. M. Endo, M. S. Strano, and P. M. Ajayan, Potential applications of carbon nanotubes, *Top. Appl. Phys.*, 111, 13–61, 2008.

61. R. H. Baughman, A. A. Zakhidov, and W. A. de Heer, Carbon nanotubes: the route toward applications, *Science*, 297, 787–792, 2002.

62. Y.-P. Sun, K. Fu, Y. Lin, and W. Huang, Functionalized carbon nanotubes: properties and applications, *Acc. Chem. Res.*, 35(12), 1096–1104, 2002.

63. M. J. O'Connel, *Carbon nanotubes: properties and applications*, Taylor & Francis, London, 2006.

64. S. M. Mukhopadhyay, A. Karumuri, and I. T. Barney, Nanotube grafting in porous solids for high surface devices, *Nanotech*, 3, 479–482, 2009.

65. O. Milton, *Materials Science of Thin Films: Deposition and Structure,* 2nd ed., Academic Press, San Diego, CA, 2002.

66. S. M. Mukhopadhyay, A. Karumuri, and I. T. Barney, Hierarchical nanostructures by nanotube grafting on porous cellular surfaces, *J. Phys. D*, 42, Sept. 2009.

67. H. T. Grahn, *Semiconductor Superlattices: Growth and Electronic Properties*, World Scientific, Hackensack, NJ, 1995.

68. R. V. Pulikollu, S. R. Higgins, and S. M. Mukhopadhyay, Model nucleation and growth studies of nanoscale oxide coatings suitable for modification of microcellular and nanostructured carbon, *Surf. Coat. Technol.*, 203, 65–72, Oct. 2008.

69. G. Decher and J. B. Schlenoff, *Multilayer Thin Films: Sequential Assembly of Nanocomposite Materials*, Wiley-VCH, New York, 2002.

70. T. Ando, A. B. Fowler, and F. Stern, Electronic properties of two-dimensional systems, *Rev. Mod. Phys.*, 54, 437–672, Apr. 1982.

71. J. R. Wagner, Jr., *Multilayer Flexible Packaging: Technology and Applications for the Food, Personal Care, and Over-the-Counter Pharmaceutical Industries*, Plastics Design Library, Elsevier, New York, 2010.

72. L. M. Falicov, Metallic magnetic superlattices, *Phys. Today*, 45(10), 46–51, Oct. 1992.

73. U. Hartmann, *Magnetic Multilayers and Giant Magnetoresistance: Fundamentals and Applications*, Springer-Verlag, New York, 2000.

74. P. Holister, C. R. Vas, and T. Harper, Nanocrystalline materials, *Cientifica*, Oct. 2003.

75. C. Suryanarayana and C. C. Koch, Nanocrystalline materials: current research and future directions, *Hyperfine Interact.*, 130, 5–44, 2000.

76. J. R. Weertman, Hall–Petch strengthening in nanocrystalline metals, *Mater. Sci. Eng. A*, 166(1–2), 161–167, July 1993.

77. P. M. Ajayan, L. S. Schadler, and P. V. Braun, *Nanocomposite Science and Technology*, Wiley, Hoboken, NJ, 2003.

78. A. Walther, I. Bjurhager, J.-M. Malho, J. Pere, J. Ruokolainen, L. A. Berglund, and O. Ikkala, Large-area, lightweight and thick biomimetic composites with superior material properties via fast, economic, and green pathways, *Nano Lett.*, 10(8), 2742–2748, 2010.

79. G. M. Luz, and J. F. Mano, Biomimetic design of materials and biomaterials inspired by the structure of nacre, *Philos. Trans. R. Soc. A*, 367(1893), 1587–1605, Apr. 28, 2009.

80. A. Solga, Z. Cerman, B. F. Striffler, M. Spaet, and W. Barthlott, The dream of staying clean: lotus and biomimetic surfaces, *Bioinsp. Biomimet.*, 2(4), 1–9, 2007.

81. T. Mueller, Biomimetics, design by nature, *Natl. Geogr.*, p. 68, Apr. 2008.

82. Z. Guo, F. Zhou, J. Hao, and W. Liu, Stable biomimetic super-hydrophobic engineering materials, *J. Am. Chem. Soc.*, 127(45), 15670–15671, Oct. 2005.

83. Y. T. Cheng, D. E. Rodak, C. A. Wong, and C. A. Hayden, Effects of micro- and nanostructures on the self-cleaning behavior of lotus leaves, *Nanotechnology*, 17, 1359, 2006.

84. C. Mao, C. Liang, W. Luo, J. Bao, J. Shen, X. Hou, and W. Zhao, Preparation of lotus-leaf-like polystyrene micro- and nanostructure films and its blood compatibility, *J. Mater. Chem.*, 19, 9025–9029, 2009.

85. K. Koch, B. Bhushan, Y. C. Jung, and W. Barthlott, Fabrication of artificial lotus leaves and significance of hierarchical structure for super hydrophobicity and low adhesion, *Soft Matter*, 5, 1386–1393, Mar. 2009.

86. J. Y. Rho, L. Kuhn-Spearing, and P. Zioupos, Mechanical properties and the hierarchical structure of bone, *Med. Eng. Phys.*, 20, 92–102, Mar. 1998.

87. S. Weiner and H. D. Wagner, The material bone: structure–mechanical function relations, *Annu. Rev. of Mater. Sci.*, 28, 271–298, Aug. 1998.

88. P. Fratzl, H. S. Gupta, E. P. Paschalis, and P. Roschger, Structure and mechanical quality of the collagen-mineral nano-composite in bone, *J. Mater. Chem.*, 14, 2115–2123, 2004.

89. T. Zimmermann, V. Thommen, P. Reimann, and H. Hug, Ultrastructural appearance of embedded and polished wood cell walls as revealed by atomic force microscopy, *J. Struct. Biol.*, 156(2), 363–369, 2006.

90. R. M. Rowell (Ed.), *Handbook of Wood Chemistry and Wood Composites*, CRC Press, Boca Raton, FL, 2005.

91. A. A. Klyosov, *Wood–Plastic Composites*, Wiley, Hoboken, NJ, 2007.

92. DoITPoMS, Wood as an engineering material, Cambridge University, http://www.msm.cam.ac.uk/doitpoms//tlplib/wood/engr_mater.php.

2

SOCIETAL IMPACT AND FUTURE TRENDS IN NANOMATERIALS

SHARMILA M. MUKHOPADHYAY

Center for Nanoscale Multifunctional Materials, Wright State University, Dayton, Ohio

Nanoscale Multifunctional Materials: Science and Applications, First Edition.
Edited by Sharmila M. Mukhopadhyay.
© 2012 John Wiley & Sons, Inc. Published 2012 by John Wiley & Sons, Inc.

1   BACKGROUND AND SCOPE

The potential applications of nanotechnology are extensive, and it is difficult to identify any type of industry that is not expected to be affected by this field. The field has therefore evoked extensive investigations and review [1–10] in relation to science, technology, and applications. Any emerging technology that can affect so many aspects of our life is bound to bring changes (good or bad) in other areas of society. Therefore, in addition to euphoria and hope for revolutionary new products, this field has also raised a significant amount of fear and skepticism, which has led to research and debates in the social, environmental, political, and ethical aspects of nanotechnology [11–20]. In this chapter we provide the reader with a brief overview of benefits and risks involved.

It is worth noting that nanomaterials are being addressed by every country on the globe that has an active research and development (R&D) program. Governmental agencies are making significant investments, and commercial sectors often surpass them. Since many materials cut across multiple products, and many sizable investments from the commercial sectors are proprietary, it is difficult to make accurate quantitative counts in terms of number of products, market value of each, percentage of the industry sector, and so on. However, some trends are constantly being monitored by nonprofit organizations, think tanks, and business bureaus. Several of these imply that the global market for nanomaterials-based products should cross well over $1 trillion by 2015. Some idea can also be obtained from published databases. Figure 1 shows the number of publications (papers and conference proceedings) related to this

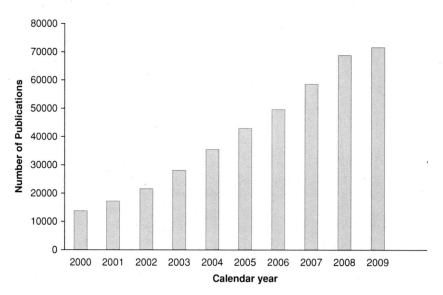

FIGURE 1   Number of publications in each calendar year within the nano-topic area. (Data from ISI Web of Science with Conference Proceedings.) The search criteria settings were kept identical for every year, in order to track only the chronological trend.

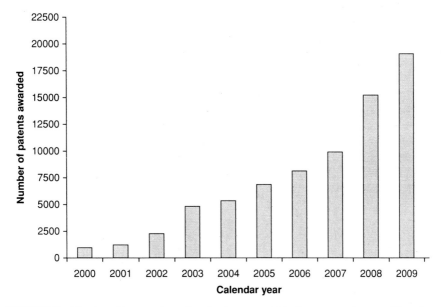

FIGURE 2   Number of patents in each calendar year within the nano-topic area. (Data from the Derwent Innovations Index.) The search criteria settings were kept identical in order to follow the chronological trend in a given database.

field in the last decade, as compiled by ISI-Web of Science. Figure 2 shows the number of patents awarded in the same period, obtained from the Derwent Innovations Index. It can be seen that during the 2000–2009 period, there has been a little over a five fold increase in the rate of publications, and over a 20-fold increase in patents awarded! This indicates that applications and commercialization options are already outpacing scientific inquiry in this emerging area.

The extent of governmental activity in the United States alone can be gauged from the fact that over 15 federal agencies participated in the National Nanotechnology Initiative (NNI) in fiscal year 2010 and the total level of research funding was about $1.8 billion. A widely cited report [21] suggests that federal funding agencies in the United States have made a cumulative R&D investment of about $14 billion between 2001 and 2011. These include not only research, development, and commercialization of nanoproducts, but also understanding of social, health, environmental, and education issues. As an example, NNI has determined that between 2005 and 2010, $480 million was spent in research related to environmental health and safety (EHS), and $260 million was invested in education, ethical, legal, and societal aspects.

The United States is not alone in this game. At least 30 other countries have organized governmental initiatives at very competitive levels. A snapshot from 2005 shows that total governmental spending worldwide was $4.6 billion, and about $1.7 billion of that was from the United States. Governmental spending is expected to have been surpassed significantly by private investments in recent years. Many of these are multinational companies with products and markets that span worldwide.

Since new knowledge and issues related to nanomaterials are unfolding every day, specific winners and losers may not be easy to forecast. However, some overall trends can be predicted. For example, if "necessity is the mother of invention," it is logical to infer that most rapid advances will come in areas where society will have the largest need. Among the most promising candidates to fulfill any need, the long-term winners will be those materials, processes, and applications that maximize benefits and minimize adverse effects.

In light of current activity and information, in this chapter we provide a condensed and simplified summary of current perspectives and future insights in this field. We do not discuss academic details and in-depth studies but, rather, focus on leaving readers from different disciplines with an impression of the wide reaches of this dynamic, multidisciplinary field.

2   NANOTECHNOLOGY FOR SOCIETAL BENEFITS

In this section, a few selected areas of societal importance are listed, along with the ways in which nanotechnology may make a positive impact. Some examples are discussed at length later in the book. Here, the goal is to provide a summary list of materials, systems, and applications that can make a global impact. It can be seen that, often, the same material occurs in multiple examples, which underscores its versatility and multifunctionality.

2.1   Clean Water

The population of the world has grown from 3 billion to almost 7 billion in the last 50 years alone. With this type of explosion, water is already scarce in many parts of the world. Moreover, freshwater supplies are becoming increasingly contaminated with industrial and other wastes. This scarcity is likely to become more global, and acutely so for future generations, unless clean water technology is improved significantly and distributed at low cost. Nanomaterials are already providing several solutions in this area, and many more are looming on the horizon [22–24]. Some examples are listed below.

- *Nanocatalysts and nanosubstrates.* Larger forms of catalytic materials, such as pellets and cartridges, have been around for many years. However, nanoparticles of these, either added to water or (preferably) supported on suitable filters, membranes, or other structures can increase catalytic activity significantly. Porous meshes made of nanotubes and nanofibers serve as excellent supports. These can provide extremely high surface activity for small amounts of nanocatalyst particles, along with the structural stability of a larger robust system. Moreover, multiple nanoparticles, each targeting a different pollutant, can be incorporated on one substrate to provide a broader spectrum of benefits. Some examples of metallic and carbon-based systems are provided in other chapters.

- *Electrodes with nanoscale fibers for deionization and desalination of water.* Carbon-based materials, discussed in several chapters in the book, play a big role in this technology, and some prototype desalination systems are available. There is the additional possibility of combining these with nanoscale photocells in order to provide this service in remote areas at minimal cost.
- *Advanced nanoporous materials for targeted filtering.* Removal of bacteria, viruses, and contaminants (organic or inorganic) is an ever-growing need. A large family of products that benefit significantly from modern nanostructures are available. Well-known nanominerals such as clays, zeolites, and gels can be combined with newer bioderived as well as synthetic nanostructures for tailored filtration. Many purification systems can be made robust and portable enough to be deployed rapidly in disaster-stricken and/or underdeveloped areas.

2.2   Food Supply

Food consumption is expected to increase in the world due to population growth. At the same time, the shortages of water discussed earlier can threaten our food production capabilities unless addressed through advanced technology. Nanomaterials can help enrich the food supply in several ways [1,8,9,25].

- *Improved greenhouse agriculture.* It is well known that conventional agricultural practices across the globe have always been inefficient and environmentally unfriendly. They consume large amounts of land, need pesticides and fertilizers, and depend strongly on weather patterns. Nanomaterials can help this by expanding the scope and capabilities of greenhouses. This can be achieved through the new generation of nano-components, such as controllers, sensors, building materials, seeds, and culture media. Such improvements will make it easier for farmers to produce fresh food locally and year round. This can reduce transportation costs, storage, and spoilage issues while reducing land consumption and pesticide use.
- *Infusion of nutrients in foods.* Nanoscale ingredients can be used to deliver vitamins and other nutrients inside natural food. This is different from the traditional (unappetizing) concept of completely artificial foods and supplements. Note that this area is expected to progress slowly, because of necessary health and safety studies (discussed later). However, once such studies have been completed and specific nanonutrients approved, future benefits can be great. There is the possibility of obtaining nanomaterial-fortified natural food that is intelligently customized for individual nutritional needs.
- *Nanomaterials for easier packaging and transportation of food.* Examples in this category include antimicrobial agents embedded in food wrapping to prevent spoilage and diffusion barriers in containers to keep food fresh for longer times. Such technologies, carefully administered so as not to contaminate the food, can reduce the need for pesticides and preservatives in the food itself.

• *Nanosensors for food safety monitoring.* The very significant impact of nanomaterials on the sensor industry is already visible, and growing every day. Among these are sensors that can detect harmful bacteria or toxins inside the food and/or in processing and storage areas. It is also possible to design sensors that can detect tampering or the use of harmful chemicals during production.

2.3   Energy Needs

This need for clean, renewable, and efficient energy has been pressing the entire world for several decades. This need will keep increasing with global growth in population and industrialization. Dependence on fossil fuels to meet our increasing energy needs can be blamed for many international conflicts. Therefore, alternative solutions can play a pivotal role in immediate geopolitical relationships in addition to long-term well-being. In this section, we provide a general list of energy-related applications [5,26–35] of nanomaterials. Selected examples of some of these are discussed in depth in Chapter 8. Energy applications have been divided into two parts: one related to applications in energy generation, storage, and transmission, and the second related to advancements for energy efficiency and savings.

Energy Generation, Storage, and Transmission

• *Stronger and lighter wind-turbine blades made from nanocomposites.* The turbine blades can be improved in many ways by using nanocomposites. Several nanocarbon-based composites are known to be stronger, tougher, lighter, and more damage tolerant. Research is also under way to make them multifunctional, such as lightning and corrosion resistant, electronically wired for automatic de-icing, and capable of storing off-peak energy.

• *Nonporous carbon and other cage-type materials.* Safe storage and transportation of hydrogen is a current bottleneck for many fuel cell and portable energy applications, and several investigations show that this can be done with fullerene and buckyball type structures. This may enable wider use of hydrogen energy devices in cars and buildings.

• *Fuel cell components.* Porous and nanostructured substrates are becoming essential for electrodes, catalysts, and other components of most fuel cells.

• *Flexible and efficient solar panels.* It is expected that future improved solar panels will use quantum structures as solar cells, luminescent nanomaterials as solar concentrators, nanowire connectors, and nanoparticles for spectrum conversion. Therefore, progressive improvements in these technologies may make solar power affordable in the near future.

• *Charge storage components* (batteries and supercapacitors). Most new designs for lithinumion and other solid batteries involve nanoscale heterostructures such as nanofiber paper (bucky-paper) and C–Si nanocomposites. Moreover, supercapacitors involve nanostructures having extremely high surface areas. These components are making rapid progress in powering energy-hungry devices, including transportation vehicles.

- *Energy transmission devices.* Many of the high-temperature superconductors developed for next-generation cables and coils are fabricated as nanolayers obtained by thin-film deposition techniques. Most of these electronic oxides also involve nanoscale defects for flux pinning. Also, it is expected that wires and cables made of copper may be replaced by weather-resistant, high-conductivity carbon nanotubes and nanofibers in the future.

Energy Efficiency and Savings   A large fraction of our energy consumption is in housing and transportation. Therefore, additional energy-saving applications are listed in the "Advanced Architecture" and "Transportation" sections.

- *Illumination devices.* Light bulbs and light-emitleing diode (LED) devices are being made from nanoscale structures such as semiconducting quantum dots or nanoparticles of photoluminescent materials. These devices already provide luminosity comparable to or higher then that of conventional lighting fixtures at fractional power and with minimal heat loss. This area is likely to show rapid improvements based on the amount of R&D activity.
- *Nanocomposites for weight reduction.* Strong, lightweight, tough, and corrosion-resistant nanocomposites can significantly reduce the weight, functionality, aesthetics, and durability of many structural components. These can enhance products for the construction, automotive, aerospace, sporting goods, and other industries.
- *Nanostructures for thermal management.* Graphitic carbon has extremely high in-plane thermal conductivity. Nanotubes and nanosheets based on controlled orientation of these planes can be powerful nanoradiators. These are being used to design superefficient cooling systems for electronics and aerospace applications.
- *Insulation and thermal barrier.* Several nanoceramic materials are used to deposit dense films and hybrid layers for thermal barrier coatings in engines. These can increase the energy efficiency of the engines. Additionally, many foams and gels with nanostructured porosity can improve insulation properties of walls and panels, thereby increasing the energy efficiency of storage containers and buildings.

2.4   Advanced Architecture

With increasing population comes increased demand for all types of buildings: residential, commercial, industrial, and others. Nanotechnology can also help in this area [1,32,33] by making many of the building materials cheaper, easier to use and faster to build, safer, and more energy efficient. Some examples are given below.

- *Aerogels and nanogels.* These nanoscale materials have superior thermal insulation, sound absorption, and an aesthetically pleasing look. They are often used to make translucent panels for modern buildings.
- *Heat-absorbing windows.* Transparent nanomaterials that can absorb and store solar heat can provide more than insulation. Advanced versions including built-in electronics can even channelize this energy for other functions.
- *Paint-on photovoltaics.* Some paints are being investigated that will have built-in optoelectronic materials. These can collect solar energy and convert it to power, which can in turn be used in basic functions of the building.
- *Electrooptic windows.* Certain materials (often as nanoscale colloids) can be made to switch between transparent and dark. These can be used as windows that respond to daylight and conserve energy.
- *Built-in sensors.* Several sensors can actually be built into a building for safety monitoring as well as for signaling the structural health of a variety of components. Alerting to upcoming maintenance issues can provide major advantages.
- *Unprecedented lighting devices.* In addition to the energy-efficient lighting mentioned earlier, unique lighting components can be built into walls and fixtures. Some are programmed to change the color and ambiance of the interior electronically; others can make multifunctional use of the same living space through such capabilities.
- *Nanocomposites and nanocoatings.* Nanoparticles are being used as reinforcements and protective coatings for building materials such as wood, glass, metal, and concrete.

2.5   Transportation

The transportation industry (automotive, railways, and aerospace) and its associated infrastructure consititute a major part of modern life, and continuous monitoring and innovation are necessary in this industry. Key drivers in this sector are expected to be issues such as reduction in weight, improved fuel efficiency, safety, service life, recyclability, reduced emission, better performance, and aesthetics. Nanomaterials can help in each of these areas [34,35]. Some examples are given below.

- *Reduction in weight.* Nanocomposites can play a big role in increasing strength/weight ratios of automotive and aerospace body parts while making them tougher and more corrosion resistant and durable. Careful tailoring of nanocomposites can make the vehicle safer, more fuel efficient, and less polluting.
- *Sensors in vehicles.* Specific to moving vehicles, critical sensors such as load balance, braking systems, air bags, and oxygen masks can be made smaller, cheaper, and more efficient by the use of nanomaterials. Smaller and cheaper sensors imply that many more such features can be embedded in vehicles.
- *Improved tires.* Carbon nanofibers added to tires can increase their wear life and friction properties more than the carbon black added to tires today.

- *Special paints and surface coatings.* Nanoparticles are added to paints and coatings for tailoring properties such as scratch and indentation protection. This concept is suitable at all levels: from mini-meteor impacts in spacecraft to minor bumps in cars.
- *Traffic lighting and signaling.* These areas can be enhanced significantly by various types of nano-LEDS and displays, as well as by lasers and signal communication tools created from nanomaterials.
- *Enhanced concrete for pavements and runways.* This may include embedded nanoparticles of special cement for self-healing after water leaks in cracks and water-repellant coatings, grooved pavements with nanoinclusions for tailored strength and durability and so on.
- *Self-monitoring infrastructures.* With the availability of carbon nanofibers and their integration with electronics, built-in sensors and triggers with multiple capabilities are possible. These may range from monitoring the structural integrity of bridges and tunnels to self-triggered de-icing.

2.6 Environment

The environment is of genuine concern to most countries around the world. Modern lifestyles involve a large number of industrial processes (which produce pollutants) as well as disposable products (which generate trash). One way of reducing these drastically would be to go back in time to a preindustrialized society, but that is neither possible not desirable. Therefore, the answer to environmental pollution is twofold: wider deployment of green products and improved remediation and cleanup efforts. Nanomaterials can make several contributions to each of these areas [24,36–41], some of which are listed below.

Green Products   These can be biodegradable and/or recyclable. These are also products that produce minimal pollutants during production or operation. Following are some that can be improved by nanomaterials.

- *Biodegradable plastics and packaging.* Due to their superior properties, plastics are preferable to paper and other degradable packaging materials in many applications. It will therefore be very desirable if some plastic and polymer products can be made biodegradable. This is possible in principle, but current processes are not economical. Recent advances in nanotechnology make it easier to adopt biological routes in synthesizing many biodegradable polymer components and provide ways of combining degradable products with plastic. These approaches together may reduce the amount of garbage created in the future.
- *Longer-lasting products for reduced trash generation.* Nanomaterials have the capability to increase the durability and usable lifespan of many consumer products and electronics, such as batteries, cell phones, remotes, and casing materials.
- *Improved recycling capabilities.* Several nanomaterials increase the potential of recycling other waste materials. Examples include nanomaterials combined

with recycled plastics for making useful by-products, and nanomaterials used to react with other industrial or agricultural by-products to form composites, gels, fillers, or fuels. It is worth noting here that although recycling of nanomaterials themselves is discussed widely in the literature, solid recycling of other components enabled by nanomaterials is at present not getting the attention it deserves.

- *Pollution reduction.* It has been mentioned that nanoscale particles are often superior as catalysts and sensors, due to their high specific surface area. These can therefore result in more efficient catalytic converters for vehicles and industry.

Environmental Cleanup and Remediation   Although minimization of trash and pollutants is essential, it is not realistic to expect that these will be reduced to zero in any industrialized society. Some amount of pollution, either routine or accidental, is to be expected. Moreover, the pollutants already added to the environment will be circulating thronghont the planet for years to come. It would be highly beneficial if our present cleanup and remediation techniques could be made more economical, effective, and widely applicable. In several respects [24,36,37,40,41], nanomaterials can provide new solutions.

- *Nanostructured membranes and filters.* Some of these can be designed to separate and/or absorb contaminants such as oil or other contaminants from water. Recent industrial accidents have shown that readily adaptable technologies should be available for unexpected situations such as chemical spills. Since hybrid filters, together with functional coatings to control their wettability and absorption, are becoming more readily available, it will be possible to keep an arsenal of deployable engineering systems containing meshes, filters, and sponges that can either separate or absorb large amounts of pollutants from water.
- *Nanoparticle slurries.* Some slurries made from nanometal suspensions can be injected into soil to change its pH. This approach can expedite the breakdown of certain harmful molecules in soil.
- *Self-assembled monolayers.* Some self-assembled monolayers supported on appropriate substrates have the ability to bind or neutralize harmful species in fluids.
- *Nanoporous filters.* Several organic–inorganic hybrid materials that have nanostructured porosity have been created for heavy metal removal from industrial wastes.
- *Photocatalytic particles for degradation of organic and inorganic components.* Semiconducting oxides such as titania have been known to degrade a large number of toxic materials in the presence of ultraviolet (UV) light. Nanoparticles of these materials can be modified to become active in visible wavelengths, which will increase their efficiency in natural sunlight.

2.7   Electronics, Sensor, and Electromechanical Devices

At present, electronics and sensors control most industrial processes and products. In the future, control of processes and products may be enhanced further by electromechanical components. Therefore, fundamental advances in these devices are bound to affect all areas of life. Many of the earlier sections have mentioned sensors and controls for specific applications, so they have not been repeated in this section. A few fundamental ways in which nanomaterials can affect future electronics, sensors, and other unique devices [42–55] are listed below.

- *Miniaturization of all electronic components by denser packing of devices.* The latest transistors and other components in very large scale integratedv (VLSI) devices are already in the 45-nm regime. This size reduction will continue as our nanofabrication capabilities increase. More dramatic change could be gradual replacement of silicon chips with completely new substrates, such as carbon nanotubes and/or graphene layers. These components will revolutionize the architecture of future electronic devices.
- *Speed of signal transfer.* Some device experts believe that in addition to the proximity of smaller components, the higher electrical mobility in graphene suggests that all electronic operations can be made faster and more efficient. This translates to reduced heat dissipation as well as increased computing speed.
- *Unprecedented electronic components through bandgap engineering.* At nanoscale dimensions, the electronic bandgap of semiconductors can be controlled by their dimensions (quantum confinement), multilayer geometries (superlattices), and core–shell structures. These can result in new types of electronics and optoelectronics for future integrated devices, well beyond current VLSI, lasers, and optical electronics.
- *Molecular electronics.* A combination of new understanding in molecular self-assembly and single-molecule devices can lead to memory, logic, and amplification functions performed by individual molecules.
- *Sensing with nanoparticles.* Semiconducting oxides and other materials capable of detecting signals always benefit from increased surface area. The microporous devices used today can be enhanced significantly if they can be deposited on higher surface-area nano-structured substrates. Such sensors, backed by additional nanoelectronic circuitry, can make giant strides in both the selectivity and the sensitivity of all sensing devices.
- *Nanoink for printable electronics.* Nanoparticles suspended in liquid or slurry with tailored rheological properties can revolutionize the field of electronic printing. A great variety of electronic devices printed on suitable substrates (i.e., plastic, paper, fabric, etc.) can add electronic functionality and precision to everyday household objects.
- *Unique nano electro mechanical structures* for future products. Biological systems have always made use of extremely efficient molecular components for electromechanical functions. Mimicking these in engineered systems has been

difficult. Many of the likely candidates are designed and synthesized around DNA molecules, peptides, carbon nanotubes, or a combination of these. Advances in nanomaterials are bringing several of these machines closer to reality [47–55]. A few examples are given below.

- *Mechanical movements based on biological motors.* Myosin is a molecular motor that can move along a track created by actin filaments. Similarly, a flagella motor allows movement in bacteria. Molecular systems also have rotary molecular motors such as ATP synthase. These chemical–electrical–mechanical energy conversion machines have been known to the biological community for decades, but nanotechnology can make it possible to use these proteins to build natural–synthetic hybrid systems suitable as mechanical devices, sensors, and actuators.

- *Nanopropellers.* When molecular end groups having specific shapes are attached to rotating molecular shafts, they may be able to push fluid forward during rotation. Such devices are being designed jointly by computational experts and chemists.

- *Tweezers.* These are molecular strands (mostly DNA molecules) that can latch on to each other at specific locations. The unlatched parts are open and dangling until another strand or molecule is introduced that can clamp these ends. Thus, "clamping" and "unclamping" of points can be controlled by switching DNA strands.

- *Nanoswitches.* These are molecules that can be cycled reversibly between two or more different states. The switching may be triggered by pH, pressure, light, current, or other stimulants.

2.8   Next-Generation Computing

Computing influences all other engineering fields by enhancing our capabilities. Computing power influences directly how well we can control devices and processes, perform modeling and simulation of complex issues, analyze data, and disseminate information. Evolutionary changes in computer speed and other characteristics are related to the advances in electronics mentioned earlier. Some revolutionary changes are believed to be possible in the field of future computing due to current advances in nanomaterials [55–59]. Some of these advances are related to the areas of electronic hardware and molecular–biological interfacing, as highlighted earlier. A few additional discussion points inherent to computing capabilities follow.

- *Information storage capacity.* Our data storage devices have been getting progressively smaller at a steady pace. Many silicon-based devices already have nanoscale components (45 nm at present), which can be reduced still further, up to a point. However, other nanomaterials are capable of creating a sudden jump in hardware design in this area.

- *Completely new species of computers.* Some unique ideas are being explored by various groups around the world. One thought is to use carbon nanotubes assembled on templates created by DNA molecules. Another is to use nanowires that can connect to, and communicate with, biological neuronal signals. It is unclear at this time which of these concepts will finally rule the computer world, but areas of neuronal computing and molecular computing are emerging.

- *Quantum computers.* This moves away from the conventional bit-by-bit data processing of today, where every unit of information is stored as binary (0 or 1) logic. Quantum computing enabled by nanomaterials can change this to "qubits," where electron spin or some other quantum mechanical state parameter can be used as data. Theoretically, these can encode information in more than two states (such as 0, 1 and superposition of 0 and 1), thus providing an explosive increase in parallel-processing power.

2.9   Health Care

Health care is an area of science of great interest to the world with a growing aging population and increasing medical needs. This is also a field expected to be very strongly affected by nanomaterials given that many biological functions in our bodies operate at these scales. The range of applications is very broad, so we provide only a few examples in the fields of internal medicine and surgery.

Internal Medicine   Nanomaterials can be functionalized with specific organic molecules and then targeted to specific tissues in the body. This is a very active field [59–64], and new developments are reported every day. Advances in nanomaterial functionalization are effectively leveraged with current progress in genetics, informatics, imaging, sensing, and robotics. A few examples are given below.

- *Nanoparticles functionalized for targeted drug delivery.* Nanoscale carriers (called *vectors*) are used in medicine to deliver functional therapeutic chemicals to specific cells inside the body. These hybrid nanoparticles can reach selected sites inside the body (e.g., cancerous sites) carrying specific drugs with them. Viruses (nanoparticles made from biomolecules) have been used for many years, but synthetic agents such as nanoparticles of gold and magnetic materials are revolutionizing this field. Gold nanoparticles can play a pivotal role in future applications because of their biocompatibility, purity, and ability to bond with multiple receptors, such as peptides, aptamers, and antibodies. This is true also for the rapidly expanding group of magnetic nanomaterials, which can be guided and detected with magnetic fields. The emerging family of nanoscale particles and colloids can improve the accuracy of delivery and localization within cells, thus minimizing adverse side effects on neighboring healthy tissues.
- *Tissue imaging and diagnostics.* Several of the nanoparticles that can be targeted to specific sites can also be used as diagnostic agents. For example, gold particles that have concentrated at certain sites can be used as a contrast agent for x-rays. This has the potential to create more location-specific images then are available using traditional barium- or iodine-based compnted tomographic (CT) imaging.

Similarly, magnetic nanoparticles targeted to specific sites can enhance the resolution and contrast of magnetic resonance images.

- *Nanomaterials for hyperthermic treatment.* Once nanoparticles, especially metallic ones, are localized at a particular site in high enough concentrations they can be activated by a magnetic field, light, ultrasound, or x-rays, depending on the metal. These allow particles to be targeted at tumors and other diseased sites, which can be cauterized selectively without damaging surrounding healthy tissue.
- *Dendritic polymers.* These are molecular chains having treelike branched architectures. Branches, which typically range from 2 to 15 nm in size, can be loaded with multiple drugs simultaneously and delivered to the tissue level in unique ways.

Surgery   The utilization of nanomaterials in surgery is expected to become a very rich field, since products can range from miniaturized and superprecision surgical tools to advanced implants and cultured tissues [59,64–69]. A few notable improvements in the horizon are listed below.

- *Adhesives and welds for tissue.* Investigations under way indicate that suspensions of nanogold-supported molecules in a suitable medium can be light-activated to join tissues. Similarly, specially formulated polymers can seal cuts on the skin without stiches. Some of these nanomaterials may be enhanced further by built-in antimicrobial or healing agents.
- *Advanced visualization and sensing tools.* This area can be truly revolutionized by the availability of nanoscale devices and sensors that may have the ability to test individual molecules within cells and to monitor different biological functions simultaneously to provide more holistic information. For example, luminescent tags made from quantum nanodots as well as other imaging tools can significantly increase the scope for endoscopic and robotic surgeries. The unprecedented optical, magnetic, and electronic capabilities of quantum structures (dots, wires, and wells), as well as advanced sensor designs and detection systems being developed today, can make it possible to take medical diagnostic tools to a completely new level.
- *Advanced lasers and cutting tools.* Cutting tools involving optical lasers and nanomechanical devices are making the surgeon's knife more and more precise. Currently, laser-based tools and nanorobots (nanobots) are being investigated which can, with the help of sophisticated guidance systems, take future surgical procedures to truly cellular levels.
- *Magnetic nanostructures for mechanical manipulation.* Magnetic particles and nanowires can be functionalized to enter the body and attach to particular cells. They can then be subjected to external magnetic fields in order to produce a controlled amount of mechanical force on the attached cells for various biological functions.

- *Biocompatible coatings on surgical tools and implants.* Examples include nonthrombogenic layers on surgical tools to prevent blood clots, and protein-attaching layers on implants to hasten tissue growth. These technologies can reduce pain and trauma during surgery while accelerating healing.
- *Nanostructures for tissue engineering.* Substrates, meshes, and composite scaffolding made from biocompatible nanofibers (carbon, peptides, DNA, etc.) can be used in implants or grafts to provide direction for cell growth and promote rapid healing. In some cases, antimicrobial agents or other drugs can be embedded in these structures for additional benefits.

2.10 Consumer Products

This is an industrial sector that is already selling well over a few hundred products involving nanomaterials [70–73]. It is worth noting that the impact of some of these materials on human health and the environment is still unclear and thus needs to be actively investigated, as discussed in the next section. Below is a list of familiar household items that may incorporate nanomaterials.

- *Sunscreens and cosmetics.* Nanoparticles of ZnO and TiO2 add reflectivity and "glow" to skin, while blocking ultra-violet rays. Therefore, they are useful in sunscreens and cosmetics. Also, some nanometallic powders have color-enhancing and flaw-concealing effects suitable for cosmetics.
- *Antibacterial fabric and appliances.* Nanoparticles of silver are being incorporated into washing machines, socks, and clothing fabrics due to their antimicrobial properties.
- *Nanocoatings for specialty properties.* Functional nanocoatings can make useful materials hydrophobic (water repellent), hydrophilic (water absorbing), oleophobic (oil repellant), and so on. These types of treatments are becoming common in fabric, household items, and food containers.
- *Nanoparticles in cleaners and surfactants.* Many types of advanced cleaners, stain removers, stain repellents, and finishing sprays used on clothes, stone, concrete, cars, and many other household gadgets already contain nanoparticles. This list is likely to increase.
- *Sporting goods.* Tennis rackets, golf shafts, golf balls, bicycle helmets, fog eliminators, goggles, and binoculars are only a few commercial products that already incorporate nanocomposites and nanocoatings.
- *Luggage and accessories.* These items need to be strong and light and have aesthetically pleasing colors and textures. It is becoming common to use polymer–matrix nanocomposites to build them and to add various nanocoatings for surface finishes.
- *Cookware and kitchen ware.* Several nanomaterials are used in high-end kitchen products to impart nonstick, antibacterial, heat-resistant, microwave-tolerant, and other properties to these products.

3   ISSUES TO WATCH

All new opportunities bring new challenges, and the final impact of anything on human society depends on our ability to balance the two sides. The possibility of adverse effects can bring fear, skepticism, and inevitable conflict between the enthusiasm of product developers and the caution of community watch groups. The cause of such growing pains in a new technology is the lack of information available. Historically speaking, lessons learned from earlier engineered materials, such as asbestos, pesticides, and other chemicals as well as natural particulates such as ash and dust, can provide some insights. However, it must be noted that modern nanomaterials come in a very wide range of compositions, shapes, structures, and surface functionalities, and can therefore pose more complex challenges.

Another special concern with nanotechnology is that the rate of new materials introduced into the market is unprecedented, and is expected to continue. This rate significantly outpaces the rate at which new studies can be carried out to understand their full impact on health, the environment, and other aspects of society. While the insufficiency of knowledge makes it a very interesting field for scholars, it does raise a red flag of caution for consumers. It is therefore very important that we, as a society, maintain our vigilance for possible adverse effects and risks at all levels. The idea of encouraging such discussions should not be to spread fear and stop R&D activity, but to encourage simultaneous investigations into possible side effects and mitigation strategies.

As each nanomaterial is ready for scale-up, discussions regarding its influence on public health, usage, fate in the environment, and economics should be encouraged. These debates bring out loopholes and blind spots that can be addressed right from the beginning, before larger problems emerge. Several journal reviews as well as market-survey and case-study types of reports have been conducted in this area [73–84], and many more are likely to emerge in the near future. In this section we touch on a few key topics. This is not meant to be an in-depth or scholarly discussion, but brief summary for general awareness by the scientific community.

3.1   Influence of Nanomaterials on the Environment and Health

As larger quantities of nanomaterials are produced, distributed, and used in industrial processes, a measurable fraction will end up as waste: in air, water, and/or soil. This can subsequently reach our bodies through several possible pathways, some of which are indicated schematically in Figure 3. These possibilities should be minimized because the long- and short-term effects of nanoparticles circulating through the body, or building up in internal organs, are largely unknown. This makes the following issues particularly important and worthy of active investigation in the coming years.

Release of Nanomaterials into the Environment   The extent of risk or exposure from each nanomaterial will depend on the products that use the material. Sometimes,

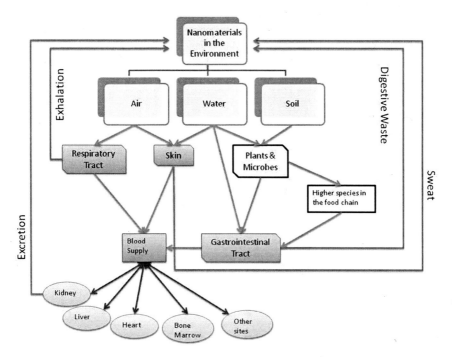

FIGURE 3   Possible pathways for exchange of nanomaterials between the environment and the human body.

multiple products may use the same material, and the risk from each may be different. The following pieces of information are essential for each product.

- The list and amounts of various nanoscale materials used to manufacture the product.
- Understanding which of these materials can escape into the environment.
- For each potential pollutant, at what stage it is likely to be released: during manufacture, during use, or after disposal. It is likely that different quantities will be released at different stages.
- Some products may be benign during a normal life cycle, but may become dangerous in an accident such as a fire or explosion.
- The surroundings where pollutants might be released is important: whether released directly into the body such as through skin, lungs, or food, or indirectly through the environment such as air, water, or soil.

Fate of Each Nanomaterial in the Environment   Once some nanomaterials are released into the environment, how they travel through the ecosystem is an essential area to be investigated. This is emerging as a major topic for research by the environmental sciences community, and is expected to grow as the list of nanomaterials in production grows. Some of the questions being addressed are:

- What are the fate and transport of materials through air, water, and soil?
- If and how are they transferred through the food chain into higher plants and animals?
- Is there potential for entry into groundwater or a change of nutrient levels in water?
- Is there a possibility for acid rain or ozone depletion?

A few of these are beginning to be addressed by some investigators, but it must be noted that the scientific community has seen only the tip of the iceberg. Many answers from these types of questions will lead into issues listed in the next section.

Possible Health Effects   Unfortunately, the very properties of nanomaterials that make them potential building blocks of future medicine (e.g., size compatibility with cell bioproteins, surface activity, possible intake into specific tissue sites) can be the properties that make them potentially damaging for health. Some major issues and concerns are:

- *Active, but unknown biokinetics.* Uptake and translocation routes for nanomaterials are many; a few obvious ones are indicated in Figure 3. For nanomaterials entering the body, the fraction that is retained versus. that which is expelled is often the first question. Of the part that stays in the body, the fraction that is absorbed by the bloodstream and transported to secondary organs and tissues is the next question. How well these can be excreted from each of the different organs poses the next level of unknowns. Those that do not clear from the body may accumulate in certain areas over time and interfere with biological processes. Ideally, each of these steps needs to be investigated for each biokinetic pathway for every nanomaterial that is expected to be produced in large quantities. Some steps are easy to determine, but others may be more complicated, and scientists often have to rely on hypothetical models.
- *Variations within the same material.* The complex pathway of each nanomaterial is further complicated by the fact that the same material may have multiple variations. Transport properties of nanoparticles are very sensitive to their surface chemistry, which depends on processing, storage, and the environment. which alter their translocation and kinetic mechanisms. Therefore, the same nanomaterial may easily enter certain tissues with one surface treatment, and be expelled harmlessly by the body with another surface treatment. This type of investigation can provide very important guidelines in future production, treatment, and disposal of nanomaterials.

Strategies for Prevention and Mitigation   While issues of nanomaterial pollution, fate, and risks continue to be addressed, very significant aid to society in general can come from R&D related to the prevention, capture, and mitigation of these materials. Fortunately, the fundamental scientific knowledge regarding the nanomaterials used for product development can be fed directly into some of these strategies. Some lines of investigation are listed below.

- *Sensors and traps.* If the size, shape, surface activity, and other parameters. of a material are known, it should be possible to design sensors that detect them, or filters and traps that capture them.
- *Surface treatments for passivation and isolation.* In relation to biokinetic studies mentioned earlier, it may be beneficial to study which types of surface treatments can make specific nanoparticles harmless by making them either difficult to absorb into the body or easy to excrete.
- *Safety guidelines and data sheets.* These have been slow to evolve. In the United States, the agency that provides such information for most materials is the Occupational Safety and Health Administration (OSHA). For nanomaterials, the OSHA web site [79] acknowledges that "the potential health effects of such exposure are not fully understood at this time." However, OSHA, along with several research groups and agencies around the world, are constantly conducting studies, exchanging data, and making progress. This progress should be monitored closely by the larger community.
- *Investigation into methods of recycling nanomaterials in creative ways.* Historically speaking, complex products (which nanomaterials are expected to be) can be difficult to recycle as is. They may be reused by incorporating them into other useful products such as composites or fillers. This field of using waste or surplus nanomaterials made for one product into secondary by-products may become a natural extension of mainstream nanomaterials research.

3.2   Management Issues in a Disruptive Technology

Nanotechnology is capable of spinning off new products at an alarming rate, and businesses communities as well as regulators are taking heed of that [85–89]. According to some recent studies, in 2007 U.S. companies used nanomaterials in products worth $59 billion. Other producers were Europe ($47 billion), the Asia–Pacific ($31 billion), and the rest of the world ($9.4 billion). The numbers for the United States and Europe are each expected to cross $1 trillion by 2015, with Asia following closely behind.

The complexity and breadth of new products are already putting a strain on intellectual property and patent management issues in the science and engineering community. Patent data shown in Figure 2 already indicates a more than 20-fold increase in the last decade. This pace is likely to increase in the near future, and managing this growth is no trivial matter.

The rapid number of products also implies that many existing or conventional products are going to be displaced or rendered obsolete. This type. of disruptive

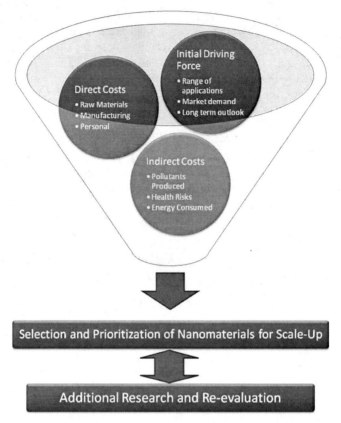

FIGURE 4 Information that should be filtered and used to make balanced regulatory decisions about individual nanomaterials.

change can result in many short-term issues, such as a shortage of trained workforce, deficiencies in supply chains, and infrastructure inadequacies. The current workforce may need some retraining, and the upcoming generation of workers needs to be educated differently, as discussed below. In many cases, supply chain management and business protocols will have to be reengineered. New infrastructure issues for a growing demand by cutting-edge technologies will need to be addressed. In addition, there is likely to be some resistance from groups that benefitted from earlier technologies.

Business ethics and social responsibility issues also become important before deciding to move forward with particular nanomaterials and products. Figure 4 is a schematic of factors that will need to be considered for nanomaterial selection and prioritization. To make the most balanced decisions, the business community, product developers, environmental scientists, toxicologists, and regulatory agencies will need to work in concert.

3.3 Educating the Workforce for Changing Technology

One of the issues that accompanies any rapidly growing new technology is a sudden need for qualified professionals. Pioneering business leaders often move very fast to create new products, leading to initial commercialization and growth, but subsequently there is a damping effect, due to a talent shortage. In such situations there can be a geographical shift of leadership toward the region that can produce or attract the most talented workforce. If no region can cope, the industry as a whole can slow down. This challenge can be especially true for nanotechnology, since this field needs a scientifically savvy workforce at all levels. Based on these thoughts and ideas [90,91], some educational aspects related to this area are highlighted below.

- *Need for technical capability.* This field is still growing, and it has not matured enough to offer assembly-line or routine operations. Therefore, workers in this area will require relatively strong analytical skills and will need to be good learners. A relatively strong foundation in science, technology, engineering, and mathematics will be required.
- *Multidisciplinary education.* In this field, students have to be able to see the link among disciplines; therefore, a more multidisciplinary training will be useful. Moreover, all students should have some awareness of all the different aspects of this field: science and engineering, together with environmental and health impacts, as well as business, societal, and ethical issues.
- *Cross-demographic and cross-cultural training.* This field currently spans the entire globe. It must be noted that some aspects of nanotechnology development will need expensive infrastructure and may focus around world-class hubs. However, many other functions are likely to be very agile, and may easily be moved to locations based on workforce, raw materials, cost, and local regulations. Therefore, it is very likely that most products will be developed in an international setting, forcing workers to interact globally rather than locally. This field may therefore become the testing ground for international collaboration and development, requiring some amount of cross-cultural experience for the workforce of any country.

3.4 Prevention of Intentional Abuse of a Very Powerful Technology

The threat of any powerful technology falling into the wrong hands is always looming [92], and this issue is all the more real for nanomaterials, given how powerful small quantities of these materials can be. While nanomaterials are looked upon as potential solutions for terror threats (i.e., better sensors, mitigation agents, ammunition etc.), they should themselves be safeguarded against terrorists. The most effective deterrent to abuse by antisocial groups may be understanding and preparedness on the side of military, public security, and law enforcement agencies. Some aspects that need to be stressed are listed below.

- Nanomaterials that can serve as effective vectors to carry therapeutic agents for healing the body can also be used to introduce toxins into the body. Therefore, widespread availability of such materials poses an enhanced threat of biowarfare unless security and controls are placed appropriately.
- Nanoparticles of some compositions are already known for their energy-dense and explosive properties, and for potential use in ammunition. Production, distribution, and transport of these materials should therefore be carefully regulated and controlled.
- Small amounts of nanomaterials can have a large impact in any application (positive or negative). Therefore, smuggling and proliferation may be easy unless sensors, controls, and strategies are put in place before these potentially powerful materials become widely available.

4   IN SUMMARY: QUEST FOR BALANCE

In summary, this rapidly growing field is opening up new possibilities along with new unknowns to be investigated. Therefore, it is important for different groups to come together proactively to find balanced answers. Few strategies in this endeavor are essential, and the society as a whole should not lose sight of them.

- The scientific community needs to work in interdisciplinary teams. This field overlaps several disciplines, and some materials can be truly multifunctional. This implies that cross-communication between disciplines can go a long way.
- Close connection between the scientific and entrepreneur communities is very important for such a rapidly growing field.
- Incentives to develop new products should go hand in hand with incentives to conduct research on the environmental impact and safety of these materials.
- Detailed understanding of the benefits versus hazards of producing and using specific nanomaterials should be developed.
- Selection and prioritization of nanomaterials for large-scale production should be made by weighing the potential benefits against not only manufacturing costs, but also societal costs in terms of energy consumed and environmental risks.
- Balanced education involving cross-disciplinary as well as cross-cultural training of the workforce will be important for the nanotechnology age.
- Collaborative teams of scientists, business leaders, citizen groups, and safety/security experts need to work together to outline priorities and regulations.

REFERENCES

1. Nanotechnology: big things from a tiny world, National Nanotechnology Initiative, available at http://www.nano.gov/Nanotechnology_BigThingsfromaTinyWorldspread.pdf.

2. Richard, E. Smalley—Nobel lecture, Nobelprize.org, July 2010, available at http://nobelprize.org/nobel_prizes/chemistry/laureates/1996/smalley-lecture.html.

3. Sir Harold Kroto—Nobel lecture, Nobelprize.org, July 2010, available at http://nobelprize.org/nobel_prizes/chemistry/laureates/1996/kroto-lecture.html.

4. J. Uldrich and D. Newberry, *Next Big Thing Is Really Small: How Nanotechnology Will Change the Future of Your Business*, Crown Business Briefings Book Series, Crown Publishing Group, New York, 2003.

5. K. E. Drexler, *The Future of Nanotechnology*, American Association for the Advancement of Science, Washington, DC, 2002.

6. G. L. Hornyak, J. J. Moore, J. Dutta, and H. F. Tibbals, *Fundamentals of Nanotechnology*, Taylor & Francis, London, 2008.

7. H. S. Nalwa, *Encyclopedia of Nanoscience and Nanotechnology*, American Scientific Publishers, Valencia, CA, 2004.

8. L. E. Foster, *Nanotechnology: Science, Innovation, and Opportunity*, Prentice Hall, Upper Saddle River, NJ, 2005.

9. C. Milburn, *Nanovision: Engineering the Future*, Duke University Press, Durham, 2008.

10. M. C. Roco, *National Nanotechnology Initiative: Past, Present, Future—Handbook on Nanoscience, Engineering and Technology*, 2nd ed., 2007.

11. Nanoscience and nanotechnologies: opportunities and uncertainties, the Royal Society, the UK National Academy of Science, the Royal Academy of Engineering, and the UK National Academy of Engineering, July 2004, available at http://www.nanotec.org.uk.

12. Public perceptions of nanotechnology, *Nat. Nanotechnol.* Focus Issue, 4(2), Feb. 2009.

13. B. H. Harthorn, Risks and benefits of nanotechnology, Nanotechnology Thought Leaders Series, AzoNano, Nov. 2009, available at http://www.azonano.com/details.asp?ArticleId=2452.

14. B. Karn, T. Kuiken, and M. Otto, Nanotechnology and in situ remediation: a review of the benefits and potential risks, Environ. Health Perspect. June 23, 2009.

15. I. Linkov and J. Steevens, *Nanomaterials: Risks and Benefits*, NATO Science for Peace and Security Series C: Environmental Security. Springer-Verlag, New York, 2009.

16. S. Morrissey, Managing nanotechnology, *Chem. Eng.* News, 34(5), 34–35, Jan. 2006.

17. G. Oberdörster, A. Maynard, K. Donaldson, V. Castranova, J. Fitzpatrick, K. Ausman, J. Carter, B. Karn, W. Kreyling, D. Lai, S. Olin, N. M. Riviere, D. Warheit, and H. Yang, Principles for characterizing the potential human health effects from exposure to nanomaterials: elements of a screening strategy, *Particle Fibre Toxicol.*, 2, 8, Oct. 2005.

18. E. Hood, Nanotechnology: looking as we leap, *Environ. Health Perspect.*, 112, 740–749, 2004.

19. T. Xia, N. Li, and A E. Nel, Potential health impact of nanoparticles, *Annu. Rev. of Public Health*, 30, 137–150, Apr. 2009.

20. P. Born, Best practices for handling nanomaterials in the workplace, *Science Daily,* Oct. 22, 2006, July 21, 2010, available at http://www.sciencedaily.com/releases/2006/10/061018150135.htm.

21. http://www.nano.gov/html/about/funding.html.

22. T. Hillie and M. Hlophe, Nanotechnology and the challenge of clean water, *Nat. Nanotechnol*, 2, 663–664, 2007.

23. Meridian Institute Document, Overview and comparison of conventional and nano-based water treatment technologies, presented at the International Workshop on Nanotechnology, Water and Development, Oct. 10–12, 2006, Chennai, India, available at http://www.merid.org/nano/waterworkshop/assets/watertechpaper.pdf.

24. G. E. Fryxell and G. Cao, *Environmental Applications of Nanomaterials: Synthesis, Sorbents and Sensors*, World Scientific, Hackensack, NJ, 2007.

25. J. Kuzma and P. Verhage, Nanotechnology in agriculture and food production: anticipated applications, Woodrow Wilson International Center for Scholars, Sept. 2006, available at http://www.nanotechproject.org/process/assets/files/2706/94_pen4_agfood.pdf.

26. L. Tsakalakos (Ed.), *Nanotechnology for Photovoltaics: A State-of-the-Art Overview*, CRC Press, Boca Raton, FL, 2010, outline available at http://www.nanoscienceworks.org/articles/nanotechnology-for-photovoltaics-a-look-at–game-changing-technologies.

27. B. Baranowski, S. Y. Zaginaichenko, and D. V. Schur (Ed.), *Carbon Nanomaterials in Clean Energy Hydrogen Systems*, NATO ASI Series, Springer-Verlag, New York, 2008.

28. M. S. A. A. Mottaleb, F. Nüesch, and M. M. S. A. Abdel-Mottaleb (guest editors), Solar energy and nanomaterials for clean Energy development, *Int. Photoenergy*, 2009.

29. P. Kamat, Carbon nanomaterials: building blocks in energy conversion devices, *Interface*, 15, 45–47, 2007.

30. M. Zachau and A. Konrad, *Nanomaterials for Lighting: Solid State Phenomena*, Trans Tech Publications, Zurich, Switzerland, 2006, pp. 99–100.

31. Nanowerk News, Quantum dots and nanomaterials: ingredients for better lighting and more reliable power, Mar. 13, 2009, available at http://www.nanowerk.com/news/newsid=9637.php.

32. G. Elvin, The nano revolution, *Architect*, May 7, 2007, available at http://www.architectmagazine.com/curtain-walls/the-nano-revolution.aspx.

33. P. J. M. Bartos, J. J. Hughes, P. Trtik, and W. Zhu (Eds.), *Nanotechnology in Construction*, Royal Society of Chemistry, London, 2004.

34. European Sixth Framework Program, *Roadmap Report Concerning the Use of Nanomaterials in the Automotive Sector*, Mar. 2006.

35. European Sixth Framework Program, *Roadmap Report Concerning the Use of Nanomaterials in the Aerospace Sector*, Mar. 2006, available at http://www.nanoroad.net/download/roadmap_as.pdf.

36. T. Masciangioli and W. X. Zhang, Environmental technologies at the nanoscale, *Environ. Sci. Technol.*, Mar. 1, 2003.

37. G. Shan, S. Yan, R. D. Tyagi, R. Y. Surampalli, and T. C. Zhang, Applications of nanomaterials in environmental science and engineering: review, *Pract. Periodi. Hazard. Toxic., Radioact. Waste Mgmt.*, 13, 110, 2009.

38. M. S. Mauter and M. Elimelech, Environmental applications of carbon-based nanomaterials, *Environ. Sci. Technol.*, 42(16), 5843–5859, 2008.

39. Nanowerk Spotlight, Turning plastic waste into a feedstock for making nanomaterials, Sept. 4, 2009, available at http://www.nanowerk.com/spotlight/spotid=12471.php.

40. S. H. Joo and F. Cheng, *Nanotechnology for Environmental Remediation*, Springer-Varlag, New York, 2006.

41. B. Karn, T. Masciangioli, W. X. Zhang, V. Colvin, and P. Alivasatos (Ed.), *Nanotechnology and the Environment: Applications and Implications*, Oxford University Press, Oxford, UK, 2005.

42. S. Das, A. J. Gates, H. A. Abdu, G. S. Rose, C. A. Picconatto, and J. C. Ellenbogen, Designs for ultra-tiny, special-purpose nanoelectronic circuits, *IEEE Trans. Circuits Syst. I*, 54(11), 2007.

43. N. Melosh, A. Boukai, F. Diana, B. Gerardot, A. Badolato, P. Petroff, and J. R. Heath, Ultrahigh density nanowire lattices and circuits, *Science*, 300(5616), 112, 2003.

44. M. Forshaw, R. Stadler, D. Crawley, and K. Nikoli, A short review of nanoelectronic architectures, *Nanotechnology*, 15, 220–223, 2004.

45. K. Tsukagoshi, N. Yoneyaa, S. Uryua, Y. Aoyagia, A. Kandad, Y. Ootukad, and B. W. Alphenaarf, Carbon nanotube devices for nanoelectronics, *Physica B*, 323(1–4), 107–114, Oct. 2002.

46. D. B. Strukov and K. K. Likharev, Defect-tolerant architectures for nanoelectronic crossbar memories, *J. Nanosci. Nanotechnol. Nanotechnol. Inf. Storage*, 7(1), 151–167, Aug. 2006.

47. H. T. Maune, S.P. Han, R. D. Barish, M. Bockrath, W. A. Goddard III, P. W. K. Rothemund, and E. Winfree Self-assembly of carbon nanotubes into two-dimensional geometries using DNA origami templates, *Nat. Nanotechnol.*, 5, 61–66, 2010.

48. R. R. Ballardini, V. Balzani, A. Credi, M. T. Gandolfi, and M. Venturi, Artificial molecular-level machines: Which energy to make them work?, *Acc. Chem. Res.*, 34(6), 445–455, 2001.

49. D. Zhong, T. Blmker, M. Wedeking, L. Chi, G. Erker, and H. Fuchs, Surface-mounted molecular rotors with variable functional groups and rotation radii, *Nano Lett.*, 9(12), 4387–4391, 2009.

50. B. Yurke, A. J. Turberfield, M. P. Mills, F. C. Simmel, and J. L. Neumann A DNA-fuelled molecular machine made of DNA, *Nature*, 406(6796), 605–608, Aug. 2000.

51. Computational sites with multiple molecular scale designs, such as: http://nanoengineer-1.com/content/index.php?option=com_frontpage&Itemid=1.

52. J. P. Collin, C. D. Buchecker, P. Gaviña, M. C. J. Molero, and J. Sauvage Shuttles and muscles: linear molecular machines based on transition metals, *Acc. Chem. Res.* 34(6), 477–487, 2001.

53. W. R. Browne and B. L. Feringa, Making molecular machines work, *Nat. Nanotechnol.*, 1, 25–35, 2006.

54. J. Wang, Can man-made nanomachines compete with nature biomotors? *ACS Nano*, 3(1), 4–9, 2009.

55. K. E. Drexler, *Nanosystems: Molecular Machinery, Manufacturing, and Computation*, Wiley, New York, 1992.

56. P. Beckett and A. Jennings, Towards Nanocomputer Architecture: Seventh Asia-Pacific Computer Systems Architecture Conference Australian Computer Science Communications, 24, 3, 2002

57. J. B. Waldner. *Nanocomputers and Swarm Intelligence*, ISTE, London, 2007, pp. 173–176.

58. P. J. Kuekes, G. S. Snider, and R. S. Williams, Crossbar nanocomputers, *Sci. Am.*, 72–80, Nov. 2005.

59. F. Patolsky, B. P. Timko, G. Yu, Y. Fang, A. B. Greytak, G. Zheng, and C. M. Lieber, Detection, stimulation, and inhibition of neuronal signals with high-density nanowire transistor arrays, *Science*, 313, 1100–1104, 2006.

60. V. Labhasetwar and D. L. Leslie-Pelecky (Eds.), *Biomedical Applications of Nanotechnology*, Wiley-Interscience, Hoboken, NJ, 2007.

61. N. A Peppas, J. Z. Hilt, and J. B. Thomas (Eds.), *Nanotechnology in Therapeutics, Current Technology and Applications*, Horizon Bioscience, Norwrich, UK, 2007.

62. W. J. Parak, D. Gerion, T. Pellegrino, D. Zanchet, C. Micheel, S. C Williams, R. Boudreau, M.A. Le Gros, C. A Larabell, and A. P. Alivisatos, Biological applications of colloidal nanocrystals, *Nanotechnology*, 14, 15, 2003.

63. S. Bellucci (Ed.), *Nanoparticles and Nanodevices in Biological Applications*, Lecture Notes in Nanoscale Science and Technology, vol. 4, Springer-Verlag, New York, 2009.

64. S. P. Leary, C. Y. Liu, and M. L. J. Apuzzo, Toward the emergence of nanoneurosurgery: III:Nanomedicine: targeted nanotherapy, nanosurgery, and progress toward the realization of nanoneurosurgery, *Neurosurgery*, 58(6), 1009–1026, June 2006.

65. A. Vogel, J. Noack, G. Hüttman, and G. Paltauf, Mechanisms of femtosecond laser nanosurgery of cells and tissues, *Appl. Phys. B*, 81, 8, Dec. 2005.

66. U. Kher, Coming up next: nanosurgery, *Time*, Dec. 4, 2000, available at http://www.time.com/time/magazine/article/0,9171,998686,00.html.

67. J. B. Elder, D. J. Hoh, B. C. Oh, A. C. Heller, C. Y. Liu, and M. L. Apuzzo, The future of cerebral surgery: a kaleidoscope of opportunities, *Neurosurgery*, 62(6), 1555–1579, 2008.

68. H. Liua and T. J. Webster, Nanomedicine for implants: a review of studies and necessary experimental tools, *Biomaterials*, 28(2), 354–369, Jan. 2007.

69. M. Berger, Improving the tools for single-cell nanosurgery, Nanowerk LLC, 2007 available at http://www.nanowerk.com/spotlight/spotid=1254.php.

70. Woodrow Wilson International Center for Scholars and the Pew Charitable Trusts, An inventory of nanotechnology-based consumer products currently on the market, 2010, Project on Emerging Nanotechnologies, available at http://www.nanotechproject.org/inventories/consumer/.

71. G. A. Kimbrell, Nanotechnology and nanomaterials in consumer products: regulatory challenges and necessary amendments, assessment presented at FDA public meeting on nanotechnology, Oct. 10, 2006, available at http://www.fda.gov/ohrms/dockets/.../06n-0107-ts00008-Kimbrell.ppt.

72. B. Ehrmann and T. Kuzma, *Real World Applications of Nanotechnology*,

73. T. Thomas, K. Thomas, N. Sadrieh, N. Savage, P. Adair, and R. Bronaugh, Research strategies for safety evaluation of nanomaterials: VII. Evaluating consumer exposure to nanoscale materials, *Toxicol. Sci.*, 91(11), 14–19, 2006.

74. K. Thomas, P. Aguar, H. Kawasaki, J. Morris, J. Nakanishi, and N. Savage, Research strategies for safety evaluation of nanomaterials: VIII. International efforts to eevelop risk-based safety evaluations for nanomaterials, *Toxicol. Sci.* 92(1), 23–32, 2006.

75. D. G. Rickerby and M. Morrison, Nanotechnology and the environment: a European perspective, *Sci. Technol. Adv. Materi*, 8(1–2), 19–24, Jan.–Mar. 2007.

76. American Bar Association Report, RCRA regulation of wastes from the production, use, and disposal of nanomaterials, section of Environment, Energy, and Resources, June 2006, available at http://www.abanet.org/environ/nanotech/pdf/RCRA.pdf.

77. C. M. Goodman, C. D. McCusker, T. Yilmaz, and V. M. Rotello, Toxicity of gold nanoparticles functionalized with cationic and anionic side chains, *Bioconjug. Chem.*, 15, 897–900, 2005.

78. T. D.Tetley, Health effects of nanomaterials, *Biochem. Soc. Trans.*, 35(3), 527–31, June 2007.

79. Occupational Safety and Health Information (OSHA) discussion on nanotechnology, available at http://www.osha.gov/dsg/nanotechnology/nanotech_healtheffects.html.

80. D. B. Warheit, P. J. A. Borm, C. Hennes, and J. Lademann, Testing strategies to establish the safety of nanomaterials: conclusions of an ECETOC workshop, *Inhal. Toxicol.*, 19(8), 631–643, 2007.

81. G. Oberdörster, E. Oberdörster, and J. Oberdörster, Nanotoxicology: an emerging discipline evolving from studies of ultrafine particles., *Environ. Health Perspect.*, 113, 823–839, 2005.

82. P. J. A. Borm, D. Robbins, S. Haubold, T. Kuhlbusch, H. Fissan, K. Donaldson, R. Schins, V. Stone, W. Kreyling, J. Lademann, J. Krutmann, D. Warheit, and E. Oberdörster, The potential risks of nanomaterials: a review carried out for ECETOC, *Particle Fibre Toxicol.*, 3(11), 2006.

83. M. Geiser and W. G. Kreyling, Deposition and biokinetics of inhaled nanoparticles, *Particle Fibre Toxicol.*, 7(2), 2010.

84. J. B. Dahmus and T. G. Gutowski. What gets recycled: an information theory based model of product recycling, *Environ. Sci. Technol.*, 41, 7543–7550, 2007.

85. J. C. Miller, R. Serrato, J. M. R. Cardenas, and G. Kundahl, *The Handbook of Nanotechnology: Business, Policy, and Intellectual Property Law*, Wiley, Hoboken, NJ. 2005.

86. Nanobusiness Alliance, http://www.nanobusiness.org/.

87. L. Gaze and J. Roderick, Nanotechnology R&D grows in beauty industry according to Thomson Reuters Intellectual Property Analysis, July 13, 2010, available at http://thomsonreuters.com/content/press_room/tlr/tlr_legal/598277.

88. D. M. Bowman, Patently obvious: intellectual property rights and nanotechnology, *Technol. Soci.*, 29(3), 307–315, 2007.

89. J. Taylor, Nanotechnology is disruptive—What this means for manufacturing sectors with reference to the UK, New Dimensions for Manufacturing: A UK Strategy for Nanotechnology, June 2002, excerpts available at http://www.azonano.com/details.asp?ArticleID=1246.

90. M. C. Roco, Converging science and technology at the nanoscale: opportunities for education and training, *Nat. Biotechnol.*, 21, 10, Oct. 2003.

91. A. E. Sweeney and S. Seal, *Nanoscale Science and Engineering Education*, Oxford, University Press, Oxford, UK. 2008.

92. Defense and national security nano, nanomaterials, and nanotechnologies, Oct. 6, 2006, available at http://nanomat.blogspot.com/.

SECTION II

PROCESSING AND ANALYSIS

3

FABRICATION TECHNIQUES FOR GROWING CARBON NANOTUBES

I. T. BARNEY

Center for Nanoscale Multifunctional Materials, Wright State University, Dayton, Ohio

1 INTRODUCTION

In 1991, Sumio Iijima published an article in *Nature* about his discovery of carbon nanotubes (CNTs) [1]. There had been earlier independent discoveries of CNTs by other researchers, but it was Iijima's work that would launch current interest in this material. Almost 20 years later, tens of thousands of articles and books have been

Nanoscale Multifunctional Materials: Science and Applications, First Edition.
Edited by Sharmila M. Mukhopadhyay.
© 2012 John Wiley & Sons, Inc. Published 2012 by John Wiley & Sons, Inc.

written studying CNTs and their uses in a wide variety of fields and applications, including electronics, composites, energy harvesting, and biomedical materials, to name just a few.

Carbon nanotubes are of great interest because of the versatility that their combined properties offer. They are electrically and thermally conductive, have a very high surface area/volume ratio, are biocompatible, possess a hollow core, have great tensile strength (especially with respect to their weight), and at present are one of the easier nanomaterials to fabricate by controlled methods and in useful quantities. Generally, it is the combination of two or more of these properties that make CNTs so valuable for specific applications.

The purpose of this chapter is to provide familiarity with the methods for growing CNTs, the advantages and limitations inherent in each method, and the primary variables controlling growth, and to offer an introduction for ongoing research in this field.

2 STRUCTURE

The properties of CNTs are determined by the structure of the bonded carbon atoms. Electrically neutral carbon has six electrons, two with the principal quantum number $N = 1$ and four with the principal quantum number $N = 2$. In the ground state of a molecule, the energy levels of the electrons in a carbon atom are $1s^2, 2s^1, 2p_x^1, 2p_y^1$, and $2p_z^1$, with the outer four electrons, also known as the *valence shell*, being involved in the bonding. When carbon atoms bond together, the valence electrons are shared according to a concept known as *hybridization*. Hybrid orbitals are formed into sp^3, sp^2, or sp hybridization by a combination of the 2s and 2p energy states. In turn, the type of hybridization that occurs determines the nature of the resulting molecule: the number of neighboring bonded atoms, the bond angle, the structure, and thus the properties.

With sp^3 hybridization, the carbon atoms have four nearest neighbors and bond angles of $109°28'$ between them, as in tetrahedral structured methane and diamond. The carbon atom has four single bonds. Next, sp^2 hybridization has one double bond and two single bonds. Bond angles are $120°$, there are three nearest neighbors, and this forms planar compounds such as benzene (C_6H_6) and graphite. With sp hybridization, a linear structure is formed having one single bond and one triple bond, the bond angles are $180°$, and the carbon atom has two nearest-neighbor atoms, as in acetylene (C_2H_2). Table 1 summarizes the types and structures of hybridized carbon bonding.

TABLE 1   Hybridization of Carbon

Hybridization	sp^3	sp^2	sp
Electron orbitals involved	s, p_x, p_y, p_z	s, p_x, p_y	s, p_x
Bond angle	$109°28'$	$120°$	$180°$
Nearest neighbors	4	3	2
Examples	Methane (CH_4), diamond	Ethylene (C_2H_4), graphite	Acetylene (C_2H_2)

The structure of CNTs is very similar to that of graphite. Graphite is made of sheets of sp^2 bonded carbon atoms arranged in a hexagonal pattern. In the graphite structure, each carbon atom has three valence electrons in covalent σ bonds that are in a plane with the atoms. A fourth electron is bonded perpendicularly to the plane (ranging above and below) in delocalized π bonds. The π bonds allow electrons to move easily within the plane of the bonded carbon atoms, which causes graphite to be electrically and thermally conductive along one plane. Interplanar van der Waals forces weakly connect the stacked sheets of graphite with each other at a distance of 0.34 nm. Electrons do not pass easily from sheet to adjacent sheet. This structure gives graphite highly anisotropic tensile, thermal, and electrical properties.

A single-walled carbon nanotube (SWNT) is nothing more than a sheet of graphite wrapped into a tube around a vector T, known as the *translational vector*. However, the curvature of the surface causes the sp^2 bonding structure to differ slightly from that of graphite. The interplanar σ bonds are confined slightly out of plane with respect to the two carbon atoms, and the perpendicular π bonds are squeezed inside the tube and are slightly less confined outside the tube. The altered structure of the bonds increases the thermal and electrical conductivity of the π bonds, improves the mechanical strength from the σ bonds, and can increase the chemical and biological activity of the surface with respect to graphite [2].

Every SWNT can be uniquely characterized by a vector C_h, which is the distance along the circumference between two equivalently positioned carbon atoms. The C_h vector is perpendicular to the translation vector T and can be described as

$$C_h = na_1 + ma_2 \tag{1}$$

where n and m are integers and a_1 and a_2 are unit vectors for graphite. By convention, it is taken that $n \geq m$. The distance between two bonded carbon atoms in a nanotube is $a_{cc} = 0.142$ nm, which means that the length of the graphite unit vectors $a = a_{cc}\sqrt{3} = |a_1| = |a_2| = 0.246$ nm.

The diameter D of a SWNT depends on the length of the vector C_h according to the equation

$$D = \frac{|C_h|}{\pi} = \frac{a(n^2 + nm + m^2)^{1/2}}{\pi} \tag{2}$$

If $n = m$, the SWNT is referred to as an *armchair tube*, and if $m = 0$, the SWNT is called a *zigzag tube*. Tubes with $n > m > 0$ are known as *chiral tubes*, and their chiral angle (θ) is described by

$$\theta = \tan^{-1}\left(\sqrt{3}\frac{m}{m + 2n}\right) \tag{3}$$

Chiral angles range from 0° (zigzag tubes) to 30° (armchair tubes). Figure 1 shows the relationship among θ, C_h, a_1, and a_2. Adding that the intertube spacing in multiwalled

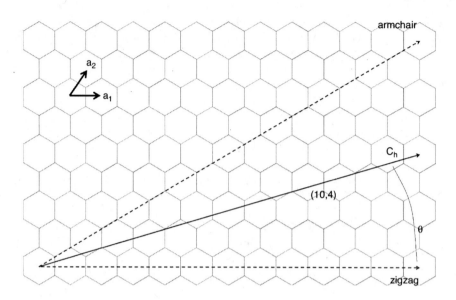

FIGURE 1   Orientation of armchair, zigzag, and (10,4) chiral nanotubes. Also, the relationship between C_h, θ, and graphite unit vectors (a_1 and a_2).

carbon nanotubes (MWNTs) is $d_t = 0.34$ nm, it becomes possible to model any desired defect-free nanotube structure.

Of course, not all CNTs are straight and defect-free. Figure 2 shows a sample of an MWNT randomly grown over silica-coated graphite using a floating catalyst chemical vapor deposition (CVD) method. Although some sections of nanotubes are

FIGURE 2   Randomly oriented MWNT grown by CVD.

straight, most have some curvature, and some even display regular helicity. A bend or "joining" between two tubes can be accomplished by inclusion of one or more pentagon–heptagon pairs [3]. Inclusions of pentagons, heptagons, and octagons into the regular array of hexagons are called *topological defects* [4–6]. One important type of topological defect, the Stone–Wales defect, is constructed by the transformation of four hexagons into two pentagons and two hexagons. Helical MWNTs are constructed by introducing uniform periodic inclusions of pentagon–heptagon pairs [7]. Additionally, MWNTs often include structural defects, such as bamboo-like cupping, cone shapes, and discontinuous walls.

A preference for SWNTs or MWNTs will depend on the application. SWNTs have higher thermal conductivity, better field emitter properties, and unique electrical properties. MWNTs are easier and cheaper to make, have higher oxidation resistance and lower reactivity with many chemicals, and can be grown in unique shapes and structures. Increasingly, double-walled carbon nanotubes (DWNTs) are of interest, as a structure that combines many of the advantages of both SWNTs and MWNTs.

3   ELECTRICAL PROPERTIES

MWNTs are all metallic (no bandgap), but SWNTs can be either conducting or semiconducting, depending on their chiral angle. All armchair nanotubes are metallic in nature. All other SWNTs are either semimetallic or semiconducting. Conducting nanotubes (semimetallic) all satisfy the condition

$$i = \frac{(n - m)}{3} \tag{4}$$

where $n \neq m$ and i is an integer. Otherwise, the nanotube is semiconducting [8]. From this it can be seen that one-third of nanotubes in an unbiased sample will be conductive and two-thirds will be semiconductors. The bandgap for semiconducting SWNTs will scale approximately with the inverse of the tube diameter:

$$E_g = \frac{2a_{cc}\gamma}{D} \tag{5}$$

E_g is the bandgap energy and $\gamma \approx 2.9$ eV is the nearest-neighbor transfer integral for graphite [9]. By measuring the optical absorption of SWNTs, which depends on the bandgap, it is possible to accurately estimate the ratio of metallic to semiconducting nanotubes in a sample [10].

The electric properties of carbon nanotubes (MWNTs and SWNTs) can be further altered by doping the structure with noncarbon atoms such as nitrogen and boron [11–13]. This creates n-type (nitrogen) and p-type (boron) conducting nanotubes, which open up possibilities for many new electronics applications. Ferromagnetic-like nanotubes have also been created by making use of iron metals or alloys trapped inside MWNT cores by using iron catalyst or ferrocene decomposition during growth

[14–16]. These are envisioned as a replacement material for current magnetic storage media.

4   FABRICATION TECHNIQUES

Fabrication methods for making carbon nanotubes can be broken down into five basic categories: arc discharge, laser ablation, high-pressure carbon monoxide (HiPCO), flame synthesis, and CVD. Each method has both advantages and limitations. Selecting a method for nanotube production should be based on the application envisioned.

The preceding section provided an introduction to how the structure of this nanomaterial can affect its final properties. Some questions to consider when choosing a method of producing CNTs for proposed applications are:

1. Are SWNTs or MWNTs preferable?
2. Would either high crystallinity or the introduction of particular types of defects improve performance?
3. What mean diameter and/or length would be best?
4. Does the application require small batches or large numbers of nanotubes? Is the price par gram a limiting factor?
5. How would the remnants of a particular catalyst affect the product?
6. What is the operating temperature and environment?

Additionally, it is important to consider the necessity to postprocess purification of the nanotubes. Arc discharge, laser ablation, and HiPCO methods require the removal of impurities produced in conjunction with the CNTs. Methods of CVD and flame synthesis may or may not require postgrowth purification.

Chemical vapor deposition and flame synthesis can also be used to grow nanotubes attached to a substrate rather than producing them loose in a vapor. This requires a substrate that can withstand the growth conditions (primarily, the temperature), which can be very limiting. However, some progress has been made toward working at lower substrate temperatures with techniques such as plasma-enhanced chemical vapor deposition (PECVD).

Regardless of the method, two things are always required to grow carbon nanotubes: a carbon source and energy. The carbon source can be some form of graphite, as is normal in the laser ablation and arc discharge techniques, a liquid such as xylene; or a gas such as CO or CH_4. Methods that primarily make use of thermal energy, such as CVD, flame synthesis, and HiPCO, generally utilize liquid- or gas-phase carbon sources.

Most methods for MWNTs and all current methods for high-volume SWNTs also make use of a metal catalyst. Many elements and combinations of elements have been utilized as catalysts (including Fe, Ni, Co, Pt, Pd, Rh, Gd, Y, La, In, and Sn [17,18]), but most processes use Fe, Ni, Co, or an alloy containing one or more of these metals. The choice of catalyst metal(s) is one of the most significant parameters for growing nanotubes and strongly influences the size, structure, and purity of the nanotubes grown. The catalyst used can also depend on the method of fabrication. Ni : Y and Ni : Co combinations are most common in arc discharge and laser ablation production of SWNTs because of the higher catalytic efficiency of these bimetal alloys. HiPCO uses Fe catalysts because of the suitability of iron pentacarbonyl [Fe(CO)$_5$] in the gas flow.

A final consideration when choosing a method is the environment necessary to produce nanotubes. Many methods require a furnace that can limit the size of substrates. Most methods require a controlled atmospheric composition, and many require pressures above or below 1 atm. Commonly used buffer gases are He, Ar, N$_2$, H$_2$, and combinations thereof. A few methods can also produce carbon nanotubes in air or liquids. Table 2 summarizes some of the benefits and limitations of the various methods for growing CNTs, as discussed in the following sections.

5 ARC DISCHARGE

Arc discharge was the method developed by Iijima [1] for controlled growth of CNTs. It is a relatively simple method of producing CNTs, but the purity and yield of the product are very sensitive to the growth parameters. Arc discharge is accomplished by producing a direct-current (dc) arc between two carbon electrodes in a controlled gas environment. Commonly used parameters set a potential difference between the electrodes of 20 to 30 V with a current between 40 and 100 A. The electrodes are cylindrical in shape, the cathode usually being of larger diameter than the anode. Anode blocks commonly range between 6 and 10 mm in diameter and are paired with cathodes between 8 and 15 mm in diameter. Both electrodes are commonly water cooled at the base to prevent damage to the apparatus when arcing occurs near the holders (at the beginning and end of the process). The evaporation chamber may also be water cooled. The electrodes can be emplaced vertically or horizontally. Carbon nanotubes are formed in the gas phase between the electrodes, using the evaporation of the anode as the source material. Deposits are formed on the cathode, chamber walls, or other collectors, depending on design, and can have varying compositions and densities of nanotubes and impurities. Understanding the composition of as-formed deposits and their geometry with respect to the growth system and method of collection is an important part of optimizing and understanding a particular arc discharge system.

Arc discharge is a batch method that always limits the maximum output of any system. Additionally, the final product requires systematic purification to remove impurities inevitably formed during the process. However, there are inherent advantages to growing both MWNTs and SWNTs by this method, as will be discussed.

5.1 Multiwalled Carbon Nanotubes by Arc Discharge

The unique advantage of growing MWNTs via arc discharge is that it is the one method for growing usable quantities that does not require a catalyst. This can lead

TABLE 2 Methods of Growing Carbon Nanotubes

Method	Single-Walled CNTs		Multiwalled CNTs	
	Advantages	Disadvantages	Advantages	Disadvantages
Arc discharge	Good balance of quality and quantity	Batch processing limits maximum output, purification required	Catalyst-free	Limited length and batch processing limits output, purification required
Laser ablation	High quality and control	Small output, purification required	Uncommon, good control of structure	Small output, purification required
HiPCO	Smallest diameter, continuous processing	Purification required, limited volume but some scalability	Not done	Not done
Flame synthesis	Not done	Not done	Cheap, easy, and fast growth on substrates possible	Poor quality and control
CVD	Longest length, highly versatile, growth on substrates possible, relatively cheap and easy	Less control over structure than some other methods, chamber size limits substrate size	Longest length, high versatility, growth on substrates, relatively cheap and easy, special structures possible (branched, helical, etc.)	Chamber size limits substrate size

to simplified purification techniques and increased suitability in several broad areas of applications. Even after repeated purification, trace amounts of transition metals remain in nanotubes grown using catalytic methods. These metal impurities can reduce the biocompatibility of the material [19].

One of the most critical parameters for growing MWNTs by arc discharge is the buffer gas. MWNTs have been grown using He, H_2, Ar, CH_4, and mixtures of these gases [14,20–24]. Helium is the most common buffer gas. It is easier to optimize a He than an H_2 system because the use of inert gas enables more uniform cathode deposits [22]. Argon is much less commonly used for arc discharge because although Ar is also inert, MWNT production is less efficient when using Ar than when using He [25]. Hydrogen has been utilized increasingly as a buffer gas, despite the difficulties in optimization. Using hydrogen instead of He can alter the mean diameter for MWNT structures in a batch [21]. Additionally, it has been shown that compared to helium, H_2 buffer gas produces high graphitization (crystallinity) in MWNTs and lower levels of impurities in the deposits collected [14,22,26]. Hydrogen is able to etch amorphous carbon selectively during the formation of nanotubes in the arc discharge plasma [20,23]. Mixing He and H_2 for the buffer gas is an effective method of combining the benefits of both, as it helps promote stability while retaining improvements in purity [23]. Compared to nanotubes grown using He in the same apparatus, MWNTs can be grown in air with similar crystallinity and greater wall thickness [24].

The plasma temperature between the electrodes is a critical parameter for arc discharge, which is why the atomic species of the buffer gas plays such a definitive role in the final product [21,22,24–27] The pressure of the buffer gas is also key in determining the state of the plasma, and thus affects both purity and the morphologies [25,27]. These two variables (pressure and gas species) are interrelated, making optimal growth pressures dependent on the buffer gas. For example, the optimal growth pressure of hydrogen is lower than that of helium [26].

Other variables also affect the state of the plasma between the electrodes. Varying the arc current and voltage between the anode and cathode will change the purity and crystallinity of MWNTs [27–29]. The gap distance, positioning of the electrodes, and diameter of the electrodes can also be used to affect the plasma formation. During production, the evaporation rate of the anode will exceed the growth rate on the cathode. Computer-controlled feed rates of the anode (often set to maintain a specific gap distance based on the voltage drop) have been shown to create more homogeneous deposits on the cathode [22,23]. Rotation of the electrodes also improves yield, but the rotation rate can depend on the system. Lee et al. reported that rotating the anode resulted in a more uniform discharge, a higher erosion rate on the graphitic anode, and thus a large yield of MWNTs [30]. These tendencies improved as they increased rotation speeds in their system. Conversely, Joshi et al. found that with cathode rotation a slower rotation rate was preferable, as the net yield in their system decreased with increasing speed, with all benefit being lost at high speeds [24].

MWNTs can also be grown by arc discharge in liquid buffer media. The advantage of a liquid environment is that it is easy to control, and the nanotubes can be collected floating on the top or resting on the bottom of the liquid, depending on the wetability. Synthesis can be carried out in liquid nitrogen [29,30] or deionized water [28,31,32] without doping the MWNTs with elements from the liquid. The range of voltages that will produce nanotubes is dependent on the liquid. LN2 (22 to 27 V) produces nanotubes within a slightly lower voltage range than that of water (25 to 30 V), but the products are fairly similar [29].

The final stage for producing MWNTs by arc discharge is purification of the soot collected. Other forms of carbon, both amorphous and graphitic based, are always produced in conjunction with the nanotubes. These nanoparticle impurities can be removed selectively with oxidative techniques (chemically, thermally, or with infrared radiation) separately or in conjunction with size-based techniques based on liquid separation or physical filtration [21,22,24].

5.2   Single-Walled Carbon Nanotubes by Arc Discharge

Arc discharge growth of SWNTs is a process with good trade-offs if you need a combination of quality and quantity. SWNTs grown by arc discharge have fewer defects than those grown by CVD methods, and higher yields than HiPCO or laser ablation. The arc discharge process for growing SWNTs is very similar to that for growing MWNTs but has one additional variable: the catalyst. Growing SWNTs requires catalysis from a transition metal nanoparticle. Not all metals are suitable. For example, manganese catalysts will grow twisted MWNTs rather than SWNTs [33]. Although many different metals and alloys have been tried, it is generally agreed that alloys containing Ni, Co, and Fe are the most productive [33–37]. Pure metal catalysts will yield SWNTs but mixed-metal catalysts are more effective [34,38]. The most commonly used alloy is Ni : Y (4.2 : 1 at%), but Ni : Co, Ni : Fe, and Ni : Co : Fe are also used [17,34,35,38–45]. The catalyst selected strongly affects the yield and can alter the diameter distribution of the nanotubes in the soot [35]. The ratio of the two metals in the alloy can be varied as a means to tune the size distribution of SWNTs [34]. A small amount of sulfur can also increase catalytic activity with some metals (most commonly with Fe and Fe-containing alloys) [38,46]. The volume of soot produced can also be increased with increasing arc current; however, this is a trade-off because although the volume of production goes up, the purity of the SWNTs in the product decreases [47]. Purity and length of SWNTs can also be increased when the arcing takes place within a magnetic field [48].

SWNT production with arc discharge is usually done using graphite electrodes much as in the MWNT process. The most common method for introducing the catalyst metal is to drill a hole in the center of the anode and fill it with graphite and metal powders in the proper atomic ratio. Metal-doped anodes are sometimes used with iron catalysts [49]. The importance of the chemical composition of the anode for SWNT production is clear, but research has also shown that the anode physical structure can be a key parameter. The output of SWNTs is affected by the grain size in the graphite: Smaller grains give higher yields [49]. Another issue with the anode material is that of cost. Price is a limiting factor for introducing nanotubes into many feasible and marketable technologies. Pure graphite anodes are already a processed, value-added product, so an alternative lower-cost carbon source would be ideal, and several possible alternatives have been presented: carbon black and coal

[43,49]. Anthracite coal anodes are of particular interest because they are cheap and plentiful while having minimal impurities with respect to other forms of coal. High-purity yields from an iron-catalyzed anthracite coal system were realized simply by surrounding the electrodes with a wire cage from which the soot could be collected [43]. These alternative carbon source methods need further refinement, but they offer a lot of promise for developing SWNTs at a lower cost.

The total yield of SWNTs is always very dependent on buffer gas and pressure [50]. The most common buffer gas for growing SWNTs is He, because it is inert and more effective than Ar [50]. Other gases that can be used (often in mixtures) include H_2, N_2, Ne, and Kr [36,37,46,51]. The choice of buffer gas will affect the size distribution of an SWNT [36,42]. Production with He is often done between 600 and 700 torr, but effective low-pressure (100 to 500 torr) growth is possible [34,38–40]. The higher pressure range is more normal because the output of nanotubes is greatly increased in He when the pressure rises from 500 torr to 700 torr [35].

The most commonly replicated method for growing SWNTs by arc discharge was developed by Journet et al. [39]. Their recipe uses a Ni : Y catalyst (4.2 at% Ni and 1.0 at% Y) with graphite electrodes in a 660-torr He atmosphere. The arc is maintained across a 3-mm gap with 100 A at 30 V dc. The diameter of the anode and cathode was 6 and 16 mm, respectively.

Process optimization is clearly important in maximizing yield and purity while minimizing cost for arc-grown SWNTs, but delivering a usable product requires a postprocessing step, purification. Arc discharge SWNTs always have some impurities inherent in the soot, collected which can include amorphous carbon, graphitic nanoparticles, fullerenes, carbon nano-anions, and both encapsulated and unencapsulated metal or metal–oxide nanoparticles. Removal of these impurities is done through two basic routes: oxidation and physical separation. Oxidation can be either gas or liquid phase; physical separation can be accomplished through filtration, centrifugation, and chromatographic methods. Important aspects to consider for developing purification processes are final purity, loss ratio of SWNTs (L_R), cost, time, scalability, and damage to the nanotube structures. The loss ratio can be defined as

$$100 \times \frac{W_0 M_S - W_F M_p}{W_0 M_S} = L_R \tag{6}$$

where W_0 and W_F are the initial and final weight percentages of SWNTs, M_S is the mass of soot before purification, and M_p is the final mass of the purified product. Purified SWNTs generally have W_F vlaues greater then 90%. Many purification methods have been developed, and research into purification of SWNTs has been as important as optimization of arc discharge parameters [36,37,40,41,45,52–54].

Liquid-phase oxidation is common to many purification methods. Four variables need to be controlled during any liquid-phase oxidation: temperature, time, concentration, and chemical oxidizer. Control of all four variables is critical to ensure that nonadvantageous carbon and various metal-based impurities are removed without significant loss or damage to nanotube structures. Common chemical oxidizers include HCl, HNO_3, H_2O_2, H_2SO_4, and HF. Gas-phase oxidation is commonly carried out in air or using custom gas blends, such as Ar–O_2 mixtures or hydrated N_2. Gas-phase oxidation is an effective method for removing amorphous carbon because it oxidizes at a lower temperature than do CNTs [45]. However, gas-phase oxidation does not remove catalyst metal remnants or graphitic carbon impurities. Carbon nanotubes have a carbon structure similar to that of the graphitic impurities, but the curvature of the tube walls makes them more sensitive to reactive oxygen. Thus, oxidation temperatures that can remove graphitic impurities will destroy the SWNTs first. MWNTs begin to react with oxygen significantly at temperatures above 450°C, and the onset temperature for SWNTs is lower due to their higher curvature [55]. Gas-phase oxidation thus is normally carried out for short periods at temperatures between 300 and 450°C [36,41,53,54]. Optimal oxidation times are roughly proportional to the inverse of the temperature setting.

The advantage of gas-phase oxidation is that it reduces the amount of chemical oxidation needed, which helps to decrease the loss ratio and the level of damage to the SWNT structure. High-temperature annealing (1200°C) of the purified nanotubes in either vacuum or hydrogen can be used to repair much of the structural damage [44,53]. A Size- and shape-based method of physical separation is necessary to remove graphitic carbon and fullerenes. Purification methods thus involve systematic combinations of oxidation and filtration often combined with centrifugation and annealing.

Figure 3 shows a simple arc discharge apparatus for growing SWNTs. The graphite electrodes are held in place by water-cooled holders (1). The anode (3) is on a computer-fed mechanical mount (5), which is able to monitor and maintain a constant gap (7) from the cathode (2). This is important because as the anode evaporates, it will decrease in length faster than the cathode will grow, and thus the gap will increase without constant feeding of the anode. The center of the anode has been drilled out and filled with Ni : Y (4.2 : 1) and graphite powder and the entire chamber has been filled with an He–H_2 gas mix.

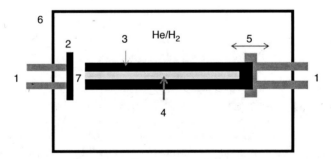

FIGURE 3   Typical arc discharge setup for growing an SWNT with (1) water-cooling inlets/outlets for electrode holders, (2) graphite cathode, (3) graphite anode with (4) catalyst and graphite powder in central hole, (5) computer-controlled anode feed, (6) chamber walls, and (7) preset arc gap maintained by (5).

6   LASER ABLATION

Laser ablation is a good method of creating small batches of carbon nanotubes with a high degree of control over batch mean structural parameters such as diameter and crystallinity. Given the batch size limitation, laser ablation is used primarily to grow SWNTs. MWNTs can be grown just as easily, but the advantages that laser ablation offers in most applications are more relevant to SWNTs. Important parameters for laser ablation are type and output of laser, processing temperature, buffer gas, pressure, chamber geometry, and target composition. Laser light on the carbon target creates a high-temperature, high-pressure bubble, which results in a shock wave and the ejection into the chamber of a vaporized plume. Growth of the nanotubes takes place in this plume within a very short time frame, from a few milliseconds to 1 s after the laser pulse [56]. Optimization of the parameters is important to control the nature of the plume and thus the interactions and growth conditions within the carbon vapor. Control of some parameters, including target composition, gas pressure, and buffer gas, not only alters the yield but can also be used to control the average characteristics of SWNTs [57].

The first parameter to address for laser ablation is the type of source, because it partially determines the ranges and optimal choices for the other parameters. This usually means a choice between Nd : YAG (primary 1064-nm or second harmonic 532-nm) lasers and CO_2 infrared (10.6-μm) lasers [56–67]. Other laser source types, including KrF (248 nm) and XeCl (308 nm), have also been shown to work, but most research focuses on the first two options [68,69].

The type of laser strongly affects the range of gas parameters in which SWNTs can be formed. With a CO_2 laser, SWNTs can be grown at room temperature without need of a furnace [63,66]. Nd : YAG laser ablation generally requires temperatures above 900°C, and reaches peak efficiency around 1200°C although this is somewhat dependent on the catalyst [63,65]. CO_2 lasers also have a lower cutoff pressure for nanotube formation, although optimal gas pressures for both systems are generally similar (200 to 600 torr) [56,57,59,63,67]. The range of operating pressures varies significantly because of dependence on other parameters. Gas pressure is critically important in laser ablation because it determines the rate of expansion for the vapor plume. If the pressure is too low, the vapor plume expands and cools too quickly, which greatly reduces the number of nanotubes grown. The purity of the SWNTs deposited is improved when the material ejected by the laser is confined within a smaller volume [64]. However, positive gains from this effect have a limited range. When the pressure is too high, carbon–carbon collisions within the plume become too frequent and the purity decreases rapidly [57].

Unlike arc discharge, helium is a poor buffer gas for laser ablation of SWNTs [57,58,66]. Both argon and nitrogen are more efficient buffer gases because they do not conduct thermal energy from the plume as quickly [57,58,61]. The cooling rate of the carbon plume in the buffer gas is inversely proportional to the square root of the molecular weight. Thus heavier gases are more efficient for laser ablation. Water vapor in the buffer gas can also be advantageous because it removes amorphous carbon during formation and promotes the growth of nanotubes [60].

Similar to arc discharge, bi- and trimetal catalysts are more efficient in laser ablation [57–61,67]. Commonly used bimetal catalysts include Ni : Co (1 : 1), Ni : Y (4.2 : 1), and Ni : Fe (4 : 1), but there is a significant variation. Metal oxide catalysts can also be used, and the additions of some oxides have been shown to lower the temperature at which SWNTs will form in Nd : YAG systems [65,66]. Metal acetates and nitrates can also be used in place of pure metals within the target [62]. Targets generally have 1 to 5 at% catalyst metal. The soot, collected from a water-cooled Cu trap downstream in the gas flow, contains metals in approximately the same proportions as in the target [58]. Metal-doped graphite is the most common target, but other types of carbon, including pitch and coke, can be used, although they are less efficient [66].

The soot collected from laser ablation has impurities of types and quantities comparable to arc discharge. Purification is equally important for final products, and the methods are similar to those discussed in Section 4.5.

7 HIGH-PRESSURE CARBON MONOXIDE

The high-pressure carbon monoxide process is used to grow loose SWNTs, and it is a continuous growth process that utilizes inexpensive precursors. HiPCO has the added advantage of being able to produce SWNTs with smaller diameters than those obtained using any other technique (0.8 nm) [70,71]. Average diameters range from 0.8 to 1.3 nm, with a mean of 1.1 nm [72]. SWNTs grown by this method have good crystallinity, which makes them ideal for many electrical applications. The precursor chemicals for HiPCO are carbon monoxide (CO) and iron pentacarbyonyl [$Fe(CO)_5$] [70]. A small amount (1.4%) of methane (CH_4) is also added to the gas flow because it has been shown to improve the yield. Increasing the methane further is found to increase the formation of amorphous carbon in SWNTs and reduce their properties [70].

The formation of SWNTs depends strongly on the heating of the gas and the pressure at which the gas is mixed [70]. Two flows of CO gas are injected into the reactor at a pressure of 30 atm. The main flow of CO is passed through a preheater which raises the temperature to 1050°C. A second room-temperature flow of carbon monoxide is passed through a bubbler of liquid $Fe(CO)_5$ before injection. Within the reactor, the hot CO is added to the room-temperature mix at a ratio of 6 : 1 using a circular showerhead. The geometry of the showerhead is particularly important to ensure optimal mixing of the gas streams [73]. The active carbon is achieved through disproportionation of the carbon monoxide by the Boudaouard reaction:

$$2CO \rightarrow C(s) + CO_2 \qquad (7)$$

The solid carbon is absorbed onto the iron catalyst to grow nanotubes. The nanotubes and nanoparticles are collected on a series of cooled filters, water and carbon dioxide are removed from the gas, and the remaining carbon monoxide is recirculated back through the reactor. The main impurities in the SWNTs collected are the remnants of the catalyst iron, which accounts for 4 to 5 at% of the product collected [72].

Due to their small diameter and high wall curvature, HiPCO SWNTs have a higher sensitivity to oxidation then that of nanotubes from other methods. Additionally, HiPCO soot does not contain the amorphous carbon and graphitic particles normal in arc discharge and laser ablation. Purification methods for HiPCO samples are concerned primarily with removal of iron oxide and thus are different from those discussed previously (iron nanoparticles exposed to air oxidize rapidly) [71,74,75]. Iron oxide also appears to have a catalytic effect on nanotube oxidation since unpurified SWNTs will oxidize at lower temperatures [74].

It was found that introducing fluorine-based compounds during oxidation can suppress the catalytic effect of iron oxide [74]. The iron can be exposed for removal without damaging the SWNTs by using a series of oxidation and deactivation steps. For this, a mix of $N_2 : O_2 : SF_6$ gases (ratio of 20 : 4 : 1) is passed over the SWNTs at 150°C for 1 h. The iron is then deactivated by passing $N_2 : SF_6$ (20 : 1) over the sample at 175°C for 30 min. Oxidation was then repeated at 175°C followed by deactivation at 200°C. This is continued up to 350°C, at which point the SWNTs are allowed to cool in nitrogen. The iron was then removed using HCl [74]. The loss ratio for this process is about 30% and the final product contains about 1 wt% Fe.

A second method for removal of the iron uses a simple bath in $Br_2(l)$ under a dry nitrogen atmosphere. The CNTs need to be stored in N_2 to exclude oxygen and water, and the bromine needs to be high purity (no chlorine contamination). The HiPCO SWNTs are soaked in $Br_2(l)$ for 30 min, washed in dilute HCl, then water, and heated to 400°C (also in nitrogen) for 1 h to evaporate the remaining bromine [75]. Bromine purification leaves 1 to 2 wt% iron, which is slightly higher than that achieved using the SF_6 gas. However, the bromine process is much faster and less expensive.

8 FLAME SYNTHESIS

Flame synthesis of carbon nanotubes has been recognized for some time. Small quantities of nanotubes are produced in the soot of ordinary fires, but producing useful quantities in a controlled manor is a relatively new and developing process. More than any other method, flame synthesis offers the chance to grow carbon nanotubes in a manner that is cost-effective and scalable enough for mass production [76–83]. Full growth of nanotubes in flames can be accomplished in about 10 s of exposure. Long exposure is not necessary and is in fact undesirable as the nanotubes can be oxidized back off the sample [79]. This is not a method for growing highly crystalline SWNTs. The nanotubes produced are MWNTs, often have large diameters, and tend to be fairly twisted, incorporating significant numbers of structural defects. Methods exist for growing nanotubes either loose or attached to a substrate, and because it is a flame environment, the nanotubes can be grown in open air [76].

The three main parameters for flame synthesis are catalyst type, flame temperature, and fuel source [82]. The fuel source provides both the energy and the carbon. Many different hydrocarbons have been shown to work, including acetylene, ethanol, methane, and butane [76–82]. There are many designs for gas outlets to control the mixing of oxygen and fuel source, but the apparatus can be as simple as a modified bunsen burner [80]. The proportions of the mixed fuel and the position of the sample in the flame (or injection point of the catalyst particles) control the temperature. Optimal temperatures depend on the catalyst and fuel type. For example, nickel-coated iron in an ethanol flame was optimized at 560°C [81]. Oxide-supported nickel catalysts in a methane flame grew nanotubes best between 750 and 800°C [82].

The burning velocity is also important because it regulates the amount of carbon available for the reaction. It was shown that too little methane would produce no growth but that too much methane in the flame would produce an amorphous carbon coating rather than nanotubes [78]. The number of walls and the diameter of the nanotubes can also be affected by the amount of carbon in the flame [79].

Substrates for flame synthesis can be simple acid-treated metals such as steel mesh or nickel plate (providing both support and catalyst) or as complex as predeposited catalyst particles supported on patterned silicon oxide [77–80]. Loose MWNTs can be produced by injecting nanoparticles into the flame or by burning metal–organic compounds such as ferrocene [$Fe(C_5H_5)_2$] along with the fuel [76,83]. Adding silicon-based compounds such as tetraethylorthosilicate, also known as TEOS [$Si(OC_2H_5)_4$], in with the ferrocene can improve the catalytic efficiency of the iron and increase the average length of the nanotubes [83].

9   CHEMICAL VAPOR DEPOSITION

CVD is generally the least expensive (after flame synthesis) as well as being the most versatile method of growing CNTs. It can be used to grow SWNTs, DWNTs, and MWNTs. Different morphologies of MWNTs are grown by this method, including aligned (straight), randomly oriented, helical, cupped (bamboo like), and iron filled. Various CVD techniques enable growth of the longest known nanotubes (centimeters in length) and can be used to grow CNTs loose or attached to a substrate. Loose nanotubes are generally collected from the walls or off the bottom of the CVD furnace. Depending on the process, this can be as aligned bundles, loose random clumps, and even felt-like mats of MWNTs. Substrates can be patterned before CNT growth to create arrays of nanotubes with desired length and geometric placement.

Two factors enable this versatility: (1) the ability to control growth rates and times and (2) the ability to grow nanotubes on a substrate. Growth times for individual nanotubes in arc discharge, laser ablation, flame synthesis, and HiPCO can range from less than 1 s up to about 10 s. In all of these other processes, the nanotubes can remain within the growth environment for only a short period of time, if for no other reason than that they are carried off to the collector. With CVD, growth processes for individual nanotubes commonly last as long as 30 min. The growth rate in a given system can be controlled over this time with the temperature and the gas composition. The use of various substrates also has a profound effect on nanotube growth rates, catalyst activity, morphologies, and patterning, as discussed later.

CVD has several primary variables that must be controlled and some optional variables that can be introduced for enhanced processing. In the simplest setups, the variables that must be controlled are substrate type or collection method, carbon source, catalyst type, method of introducing the catalyst, furnace or experimental chamber geometry, gas composition, and growth temperature. Optional variables to control include pressure, electromagnetic fields or radiation, and post processing techniques.

The two biggest inherent limitations on CVD processes are substrate size and that SWNTs usually have lower crystallinity than those made with arc discharge, laser ablation, or HiPCO. Some methods of CVD also require purification to remove excess catalyst metal [84]. Crystallinity can be improved through postgrowth annealing [85,86]. This can often be done in the same furnace chamber in which CNTs were grown, although annealing temperatures are higher and the gas composition will probably be different from the growth conditions. Unlike flame synthesis, CVD substrates are limited in size by the hot zone (or growth region) of the process chamber. Quartz or mullite tube furnaces are common growth apparatus, so the hot zone in these systems is less than a foot long and a couple of inches wide.

CVD catalysts are also generally Co, Ni, or Fe metals and alloys. Introduction of catalyst metals comes in two main groups: predeposited or floating catalyst. Predeposition methods always have a substrate and can be done physically or chemically. Chemical methods include various forms of solgel, reduction, precipitation, and thermal decomposition reactions [18]. Physical deposition is commonly done with lithography, printing methods, ion or electron beam evaporation, impregnation, or magnetron sputtering [18,87–92]. Catalysts are often deposited in thin films and the thickness of the film helps determine the size of nanoparticles formed during reaction and thus influences the diameter of the nanotubes [87]. Predeposited catalysts can be patterned on the substrate to limit growth of CNTs to the patterned areas [93]. The substrate under the catalysts can also be patterned using a variety of techniques to prevent catalytic reactions over specific areas.

Unlike predeposition systems, floating catalyst methods are based on introducing metal–organic compounds into the gas flow. The most common precursor is ferrocene $[Fe(C_5H_5)_2]$, but other chemicals can be used, including cyclooctatetraene iron tricarbonyl $[(C_8H_8)Fe(CO)_3]$, cyclopentadienyliron dicarbonyl dimer $[(C_5H_5)_2Fe_2(CO)_4]$, and iron pentacarbonyl $[Fe(CO)_5]$ [94]. The metal–organic precursor is usually dissolved in a liquid carbon source such as toluene, xylene, or benzene [15,86,94–102]. Ferrocene can also be evaporated by itself or with methane to grow nanotubes [103,104].

Oxides are the most common substrates. Some examples are SiO_2, Al_2O_3, MgO, and TiO_2 [15,87–89,91–94,105–110]. Various forms of silicon oxide are the most common. MWNTs generally show good adhesion with silica [87]. Growth is highly substrate dependent [110–114]. Interactions between the substrate and catalyst can be physical, electronic, and/or chemical in nature. For example, a layer of oxide is necessary to grow nanotubes on silicon because the substrate will interact chemically and form metal—silicides, which poison the catalysts. A silica layer additionally acts to promote catalyst activity, and this benefit increases significantly with the thickness of the silica up to about 30 nm [107]. A layer of silica also promotes catalyst activity on graphitic substrates [114]. CNT can be grown on conductive substrates, but this is generally less effective than growth on insulating ceramics [94,103,115]. An underlayer of Ti or Al under an iron or nickel catalyst layer can also increase nanotube yield [88]. Metallic underlayers can help protect the catalyst, improve adhesion, and increase electrical and thermal transport to and from the substrate [87]. Again, the effectiveness of an underlayer is strongly substrate and catalyst dependent.

A wide selection of carbon sources have been used for CVD. Some examples are acetylene, ethylene, methane, toluene, xylene, benzene, and propylene. Optimal growth and injection temperatures can vary by source and catalyst [111]. It is known that the type of source affects growth rates and thus MWNT structures, but Qingwin et al. found that the structure of the source could also affect whether SWNTs or MWNTs were preferential. Aromatic molecules were found to favor SWNT growth, while aliphatic molecules favored MWNTs [116].

There is a lot more variation in gas compositions for CVD methods than is commonly found in other processes for growing nanotubes. The primary factors affecting gas composition are the carbon source, the catalyst source, and the structure of CNT desired. Common neutral carrier gasses are Ar and N_2. Hydrogen and sometimes ammonia are used as reducing agents. Water or occasionally small amounts of oxygen are used to remove amorphous carbon buildup and prevent catalyst deactivation. This can both improve purity and crystallinity and enable longer growth periods and thus increased CNT length. Methane and acetylene can be used by themselves or more commonly with a reducing agent. The optimal gas balance is also affected by the operating temperature and pressure.

For purely thermal CVD systems, the growth temperature for MWNT is usually between 700 and 850°C. SWNT are generally grown above 900°C. Substrates such as graphite would oxidize at these temperatures in the atmosphere but can be used here as long as oxygen is not present in significant concentrations. However, not all substrates can withstand these temperatures even in neutral environments. The primary tactic for lowering growth temperatures is to use either radio frequency or microwave power to form plasma in the reactor [90,93,103,105]. This is appropriately known as *plasma enhanced chemical vapor deposition* (PECVD). A voltage bias can also be used to increase growth and affect growth direction of the nanotubes [105]. PECVD can lower substrate temperatures by several hundred degrees Celsius, which opens up the window for many new substrate materials and potential applications. PECVD is a large field of research all on its own, and the references given are only meant to serve as a starting point for readers interested in further study of this very important technique.

Chemical vapor deposition techniques are extremely varied and thus findings about growth in one system are not necessarily applicable to every other CVD setup. For example, some CVD systems will not produce CNT without hydrogen gas, but others will. The variables in any CVD system are highly interdependent. However, there are some trends that hold across systems. Settings that increase growth rates (i.e., high temperature, high carbon input, high pressure, etc.) generally cause an increase in defects and lead to the formation of bamboo structures near the upper

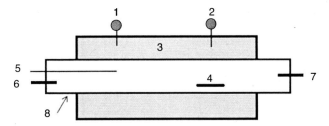

FIGURE 4 Floating catalyst thermal CVD system with (1) injection temperature, (2) growth temperature, (3) tube furnace, (4) silicon substrate, (5) xylene–ferrocene injection pump, (6) Ar–H_2 gas inlet, (7) gas outlet, and (8) quartz furnace tube.

limits at which MWNT are formed [98,117–119]. This is one reason that it is very important to consider the transport mechanisms of carbon in any CVD system. There are two related rates of carbon transport to consider: transport through the reactor and transport through the catalyst [119]. Restricted transport through the reactor can, of course, limit transport through the catalyst. Too much carbon at the catalyst can lead to catalyst deactivation and the formation of amorphous products, whereas too little will limit growth [92]. Formation of amorphous carbon leads to catalyst deactivation and decreasing growth rates with time [120]. This decrease in growth rates can be partially countered by adding water to the gas.

Figure 4 shows a relatively simple floating catalyst thermal CVD system. The reaction is caused by the decomposition of a ferrocene–xylene solution in an Ar–H_2 atmosphere injected at (5) and (6), respectively. The system is set up in a multizone quartz tube furnace with two important temperature regions. The first region (1) is the injection temperature for the ferrocene–xylene mixture, and the second region (2) is the hot zone of the furnace, where an oxidized silicon substrate is placed (4). The injection temperature is a critical variable because it governs the decomposition of the catalyst–source solution. Having too high of an injection temperature causes increased levels of impurities but too low of an injection temperature limits formation of the nanotubes.

10   SUMMARY

There are five primary methods for growing CNTs: arc discharge, laser ablation, HiPCO, flame synthesis, and CVD. Each method has strengths and weaknesses in providing these important nanostructures to a growing field of applications. Arc discharge, laser ablation, and HiPCO are more often (always in the case of HiPCO) used for SWNT growth. Flame synthesis is used only to grow MWNTs. CVD is commonly used for both SWNTs and MWNTs. Both of these methods can grow nanotubes attached to a substrate. Choosing a particular production method for any application should be based on the size, structure, and quantity of CNT required, as well as the appropriateness of the method in question.

REFERENCES

1. S. Iijima, *Nature* **354** (1991) 56.
2. M. Meyyappan (Ed.), *Carbon Nanotubes: Science and Applications*, CRC Press, Boca Raton, FL, 2005.
3. B. I. Dunlap, *Phys. Rev. B* **46** (1992) 1933.
4. A. L. Mackay and H. Terrones, *Nature* **352** (1991) 762.
5. T. Lenosky, X. Gonze, M. Teter and V. Elser, *Nature* **355** (1992) 333.
6. R. Saito, G. Dresselhaus, and M. S. Dresselhaus, *Chem. Phys. Lett.* **195** (1992) 537.
7. P. R. Bandaru et al., *J. Appl. Phys.* **101** (2007) 094307.
8. M. J. O'Connell (Ed.), *Carbon Nanotubes: Properties and Applications*, CRC Press, Boca Raton, FL, 2006.
9. J. Jiang et al., *Chem. Phys. Lett.* **392** (2004) 383.
10. Y. Miyata et al., *J. Phys. Chem. C* **112** (2008) 13187–13191.
11. K. C. Mondal et al., *Chem. Phys. Lett.* **437** (2007) 87–91.
12. P. Ayala et al., *J. Mater. Chem.* **18** (2008) 5676–5681.
13. R. Czerw et al., *Nano Lett.* **1** (2001) 457–460.
14. B. Ha et al., *Physica B* **404** (2009) 1617–1620.
15. C. T. Kuo et al., *Diamond Relat. Mater.* **12** (2003) 799–805.
16. R. Kozhuharova et al., *J. Mater. Sci. Mater. Electron.* **14** (2003) 789–791.
17. M. Yadasaka et al., *Appl. Phys. A* **74** (2002) 377–385.
18. A. C. Dupuis. *Prog. Mater. Sci.* **50** (2005) 929–961.
19. K. Pulskamp, S. Diabate, and H. F. Krug. *Toxicol. Lett.* **168** (2007) 58–74.
20. X. Zhou et al., *Carbon* **35** (1997) 775.
21. X. Zhao et al., *J. Cryst. Growth* **198–199** (1999) 934.
22. Y. Ando et al., *J. Cryst. Growth* **237–239** (2002) 1926.
23. T. Suzuki et al., *J. Nanopart. Res.* **8** (2006) 279.
24. R. Joshi et al., *Diamond Relat. Materi.* **17** (2008) 913.
25. T. W. Ebbesen, and P. M. Ajayan, *Lett. Nat.* **358** (1992) 220–222.
26. X. K. Wang et al., *J. Mater. Res.* **10** (1995) 1977.
27. M. Cadek et al., *Carbon* **40** (2002) 923.
28. J. Guo et al., *Mater. Chem. Phys.* **105** (2007) 175.
29. M. V. Antisari et al., *Carbon* **41** (2003) 2393.
30. S. H. Jung et al., *Appl. Phys A* **76** (2003) 285.
31. S. J. Lee et al., *Diamond Relat. Mater.* **11** (2002) 914.
32. H. Lange et al., *Carbon* **41** (2003) 1617.
33. J. M. Lambert, P. M. Ajayan, and P. Bernier, *Synth. Met.* **70** (1995) 1475–1476.
34. H. Takahashi, M. Sugano, A. Kasuya, Y. Saito, T. Koyama, and Y. Nishina, *Mater. Sci. Eng. A* **217–218** (1996) 48–49.
35. Z. Shi, Y. Lian, F. H. Liao, X. Zhou, Z. Gu, Y. Zhang, S. Iijima, H. Li, K. T. Yue, and S. L. Zhang, *J. Phys. Chem. Solids* **61** (2000) 1031–1036.
36. Y. Ando et al., *Diamond Relat. Mater.* **14** (2005) 729–732.

37. X. Zhou et al., *Diamond Relat. Mater.* **15** (2006) 1098–1102.
38. Y. S. Park et al., *Synth. Met.* **126** (2002) 245–251.
39. C. Journet, W. K. Maser, P. Bernier, A. Loiseau, M. Lamy de la Capelle, S. Lefant, P. Deniard, R. Lee, and J. E. Fischer, *Lett. Nat.* **388** (1997) 756–758.
40. Z. Shi, Y. Lian, X. Zhou, Z. Gu., Y. Zhang, S. Iijima, L. Zhou, K. T. Yue, and S. Zhang, *Carbon* **37** (1999) 1449–1453.
41. Z. Shi, Y. Lian, X. Liao, X. Zhou, Z. Gu, Y. Zhang, and S. Iijima, *Solid State Commun.* **112** (1999) 35–37.
42. S. Farhat, M. Lamy de La Chapelle, A. Loisseau, S. Lefrant, C. Journet, and P. Bernier, *J. Chem. Phys.* **115** (2001) 6752.
43. J. Qui et al., *Carbon* **41** (2003) 2170.
44. S. R. C. Vivekchand and A. Govindaraj, *Proc. Indian Acad. Sci. (Chem. Sci.)* **115** (2003) 509–518.
45. S. Gajewski et al., *Diamond Relat. Mater.* **12** (2003) 816–820.
46. X. Sun et al., *Materi. Lett.* **61** (2007) 3956–3958.
47. D. He et al., *Diamond Relat. Materi.* **16** (2007) 1722–1726.
48. M. Keidar et al., *J. Appl. Phys.* **103** (2008) 094318.
49. H. Lange et al., *Diamond Relat. Mater.* **15** (2006) 1113–1116.
50. I. Hinkov et al., *Carbon* **43** (2005) 2453–2462.
51. V. V. Grebenyukov et al., *Fullerenes Nanotubes Carbon Nanostruct.* **16** (2008) 330–334.
52. L. A. Montoro and J. M. Rosolen, *Carbon* **44** (2006) 3293–3301.
53. C. Wu et al., *J. Phys. Chem C* **113** (2009) 3612–3616.
54. H. G. Cho et al., *Carbon* **47** (2009) 3544–3549.
55. C. P. Deck, G. S. B. McKee, and K. S. Vecchio, *J. Electron. Mater.* **35** (2006) 211.
56. F. Kokai et al., *J. Phys. Chem. B* **104** (2000) 6777–6784.
57. E. Munoz et al., *Appl. Phys A.* **70** (2000) 145–151.
58. W. K. Maser et al., *Opt. Mater.* **17** (2001) 331–334.
59. Y. Zhang et al., *Appl. Phys. Lett.* **73** (1998) 3827.
60. M. Bystrzejewski et al., *Chem. Mater.* **20** (2008) 6586–6588.
61. E. Munoz et al., *Synth. Met.* **103** (1999) 2490–2491.
62. X. Lin et al., *Carbon* **45** (2007) 196–202.
63. M. Yudasaka et al., *J. Phys. Chem. B* **103** (1999) 3576–3581.
64. A. A. Puretzky et al., *Appl. Phys. A* **93** (2008) 849–855.
65. M. H. Rummeli et al., *Nano Lett.* **5** (2005) 1209–1215.
66. W. K. Maser et al., *Nanotechnology* **12** (2001) 147–151.
67. H. Zhang et al., *J. Phys. Chem. Solids* **62** (2001) 2007–2010.
68. G. Radhakrishnan et al., *Thin Solid Films* **515** (2006) 1142–1146.
69. M. Kusaba and Y. Tsunawaki, *Thin Solid Films* **506–507** (2006) 255–258.
70. P. Nikolaev et al., *Chemi. Phys. Lett.* **313** (1999) 91–97.
71. V. A. Karachevtsev et al., *Carbon* **41** (2003) 1567–1574.
72. I. W. Chiang et al., *J. Phys. Chem. B* **105** (2001) 8297–8301.
73. M. J. Bronikowski et al., *J. Vac. Sci. Technol. A* **19** (2001) 1800.

74. Y.-Q. Xu et al., *Nano Lett.* **5** (2005) 163–168.
75. Y. Mackeyev et al., *Carbon* **45** (2007) 1013–1017.
76. R. L. Vander Wal et al., *Appl. Phys. A* **77** (2003) 885.
77. C. Li et al., *Diamond Relat. Mater.* **17** (2008) 1015–1020.
78. Q. Zhou et al., *J. Alloys Compounds* **463** (2008) 317–322.
79. L. Wang et al., *Physica B* **398** (2007) 18–22.
80. C.-C. Hsieh et al., *J. Phys. Chem C* **112** (2008) 19224–19230.
81. Y. X. Liu et al., *Appl. Surf. Sci.* **255** (2009) 7985–7989.
82. T. X. Li et al., *Proc. Combust. Inst.* **32** (2009) 1855–1861.
83. C. J. Unrau and R. L. Axelbaum., *Carbon* **48** (2010) 1418–1424.
84. X. H. Chen et al., *Mater. Lett.* **57** (2002) 734–738.
85. L. Ci et al., *J. Cryst. Growth* **233** (2001) 823–828.
86. E. Kowalska et al., *J. Mater. Res.* **18** (2003) 2451.
87. J. K. Radhakrishnan et al., *Appl. Surf. Sci.* **255** (2009) 6325–6334.
88. L. Delzeit et al., *J. Phys. Chem B* **106** (2002) 5629–5635.
89. K. Bartsch and A. Leonhardt, *Carbon* **42** (2004) 1731–1736.
90. H. Lee et al., *J. Alloys Compounds* **330–332** (2002) 569–573.
91. O. A. Nerushev et al., *J. Appl. Phys.* **93** (2003) 4185.
92. M. Jung et al., *Diamond Relat. Mater.* **10** (2001) 1235–1240.
93. H. Wang et al., *Appl. Surf. Sci.* **181** (2001) 248–254.
94. J. D. Harris et al., *Mater. Sci. Eng. B* **116** (2005) 369–374.
95. A. Barreiro et al., *Appl. Phys. A* **82** (2006) 719–725.
96. H. Liu et al., *Diamond Relat. Materi.* **17** (2008) 313–317.
97. X.-Y. Liu et al., *Fullerenes Nanotubes Carbon Nanostruct.* **10** (2002) 339–352.
98. H. Zhang et al., *Physica B* **337** (2003) 10–16.
99. C. Singh et al., *Carbon* **41** (2003) 359–368.
100. B. Xiaodong et al., *Tsinghua Sci. Technol.* **10** (2005) 729.
101. H. Kim and W. Sigmund, *Carbon* **43** (2005) 1743–1748.
102. Bhalchandra A Kakade et al. *Carbon* **46** (2008) 567–576.
103. Y. Shimizu et al., *Chem. Vap. Depos.* **11** (2005) 244–249.
104. L. C. Qin, *J. Mater. Sci. Lett.* **16** (1997) 457–459.
105. T. M. Minea et al., *Surf. Coat. Technol.* **200** (2005) 1101–1105.
106. Y. H. Mo et al., *Synth. Met.* **122** (2001) 443–447.
107. K. Kordas et al., *Appl. Surf. Sci.* **252** (2005) 1471–1475.
108. H. C. Choi et al., *Chem. Phys. Lett.* **399** (2004) 255–259.
109. Y. Soneda et al., *Carbon* **40** (2002) 955–971.
110. R. L. Vander Wal et al. *Carbon* **39** (2001) 2277–2289.
111. K. Hernadi et al., *Carbon* **34** (1996) 1249–1257.
112. A. Cao et al., *Appl. Phys. Lett.* **84** (2004) 109.
113. P. J. Cao et al., *Materi. Lett.* **61** (2007) 1899–1903.
114. S. M. Mukhopadhyay et al., *J. Phys D* **42** (2009) 195503.

115. S. Zhu et al., *Diamond Relat. Mater.* **12** (2003) 1825–1828.
116. Q. Li et al., *Carbon* **42** (2004) 829–835.
117. W. Z. Li et al., *Appl Phys A* **73** (2001) 259–264.
118. W. Z. Li, *Appl Phys A* **74** (2002) 397–402.
119. G. Gomes et al., *Carbon* **42** (2004) 1473–1482.
120. C. Singh et al., *Physica B* **323** (2002) 339–340.

4

NANOPARTICLES AND POLYMER NANOCOMPOSITES

GUILLERMO A. JIMENEZ, BYOUNG J. LEE, AND SADHAN C. JANA

Department of Polymer Engineering, University of Akron, Akron, Ohio

1   INTRODUCTION

The advent of nanotechnology research in academic and industrial laboratories has spurred significant growth opportunities for nanocomposites for the consumer, defense, aerospace, and health industries [1–4]. The number of journal papers reporting scientific work on nanotechnology grew rapidly from 6422 in 1994 to 36,865 in 2004 [1]. More than 1,9000 patents issued by the U.S. Patent and Trademark Office (PTO) in 30 years between 1976 and 2006 carried the term *nanotechnology* in their titles and claims [2]. Between 2005 and 2006 the U.S. PTO issued 4081 nano patents involving 9491 inventors representing 34 countries [2]. These activities endorsed how

Nanoscale Multifunctional Materials: Science and Applications, First Edition.
Edited by Sharmila M. Mukhopadhyay.
© 2012 John Wiley & Sons, Inc. Published 2012 by John Wiley & Sons, Inc.

widespread the research activities on nanotechnology have been across several disciplines. In this chapter we present highlights on a branch of nanotechnology research involving only the polymers, specifically on polymer nanocomposites.

The central objective of polymer nanocomposites research is to find ways to achieve large enhancements of a set of properties over those of the matrix polymers using only small fractions of nanofillers. The nanofillers are defined to have at least one dimension less than 50 nm [5]. The nanofillers are expected to bring out new chemical and physical properties otherwise not achievable from the matrix polymers. It is known from decades-long research that some nanofillers improve only one type of property (e.g., electrical, mechanical, or thermal), while a select few nanofillers can contribute to improvement of multiple properties at the same time. Examples in the latter category include enhancements in both mechanical and electrical properties, or electrical and thermal properties, or mechanical and barrier properties, and so on.

The challenges in this field of research are threefold. First, proper nanofiller–polymer pairs must be identified. Otherwise, desired improvements cannot be obtained. Second, the nanofillers must be dispersed to nanoscale in the matrix polymers; otherwise, properties improvements cannot be fully realized. Note that true nanocomposites result when nanofillers are dispersed to nanoscale in the matrix materials. Third, all true nanocomposite materials may not be cost-effective. This cost-constraint largely governs the scope of current nanocomposite research and commercialization in industrial laboratories. In the same vein, some novel, innovative nanocomposite research carried out in academic laboratories does not see the light of commercial usage, due to the prohibitive cost of nanoparticles, time-consuming processing and fabrication steps, and sometimes, lack of a proper market. This is supported by a large difference in the number of peer-reviewed journal papers published: 213,847 between 1976 and 2004 [1] compared to a total of 4788 patents published during the same period in top 10 countries, including the United States (3450), Japan (517), and Germany (204) [2]. Of the three challenges presented above, academic research is concerned with the first two, whereas industrial research is focused on preparation and commercialization.

The definitions of nanofillers and nanocomposites widely used in scientific literature are varied. Some guidance on the definition of nanoparticles is provided in Ref. 5. Nanoparticles should have at least one dimension of less than 50 nm. We will term this the *defining dimension*. A platelet type of nanoparticle has length, width, and thickness, of which only the thickness is less than the defining dimension. The length and width of platelet nanoparticles can range from a few hundred nanometers to up even a micrometer. In view of this, a particle with 100 individual platelets each 1 nm thick cannot be termed a nanoparticle. A rodlike nanoparticle has a diameter less than the defining dimension, although the length can be several hundred nanometers. Consequently, an agglomerate of rodlike nanoparticles with about 100 rods cannot be termed a nanoparticle, as the agglomerate diameter may be larger than the defining dimension. All spherical nanoparticles less than 50 nm in diameter can meet the defining dimension of nanoparticles, but their agglomerates with effective diameter greater than 50 nm cannot be termed nanoparticles. In some published work, the spherical, rodlike, and platelet-type nanoparticles are termed zero-, one-, and two-dimensional, respectively.

In physical form, these nanoparticles are usually available as agglomerates. In the case of platelet-type nanoparticles, a large number of individual particles are held together as microscopic agglomerates by attractive forces originating from electrostatic, ionic, and van der Waals forces. The state of agglomeration of nanoparticles is usually dictated by the synthesis methods and, accordingly, the particle agglomerates serve as the starting point in the preparation of polymer nanocomposites. The agglomerates are combined with the matrix polymer in solution or in melt state and are subjected to several dispersion forces, including shear, ultrasound, and gas expansion in efforts to render individual particles dispersed to the nanoscale in the matrix. The transformation from initial microscopic particle agglomerates to individual nanoparticles by dispersion forces poses severe challenges to engineers and scientists, as outlined in this chapter.

In this chapter we present some general trends in the field of research and provide a detailed account of polymer nanocomposites research involving two specific nanofillers: carbon nanotubes (CNTs) and carbon nanofibers (CNFs) and polyhedral oligomeric silsesquioxanes (POSSs).

2   NANOPARTICLES

Nanoparticles are available in primarily three simple geometric forms: spheres, cylinder or rodlike, and platelet type. The primary particles of carbon black, fumed silica, and precipitated silica are available as spheres. However, the electrostatic forces in carbon black and in cases of fumed and precipitated silica, hydrogen-bonding interactions render a fractal nature to the particle agglomerates. Rodlike nanofillers include carbon nanofibers, carbon nanotubes, halloysite-type nanoclay [6–11], copper, and silicon carbide nanorods [12–14]. Layered silicate clays, platelet-type nanofillers, have been used overwhelmingly to produce polymer nanocomposites, although graphene nanofillers recently emerged as more attractive platelet-type nanofillers.

Carbon black and silica have received considerable attention in the literature in conjunction with the development of rubber compounds, although interest in nanometer-size carbon black particles is recent [15,16]. Layered silicate clay particles have also received much attention from clay mineralogist and polymer researchers, due to their abundance in nature, relatively simple modification chemistry, and ease of handling. Several review articles and research monographs addressed important attributes of nanoclays and their polymer composites [17–25]. Research on graphenes is relatively new, although several review articles have been published delineating the major achievements of chemical modifications of graphenes [26–31]. In view of this, we present a detailed account of two nanoparticle: CNT–CNF and POSS and associated composites.

2.1   Carbon Nanotubes and Nanofibers

Carbon is known to be available in several allotropic forms, such as diamond, graphite, fullerenes, and carbon nanotubes and nanofibers. Of these, graphite, fullerenes, and carbon nanotubes and nanofibers have been found attractive for the development

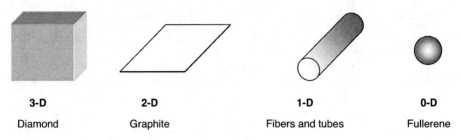

| 3-D | 2-D | 1-D | 0-D |
| Diamond | Graphite | Fibers and tubes | Fullerene |

FIGURE 1   Dimensionality of carbon materials.

of polymer nanocomposite materials. Graphite is available in the form of two-dimensional hexagonal arrangement of carbon atoms in sp^2 hybridization, whereas fullerenes are considered to be zero-dimensional and carbon nanotubes and nanofibers are known to be in one-dimensional forms [32]. The dimensionality of carbon nanomaterials mentioned above is presented in Figure 1. A carbon nanotube can be visualized as a cylinder made up of a rolled graphene layer capped by hemispheres of fullerenes (Figure 2). The curvature in the graphene layers increases the energy of the tube, although the lack of dangling bonds at the edges of the graphene layers lowers the total energy. Carbon fibers are synthesized from several precursors, such as hydrocarbons via decomposition, graphite via chemical vapor deposition, mesopitch fibers via graphitization, and polymer fibers via graphitization, each precursor producing fibers with different morphologies. The mechanical strength of these fibers is based on the fiber axes lying close to the in-plane direction of a graphene layer [33]. Vapor-grown carbon fibers (VGCFs) have hollow cores similar to those of CNTs and have a wide range of diameters, from less than 1000 Å (nanofibers) to more than 100 μm (microfibers) with an aspect ratio (i.e., length/diameter ratio) varying from

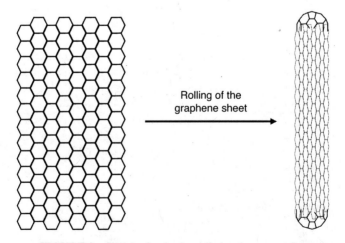

Rolling of the graphene sheet

FIGURE 2   Sketch of a single-walled carbon nanotube.

1000 for nanofibers to 10,000 for microfibers [34]. Carbon nanotubes have aspect ratios comparable to those of VGCFs, while fullerenes have aspect ratios close to unity. Compared to VGCFs, carbon nanotubes are usually composed of smaller numbers of graphene cylinders. For example, multiwalled carbon nanotubes (MWNTs) show several concentric graphene tubules. Compared to conventional carbon microfibers, VGCFs shows high-purity graphene layers with a smaller number of active sites on the fiber surfaces, which makes them resistant to oxidation and less prone to bond to matrix materials [35].

Vapor-grown carbon fibers obtained by decomposing hydrocarbons at temperatures between 700 and 2500°C may grow on substrates (fixed catalyst method) or without a substrate (floating catalyst method). The diameter of the fibers is related to the diameter of the catalyst particles in a fixed catalysis dehydrogenation reaction of a hydrocarbon. The structural morphologies of VGCFs depend on the catalytic methods. For example, micrometer-sized continuous fibers grow with the fixed catalyst, while the floating catalyst method yields nanosized discontinuous filaments [34].

The electrical conductivity of carbon materials depends on lattice perfection: Higher crystal perfection leads to higher electrical conductivity. In this context, thermal treatment can improve the graphitic character of carbon fibers and significantly improve the electrical resistivity [36]. This feature can be observed in Figure 3, which depicts the electrical resistivity, ρ, of various carbonaceous materials compared against that of copper.

Carbon nanofibers are composed of approximately 90 wt% elemental carbon; the rest is contributed mainly by oxygen and hydrogen. Surface defects such as vacancies, dislocations, edges, and steps are present on the surfaces of CNFs [37]. The surfaces

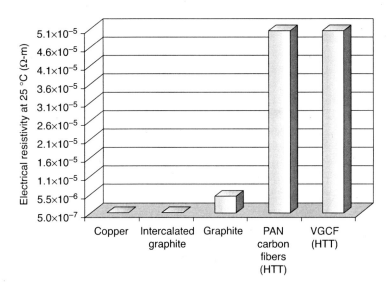

FIGURE 3 Electrical resistivity of various forms of carbons compared to that of copper (HHT, heat treatment temperature). (Adapted from [36].)

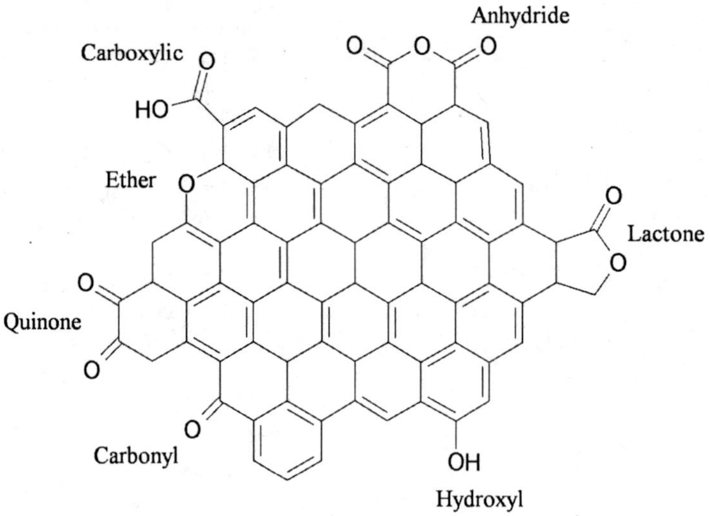

FIGURE 4 Functional surface groups containing oxygen in carbon nanofibers. (Adapted from [38].)

of the carbon fibers are characterized as active sites or as chemical functionality the latter relating to the functional groups containing mainly oxygen and other atoms, such as nitrogen or sulfur. Two categories of functional groups or surface oxides are encountered on CNFs: acidic and basic surface groups. Although these two groups are present on carbons simultaneously, the first ones are found most commonly on the fiber surfaces. Among the most important oxides with acidic character are the carboxyl and phenolic —OH groups. In addition, anhydrides, lactones, and lactols are also found on the carbon surface [38]. The acid dissociation constants of the various groups, such as carboxyls, phenols, and lactones, differ by several orders of magnitude. Therefore, an estimate of their relative amounts can be obtained by titration with bases of different strength [39]. Carbonyl groups are also present on the carbon surface as isolated or conjugated structures (e.g., quinones). Potentiometric techniques and cyclic voltammetry are suitable for the determination of weak acidic groups and quinones. Figure 4 shows a scheme of typical functional groups having oxygen in their structures that are possible to find on the surface of CNFs.

Characterization of surface groups can also be carried out by x-ray photoelectron spectroscopy (XPS) method. This technique has been used extensively to characterize carbon materials: and in particular, surface-treated carbon nanofibers [38–42]. XPS data can reveal chemical information up to a 10-nm depth from the surface [43]. Normally, carbon materials show two main XPS peaks, assigned to carbon and oxygen atoms, respectively, denoted as C(1s) and O(1s). C(1s) shows the following peaks: graphitic peak at 284.6 eV, —C—OH or —C—O—C— peak at 286.3 eV, —C=O peak at 287.7 eV, —COOH or —COOR peaks at 289.4 eV, and —COO— and the π–π^* shake-up satellite at 290.6 eV [39,40,42,44,45]. A number of other surface characterization techniques are also applied to carbon fibers, including Auger spectroscopy, secondary ion mass spectroscopy (SIMS), infrared spectroscopy (IR), Raman spectroscopy and surface-enhanced Raman spectroscopy (SERS), surface energy through wetting experiments, inverse gas chromatography, surface area and pore structure by gas or liquid adsorption, and scanning tunneling spectroscopy to obtain information on the surface roughness [43]. The surface modification of CNFs and CNTs plays an important role in the preparation of composites, providing strong fiber–matrix bonding and thus improving the mechanical properties of the materials.

Two main paths are usually followed for functionalization of CNFs and CNTs: (1) attachment of organic moieties to carboxylic acid groups formed previously by oxidation, and (2) direct covalent bonding of organic molecules (e.g., oligomers and polymers) [46] to the surface double bonds via chemical or electrochemical reactions. Oxidation, fluorination, amidation, and esterification, primarily, are used for chemical functionalization of CNTs and CNFs [47]. Polymers or oligomers are bonded chemically to CNTs and CNFs by two grafting approaches: grafting from and grafting to [48]. A monomer chemically attached to the surface of a CNF or CNT is polymerized in the "grafting from" approach. Polymers with reactive end groups react with the functional groups on the nanofiber or nanotube surfaces in the "grafting to" approach and produce low grafting density due to slow diffusion of the polymer chains to the fiber surface. Chemical modifications of CNTs and CNFs using surface carboxyl groups are not efficient, due to low concentration and low chemical reactivity of the carboxyl groups. In view of this, carboxylic acid groups are often converted into a much more reactive acyl chloride using thionyl chloride, $SOCl_2$, which in turn allowed grafting of poly(ethylene oxide) [49], poly(propionylethylenimine-*co*-ethylenimine) [50], and polyurea [51] onto MWNTs. Wei et al. [52] grafted a poly(ethylene glycol) (PEG) onto the surface of VGCFs by using N,N'-dicyclohexylcarbodiimide (DCC) as a condensation agent (Figure 5) and determined using thermogravimetrical analysis that 11 mol% of grafting was equivalent to 1 mol% of COOH used in the reaction.

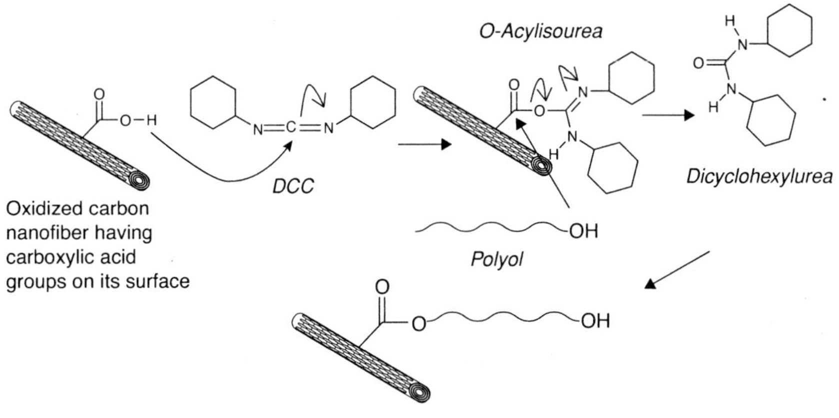

FIGURE 5   Reaction mechanism of the DCC-aided condensation of a polyol onto the surface of CNFs.

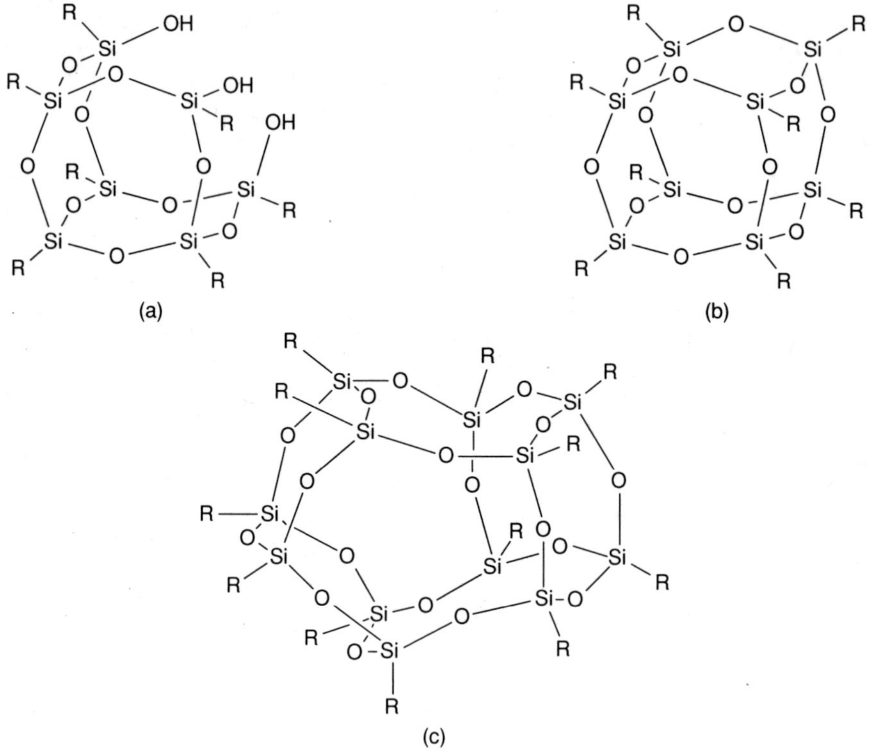

FIGURE 6   Structures of various types of silsesquioxanes: (a) partial cage structure; (b) cage structure; (c) random structure.

2.2 POSS

Polyhedral oligomeric silsesquioxanes (POSSs) are a family of polycyclic compounds of silicon and oxygen (Figure 6) [53] with the generic formula $(RSiO_{3/2})_{2n}$, where n is an integer and R is an organic group. The silsesquioxanes were first discovered in 1946 in the form of completely condensed methyl-substituted POSSs [54] of general formula $(CH_3SiO_{1.5})_{2n}$. Barry et al. [55] were first to reveal the structures of methyl-, ethyl-, n-propryl-, n-butyl-, and cyclohexyl-POSSs. Brown and Vogt published a method of synthesis of POSSs in 1965 [56] based on polycondensation of cyclohexyltrichlorosilane in an acetone–water mixture. The oligomers containing silanol groups, such as incompletely condensed dimers $[RSi(OH)_2OSi(OH)_2R$, R = cyclohexyl] and higher oligomers, were found at the beginning of the reaction. The incompletely condensed cubic silsesquioxane trisilanol $(R_7Si_7O_9)(OH)_3$ and the completely condensed $R_6Si_6O_9$ without the silanol groups were found after long reaction times, spanning about a month. Feher et al. [57,58] reported hydrolytic polycondensation of cyclopentyltrichlorosilane, leading to the formation of cyclopentyltrisilanol

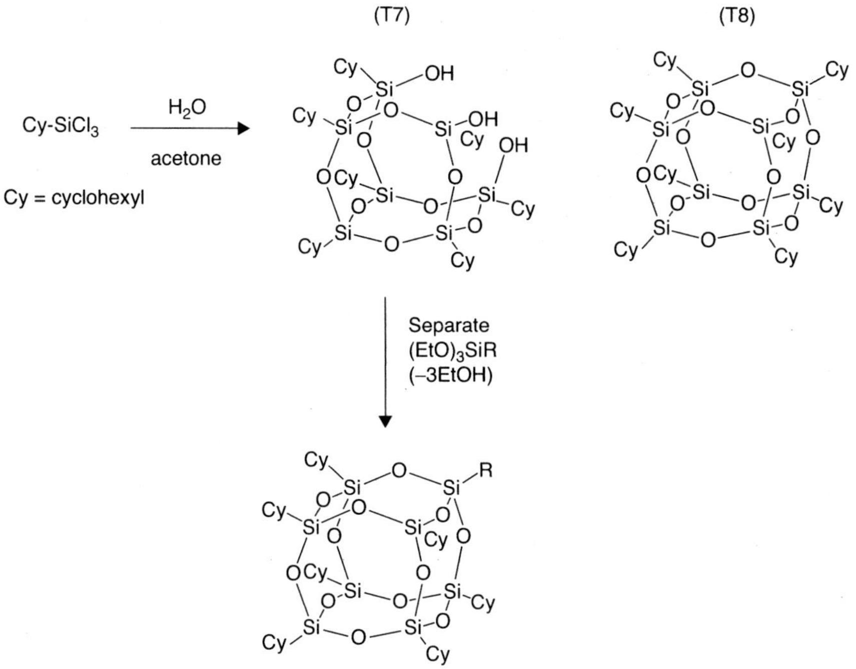

FIGURE 7   Synthesis of monofunctional POSS molecules. (Adapted from [53].)

$[(C_5H_9)_7Si_7O_9(OH)_3]$ and polycondensation of cycloheptyltrichlorosilane, leading to a mixture of cycloheptyltrisilanol $[(C_7H_{13})_7Si_7O_9(OH)_3]$ and cycloheptyltetrasilanol $[(C_7H_{13})_6Si_6O_7(OH)_4]$. Multistep hydrolysis–condensation reactions lead to the formation of POSS molecules (Figure 7) [59]. First, organopolysilanol compounds are formed from the hydrolysis of alkylsilane with water (step 1). The polysilanol compounds condense with each other, leading to the formation of POSS, depending on the concentration of water or solvent, and the pH of the medium (step 2). The resulting mixture contains oligosiloxane dimers or tetramers, POSSs, such as the tetra- and trisilanol silsesquioxanes, depending on the kinetics and solubility of the final compounds.

Lichtenhan et al. [60] obtained a patent on the preparation of completely condensed and incompletely condensed POSSs with isobutyl and isooctyl organic sidegroups that can be prepared on large scales in a short time using base-catalyzed polycondensation reactions. This alleviated a major barrier in the use of POSSs. This has led to the development of large-scale synthesis processes [61] of a variety of POSS chemicals: for example, covalently bonded reactive functional groups that are appropriate for polymerization, grafting, or other transformations. POSS molecules are currently available commercially in solid or liquid oil form from Hybrid Plastics Company (Hattiesburg, MS).

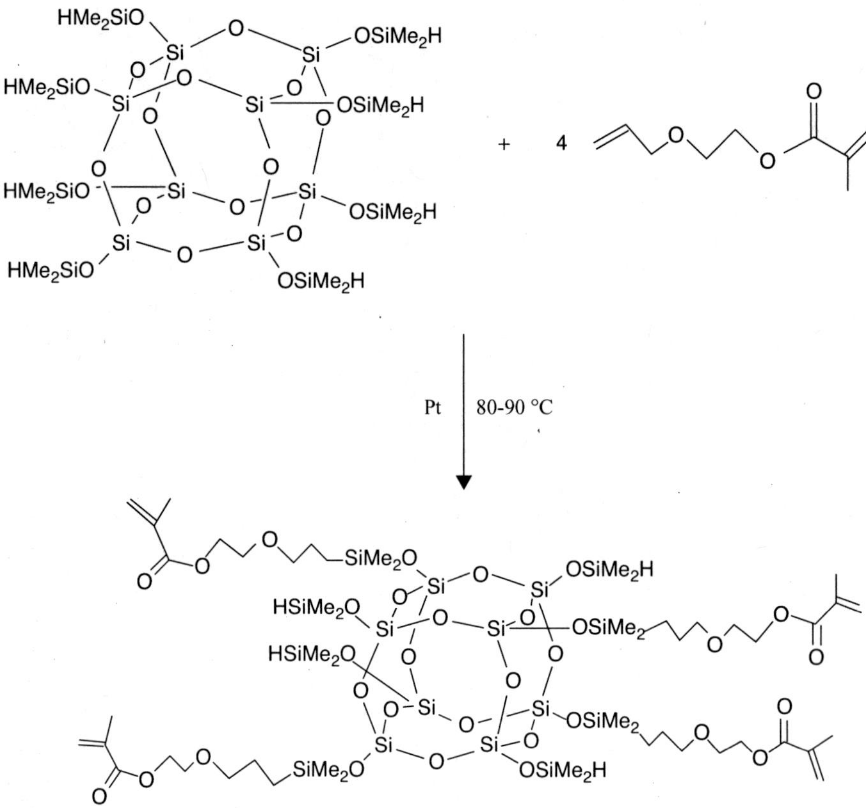

FIGURE 8 Synthesis of multifunctional methacryl (MA)–POSS.

Reactive Monofunctional POSSs Reactive monofunctional POSS molecules with R as the reactive functional group (Figure 8) can be synthesized from controlled hydrolysis and condensation of commercially available organotrichlorosilanes and R—Si(OEt)$_3$ or a related trialkoxysilane [62,63]. Fully condensed POSS cages, with a single specific reactive function are obtained. Examples of specific reactive groups attached are hydride, chloride, hydroxide, nitriles, amines, isocyanates, acrylics, epoxides, norbornyls, alcohols, and acids. These monofunctional POSS molecules are used for synthesis of grafted or copolymerized nanostructured composites.

Reactive Multifunctional POSSs Reactive multifunctional POSS molecules are synthesized from condensation of R—Si(OEt)$_3$; where R′ is a reactive group. The condensation of R′—Si(OEt)$_3$ usually produces an octa-functional POSS, R′$_8$(SiO$_{1.5}$)$_8$. Other multifunctional POSS molecules are produced by functionalizing fully condensed POSS cages: for example, via hydrosilylation of alkenes or alkynes with octahydro-POSS (HSiO$_{1.5}$)$_8$ and octasilane-POSS (HMe$_2$SiOSiO$_{1.5}$)$_8$ cages in the presence of Pt catalysts (Figure 8) [64,65]. Multifunctional POSS derivatives are synthesized from simple hydrolytic condensation of modified aminosilanes [66].

Applications of POSSs   Introduction of POSSs in polymers leads to significant improvements in a variety of physical and mechanical properties, due to reinforcement at the nanoscale and the silica-like properties of the POSS cage. Recent work showed that physical cross-links formed by the POSS aggregates significantly retarded thermal motion of the polymer chains [67,68]. The silicon oxide cage was found to react with atomic oxygen and to protect the surfaces by rapidly forming a protective silica (SiO_2) layer throughout the organic matrix [53]. The self-assembly of POSS molecules in immiscible polymer blends leads to some degree of compatibilization. Other applications for POSS molecules include formation of dendritic macromolecules [69,70]. A phosphine-substituted POSS can be used as a core in the synthesis of dendritic macromolecules [69]. A new metallodendrimer containing this POSS and ruthenium(II)-based chromophores has been synthesized [70].

3   NANOCOMPOSITES

Several methods have been used in literature for the preparation of nanocomposites of carbon nanotubes, carbon nanofibers, and POSSs. In cases of CNTs and CNFs, the research effort can be divided into three focus areas: (1) unentangling nanotubes from agglomerates without fiber length reduction, (2) producing uniform dispersion, and (3) aligning nanotubes so as to obtain high electrical conductivity at the lowest possible loading of CNTs. Research on the first focus area utilizes low-viscosity solvent as the medium for producing uniform suspensions. Good dispersion in solvent is achieved using ultrasound and mechanical mixers. The suspension and the polymer solution in the same solvent are mixed and the composite is recovered after removal of the solvent. Attrition of fiber lengths is limited due to the mild shear conditions of the process. The nanocomposites of low-viscosity resins such as epoxy and polyester are prepared in the same manner.

In the second focus area, the uniform dispersions of CNTs and CNFs are usually obtained by melt mixing of the master batches of CNTs and CNFs with the polymer. The master batches are usually prepared by mixing CNTs with the polymer in solution. This promotes proper wetting of CNTs by the polymer. Some researchers also mix CNTs and polymers in a powder state before melt extrusion. A high shear stress in the extruder can still cause substantial fiber breakage.

Research in the third focus area uses some ingenious fabrication methods whereby the nanotubes are given alignment in a preferential direction so as to produce composites with anisotropic electrical conductivity and, sometimes, enhanced mechanical performance. Solution mixing and melt extrusion processes are avoided in this research, as they produce random mixtures of CNTs. Instead, alignment is achieved during fabrication of the articles: for example, by injection molding or extrusion drawing. The alignment of CNTs in injection-molded articles is concentrated primarily in the skin layer, while the core remains random. The uniaxial drawing process in conjunction with film casting and fiber spinning produces substantial alignments. Some researchers use secondary stretching methods such as calendaring for the simultaneous alignment of polymer chains and CNTs.

3.1   Polymer–Carbon Nanofiber and Nanotube Composites

Both thermosetting resins and thermoplastic polymers are used as matrices for fabrication of polymer–CNF and CNT composites with anticipated improvements in mechanical and electrical properties. In addition, these graphitic materials contribute to higher thermal conductivity and slower thermal degradation of polymers [71–75]. Epoxy and phenolic esters are the primary thermosetting resins, and polypropylene, polycarbonate, polystyrene, and polyamides are popular thermoplastic polymers for the development of polymer nanocomposites. Adequate interfacial adhesion is a prerequisite for development of synergistic polymer–CNF or polymer–CNT composites [37]. This occurs when the molten thermoplastic polymer or the liquid precursor of a thermosetting resin wets a CNF or a CNT. The wetting behavior, on the other hand, is determined from the contact angle and viscosity.

Tensile properties   The tensile strength (σ_C) of an ideal composite, where all fibers are aligned in the direction of uniaxial tension, can be calculated using the fiber strength (σ_f), the strength of the matrix at ultimate fiber deformation ($\sigma_{m\varepsilon}$), and the volume fraction of the fiber (V_f) by using the rule of mixture [76]:

$$\sigma_C = \sigma_f V_f + \sigma_{m\varepsilon}(1 - V_f) \tag{1}$$

The potential of discontinuous short fibers such as CNFs and CNTs on structural reinforcements can be supported by the Cox model [77], which was later extended by Baxter [78]. In this model, the modulus of composite (E_c) is determined from the fiber (E_f) and matrix moduli (E_m) and the volume fraction of fiber (V_f) and matrix (V_m):

$$E_C = E_m V_m + E_f V_f g \left(\frac{l}{d}\right) f(\theta) \tag{2}$$

Functions f and g in equation (2) can take values between 0 and 1. The function g is small for low-aspect-ratio particles, but increases rapidly as the aspect ratio increases. The function f depends on the orientation angle of fibers, θ. Equation (2) implies that the largest improvement in modulus is obtained when high-aspect-ratio particles are oriented.

Thermal Management   Polymer composites of CNTs and CNFs are suitable for high-performance thermal management applications such as electronic heat sinks and radiator fins. Composites with fibers oriented in preferred directions can offer desired combinations of thermal conductivity and coefficient of thermal expansion. On the other hand, composites prepared with randomly oriented short fibers offer

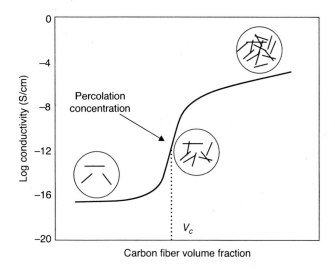

FIGURE 9   Dependence of the electrical conductivity on the filler content.

near-isotropic thermal properties and are more suitable for applications involving carbon–carbon composites for pistons, brake pads, and heat sink applications.

Electrical Conductivity   Electrical conductivity is a strong function of filler volume fraction. At low filler loadings, the conductivity of the composite is close to that of the electrically insulating polymer matrix. At or above some critical loading (i.e., percolation threshold), the conductivity increases by several orders of magnitude with very little increase in the filler amount. At the percolation threshold the filler particles begin to form continuous conductive networks in the composite (Figure 9) [79]. The conductivity levels off and approaches that of the filler material at concentrations higher than the percolation threshold.

Several models predict the electrical conductivity behavior of composites based on structural properties, interfacial properties, and constituent conductivity of both the filler and the polymer. The type, shape, and orientation of the filler particles within the matrix determine the electrical conductivity of the composites. In the case of spherical conductive particles, the smaller the particles, the lower the percolation threshold. For fillers with an aspect ratio greater than unity, a lower percolation threshold results for higher-aspect-ratio fillers with broader distribution of lengths [79]. Most models are of the statistical percolation type and predict the conductivity based on the probability of particle contacts within the composite. One of basic statistical models follows a power-law equation:

$$\rho = \rho_0 (V - V_C)^s \tag{3}$$

where ρ is the conductivity of the mixture, ρ_0 the conductivity of the filler, V the volume fraction of the filler, and V_C the percolation volume fraction. The critical exponent, s, depends on the dimension of the lattice. More accurate statistical models include polymer gelation and a general effective media equation that considers the resistivity of both components.

3.2 POSS–Polymer Composites

Several approaches are followed in the literature for preparation of POSS–polymer composites. One approach is to form covalent bonds, whereby POSS molecules are chemically tethered to one or more polymer chains. In another approach, POSS molecules are linked covalently to monomers and the monomer is later polymerized. A third approach utilizes mixing of the polymer with POSS in solution or in the melt state. In the latter approach, researchers exploit noncovalent interactions such as hydrogen bonding to achieve acceptable dispersion. In the following, several examples of each approach are discussed.

Chemical Tethering of POSS with Monomer and Subsequent Polymerization In this approach, POSS molecules are chemically attached to the monomer and the monomer is polymerized. A monofunctional POSS molecule can form only one attachment to the monomer (e.g., with methacrylate). In this case, POSS cages are found on alternative carbon atoms in the backbone of the linear polymer. Alternatively, these POSS-tethered monomers can be copolymerized with another vinylic monomer. The molar ratio of monomers can be used to control the proportion of POSS in the polymer chains. On the other hand, POSS molecules tethered to two or more vinylic monomers may serve as cross-link sites and help produce fully cross-linked polymers with substantial stiffness and mechanical and thermal stability. POSS-tethered vinylic monomers (e.g., methacrylates) are polymerized by a free-radical method, while POSS-tethered epoxies and polyurethanes are obtained by condensation polymerization techniques. A novel method based on ring-opening metathesis polymerization is also used for the synthesis of copolymers of olefins containing POSS. An example has been found where cyclooctene and POSS–norbornylene monomers are copolymerized and random copolymer is obtained [80].

Chemical Tethering of POSS with Presynthesized Polymers POSS molecules can be tethered to chain ends or side chains of presynthesized polymers. For example, polyimide–POSS composites are synthesized from mixtures of diamine functionalized POSS, acid dianhydride, and aromatic diamines. A significant portion of the research and development efforts for POSS–polymer composites has been focused on the synthesis and characterization of copolymerized POSS–polymer systems. Table 1 lists the POSS-related activities of various research groups with active interests in POSS–polymer composites. The studies included in this survey focused on self-assembled morphology of POSS and thermal and mechanical properties of POSS–polymer nanocomposites.

　　Mather and co-workers studied structure–property relationships of POSS–polymer systems [80] and used POSS–norbornyl polymer compound to demonstrate how the self-assembled structures of POSS molecules were formed via POSS–polymer

TABLE 1   Summary of Research Focus of Major Research Groups on POSS–Polymer Composites

Research Group	Major Contributions	Polymer System
P. Mather (University of Connecticut, Case Western Reserve University)	Synthesis and characterization of POSS–polymer and ABA triblock copolymer composites	Polynorbornyl, polystyrene, ABS (acrylonitrile–butadiene–styrene), polyurethane, epoxy, poly(methyl methacrylate)
B. Coughlin (University of Massachusetts)	Synthesis and characterization of POSS–polymer composites	Polybutadiene, polystyrene, polyolefins
A. Lee (Michigan State University)	Synthesis and characterization of POSS–polymer and thermoset composites	SBS (styrene–butadiene–styrene), epoxy
D. Schiraldi (Case Western Reserve University)	Synthesis and characterization of POSS–polymer composites Characterization of nonreactive POSS–polymer blend	Poly(ethylene terephthalate), polyamide, polycarbonate
M. Capaldi (MIT)	Modeling of POSS–polymer composites	Polyethylene

interactions at the molecular level [81]. These authors also observed block and random POSS–norbornyl copolymers by varying the concentration and the nature of side groups such as cyclopentyl and cyclohexyl on POSS molecules. These authors showed that slight changes in domain size (10 to 20 nm) can result in significant changes in glass transition behavior. Mather [82] synthesized ABA triblock copolymers with a soft middle poly(butylacrylate) (pBA) segment and POSS-containing outer blocks and identified several separated microstructures as a function of composition and the degree of polymerization of the two segments. One example is a morphology consisting of pBA cylinders in a continuous POSS phase. The thermal and linear rheological behavior of random styrene–styrene-based butyl (Bu) POSS copolymers was recently reported [83]. A reduction in T_g with increasing (Bu)POSS concentration in the molecularly homogeneous copolymers was attributed to increased free volume. The favorable (Bu)POSS–polystyrene interaction was believed to produce small POSS particles, 1.5 to 3 nm in size, for loading of POSS copolymer between 6 and 30 wt%.

The cyclopentyl–POSS/poly(methyl methacrylate) (PMMA) copolymer was found to compatibilize the blend of polystyrene (PS) and poly(methyl methacrylate)

(PMMA) homopolymer [84] and led to reduced domain size, increased interfacial width, and improved fracture toughness in the blends. This was attributed to favorable interactions of the cyclopentyl group on POSS with a PS phase. Mather and co-workers used POSS as the building blocks to improve the toughness of epoxy-based thermosets [85]. These authors argued that the main toughening mechanism in the POSS-reinforced epoxy thermoset system was void formation at the nanometer scale (8 to 12 nm), possibly templated by limited POSS aggregation. In their studies on POSS telechelics [86,87], they synthesized amphiphilic telechelics with POSS end caps, which allowed control of the molecular architecture and resulted in a dramatic decrease in the crystallinity of the PEO segments.

Zhang et al. [88] prepared hybrid norbornylene-substituted POSS–polyolefin copolymers using single-site catalysts. These copolymers exhibited improved thermooxidative stability due to incorporation of a wide range of POSS amounts in these polyolefin–POSS copolymers. Coughlin and co-workers [89–91] synthesized POSS–polybutadiene and POSS–polyethylene copolymers by ring-opening metathesis copolymerization and single-site catalysts and argued that POSS–copolymer can adopt two-dimensional networks as the highest possible architecture, due to the constraints of the polymer chains; on the other hand, three-dimensional structures are expected for most POSS–polymer blends. Iyer and Schiraldi [92] prepared copolymers of polyesters and polyamides with 5 wt% glycidyl phenyl-POSS and aminoethyl aminopropyl isobutyl-POSS respectively and compared the properties with those prepared by blending, i.e., without chemical reactions with the polymer chains. Blends of PET with glycidyl phenyl-POSS resulted in little or no improvement in the thermomechanical properties under most conditions. Polyamides exhibited stronger potential for interactions due to hydrogen bonding with the amino groups in aminoethyl aminopropyl isobutyl-POSS. Cardoen and Coughlin [93] prepared hemi-telechelic POSS–polystyrene copolymers from the reactions between the POSS molecules with NCO side groups and hydroxyl-terminated polystyrenes and found that the presence of POSS tethered at the end of the polystyrene chains did not change the glass transition temperature, suggesting that the POSS moieties and polystyrene chains were secluded from each other. The POSS molecules crystallized and did not mix with the polystyrene phase. Consequently, improvements in mechanical properties were not observed.

Lee and co-workers found that T_g increased with an increasing weight fraction of the monofunctional POSS–epoxy molecules in epoxy thermosets [94,95] and that the monofunctional POSS–epoxy nanoreinforcements to such networks did not alter the shape of the viscoelastic spectrum, despite their ability to shift it to higher temperatures [94]. These investigators found that the phenyltrisilanol POSS exerted catalytic activity in the curing of epoxy–amine networks and promoted a higher degree of curing of epoxy networks [95]. Lee and co-workers [96,97] found in copolymers of POSS-grafted SBS and styrene-*co*-vinyl diphenylphospine oxide that POSS enhances the T_g of the butadiene segment and improves the storage modulus at temperatures where the styrene begins to soften. Schiraldi and co-workers worked on the properties of POSS-incorporated poly(ethylene terephthalate) (PET) fibers [98,99]. They prepared the fibers by melt spinning of a POSS–PET composite with nonreactive POSS, trisilanol POSS, and reactive POSS. The reactive POSS was obtained by in situ polymerization with 2.5 wt% propanediol isobutyl–POSS or triepoxy–POSS. Significant increases in tensile modulus and strength were achieved in PET fibers with nonreactive POSS at room temperature. Also, PET/trisilanol–POSS fibers exhibited better high-temperature modulus retention [98]. The tensile strength and modulus of the melt-mixed composite fibers decreased compared with PET, due to phase separation of epoxy–POSS from melt-mixed PET/epoxy–POSS blends [99].

Tamaki et al. [100] reported that trisilanol–POSS undergoes phase changes when held for prolonged periods of time at higher temperatures [e.g., at PET processing conditions ($\sim$300°C)]. However, Zeng et al. [101] studied the structural changes of trisilanol–POSS with isooctyl side groups as a function of time, temperature, and air–nitrogen atmosphere and observed no differences in thermomechanical properties between samples produced using trisilanol–POSS or precondensed POSS. Zhao and Schiraldi [102] studied polycarbonate (PC)–POSS composites prepared by melt blending and compared the thermal and mechanical properties as the nature of POSS molecules was changed, such as fully condensed cage–POSS with a phenyl substituent or trisilanol–POSS with a phenyl or isooctyl substituent. POSS derivatives exhibit different compatibilities with PC, depending on the specific structure of the POSS used. Trisilanol POSS molecules generally provide better compatibility with PC than other POSS derivatives tested in their study. Unfortunately, the tensile properties were found to be lower than PC and decreased monotonically with POSS content.

Capaldi et al. [103] performed lattice dynamics simulations of various crystal forms of POSS with an octacyclopentyl side group as well as molecular dynamics simulations of POSS particles dispersed in an organic matrix at temperatures from 300 to 500 K. They focused on the dynamics of the POSS particles as they interact with the matrix material and observed interesting aggregation properties at lower temperatures (300 K), in accordance with prior experimental observations. In another publication [104], they determined the anisotropic elastic constants of crystalline octacyclopentyl–POSS (CpPOSS) using molecular dynamics simulation. They used the force field to calculate the elastic constants based accurately on the triclinic crystal structures of CpPOSS, as well as the vibrational frequencies of CpPOSS. The moduli for CpPOSS are anisotropic, with a Reuss-averaged bulk modulus of 7.5 GPa, an isotropic averaged Young's modulus of 11.78 GPa, and an isotropic averaged shear modulus of 4.75 GPa.

Ricco et al. [105] and Baldi et al. [106] studied POSS–polyamide 6 (PA6) nanocomposites using an in situ polymerization method. In their first publication [105], poor dispersion of nonpolar POSS in the matrix was observed due to interaction among highly polar PA6 chains. These authors next studied the behavior of POSS-modified PA6. In this case, PA6 chains terminated at one end with aminopropyl heptaisobutyl–POSS were used [106]. They showed that the presence of POSS resulted in decreased stiffness relative to that of PA6 neat polymers, which was attributed to a reduction in the degree of crystallinity. Liu and Lee [107] prepared POSS tethered aromatic polyamide nanocomposites with various POSS fractions through Michael addition between maleimide-containing polyamides and amino-functionalized POSS.

TEM images revealed that self-assembled POSS molecules were of submicrometer size and were homogeneously dispersed in the polymer matrix. While the tensile strength of these composites decreased, POSS modification increased the storage modulus and Young's modulus. In addition, a slight decrease in T_g from 312°C to 305°C was observed.

Self-Assembly of POSS in Polymers

Self-Assembled Structure of POSS in polymers POSS has the capability to build higher dimensionality in the polymer matrix through aggregation called *self-assembly*. The ability to self-assemble is a key attribute of POSS nanostructured materials. This is aided by the versatility of functional groups that can be incorporated into POSS molecules. Consequently, many researchers attempted incorporation of POSS molecules into polymer matrices via two common routes: grafting and copolymerization.

Mather and co-workers [80] worked on structure–properties relationships of POSS–polymer systems considering composites of POSS–norbornyl copolymers. These authors showed that changing the organic side groups on the POSS cage can alter the diameter of cylindrical assembled domains of POSS aggregates [81]. They studied the microstructure and shape-memory properties of these copolymer composites having either cyclohexyl corner groups (CyPOSS) or cyclopentyl side groups (CpPOSS). From the TEM image of a polynorbornyl composite with 50 wt% of CyPOSS (50CyPN), cylindrical aggregates with average length and diameter approximately 62.5 and 12 nm, respectively, were identified. The size of aggregated domains decreased, however, to about 36 nm in length and about 6 nm in diameter for the case of polynorbornyl composite with 50 wt% CpPOSS (50CpPN). The mechanism of aggregation of POSS molecules into a cylindrical shape did not become apparent from only two TEM images at 50 wt% loading of POSS. More recently, the same group reported self-assembled structures for random copolymers of polystyrene (PS) and styryl-based POSS [83]. It was demonstrated by TEM that the butyl-POSS dispersed in the polymer matrix at a molecular level (e.g., the size of dispersed domain is between 1.5 and 3 nm, approaching the size of single POSS molecules).

A visual model of a POSS–polymer assembly has been suggested by Coughlin and co-workers [89–91] based on the TEM images of POSS–polybutadiene (PBD) and polyethylene (PE) copolymers. It was seen that the size of POSS particles in PE copolymer was small and the nature of self-assembled structure in POSS–PE copolymer composite was not revealed. The POSS–PB composites were prepared by the method of ring-opening metathesis copolymerization or single-site catalysis. In this case, a bottom-up polymer nanocomposite was obtained due to self-assembly of POSS particles caused by the affinity between POSS units. Crystallization of the polymer into a two-dimensional lattice was obtained due to restrictions of POSS molecules bonded covalently to polymer chains. A schematic model was proposed by the authors as presented, which was obtained in relation to TEM data containing 10 and 40 wt% POSS with cyclohexyl side groups, displaying the "raft-like" structure of POSS [89]. From these studies we learn that only amorphous polymer (e.g., a PBD with POSS system) can offer a well-ordered self-assembled structure, while in the case of crystallizable polymer (PE), the POSS assembled structures can be too small and are randomly distributed. In the latter case, crystallization of polymer chains is a dominant phenomenon, and POSS molecules attached to the backbone of polymer chains produce random dispersion.

Leu et al. [108] investigated polyimide–POSS copolymer prepared by copolymerization reaction between POSS–diamine and pyromellitic dianhydride (PMDA). This polyimide–POSS nanocomposite exhibited self-assembled structures when the amount of POSS exceeded 10 mol% (26.6 wt%), as evident from TEM images. Domains of parallel dark lines consisting of POSS molecules about 2 nm thick were formed. This self-assembly was caused by much stronger polar–polar interactions between the imide segments than the van der Waals interaction between POSS–POSS segments. From Leu et al.'s study [108], it is evident that associative polymer with copolymerized POSS behaves differently from amorphous polymer such as PBD–POSS, as in the work reported by Coughlin et al. [89].

Lee et al. [109] investigated the miscibility of a phenolic resin with isobutyl-POSS by preparing the materials in solution in tetrahydrofuran (THF). A POM image of a phenolic–POSS hybrid material (20 wt% POSS) showed that the crystals of POSS were arranged evenly in the polymer matrix and the self-assembled aggregates of POSS molecules were confirmed by AFM. The sizes of the POSS domains in a composite with 20 wt% POSS crystals fell in the range 1.5 to 2.0 μm, which was considered to be due to hydrogen-bonding interactions between POSS and phenolic phases, respectively via siloxane Si—O—Si group and phenyl—OH groups. The AFM image confirms the self-assembly behavior of POSS. Note that the assembled size of POSS molecules in this case was much greater than the few nanometers in the case of copolymerized POSS composites.

Although previous studies show that one can obtain POSS self-assembled nanoparticles ranging in size from 1.5 to 100 nm (anisotropic) in various polymer composites (Table 2), the only copolymer systems that have been studied had relatively high loadings of POSS. On the other hand, nanoscale self-assembly of POSS molecules via melt blending has not been studied extensively. In this context, self-assembly between polymer and nonreactive POSS molecules can be considered an attractive area of research.

Self-Assembly of POSS Nanoparticles Owing to the complex chemistry of POSS and polymer, neither the microstructure of POSS–polymer composites nor the mechanism of reinforcement is well understood. Such reinforcement may arise either from isolated POSS nanoparticle units or from aggregates of these units into larger POSS clusters. The degree of aggregation by self-assembly may be expected to depend on the fraction of POSS nanoparticles, the degree of compatibility between POSS and the host polymer, and the tendency for the host polymer to form a separate crystalline phase.

Capaldi et al. [103] simulated the aggregation characteristics of a POSS–PE blend system. The evolution of structure was monitored over the time scales accessible within their simulations. They showed snapshots of the system with 25 wt% of POSS

TABLE 2 Summary of Self-Assembled Structures Observed by Previous Researchers

Authors	Polymer/Composite	POSS Type (R = alkyl, X = reactive)	Self-Assembly Shape and Size (L = length, D = diameter, T = thickness)
Jeon et al. [81]	Polynorbornyl/ copolymer–POSS	R = cyclopentyl (Cp) and cyclohexyl (Cy) X = monofunctional	Cylinder ($L \times D$) 50% Cp = 36 × 6 nm 50% Cy = 63 × 12 nm
Wu et al. [83]	Polystyrene/ copolymer–POSS	R = isobutyl X = mono–functional	Sphere (D) 6–30% = 1.5–3 nm
Leu et al. [108]	Polyimide/ copolymer–POSS	R = cyclopentyl (Cp) X = monofunctional	Plate ($L \times T$) 27–40% = 100 nm × 2–4 nm
Zheng et al. [89]	Polybutadiene/ copolymer–POSS	R = cyclopentyl X = monofunctional	Plate ($L \times T$) 10% = 50 nm × 3–5 nm 40% = 3–5 μm × 3–5 nm
Lee et al. [109]	Phenolic resin/ solvent mixing–POSS	R = isobutyl (8 corner) Nonreactive	Sphere (D) 20% = 1.5–2 μm

at 500 and 300 K. In this case, the simulation was carried on for 5 ns at 500 K, holding the number of atoms, the pressure, and the temperature constant. In the case of 300 K, simulation was continued for 10 ns at 300 K, holding the number of atoms, the stress tensor, and the temperature constant. POSS particles are initially distributed randomly within the matrix at 500 K. Upon cooling to 300 K, the particles aggregate further within the polymer matrix, with the aggregates now developing into an identifiable internal structure.

Based on the work of Capaldi et al. [104], Jana and co-workers developed a new method for development of PP–POSS nanocomposites using a soribitol-type nucleating agent as a dispersion aid [110,111]. They hypothesized that POSS molecules containing hydroxyl or silanol groups should disperse well in molten PP, due to POSS–sorbitol hydrogen-bonding interactions and due to the solubility of sorbitol in molten PP. The sorbitol then forms three-dimensional fibrillar networks upon cooling of the melt and provides templates for self-assembly of POSS molecules into nanoparticles. In the latter step, the self-assembly of POSS molecules is promoted by hydrogen bonding, which is reestablished as the melt cools down. Perilla et al. [110] showed that hydrogen bonding interactions between sorbitol and POSS molecules interferes with sorbitol fibrillation. A schematic diagram of the self-assembly of POSS molecules during mixing and processing, such as fiber spinning and film casting, is shown in Figure 10. For example, in the case of fiber spinning and film casting, the final stage of self-assembly is expected to occur during nonisothermal cooling. This scenario was captured by Lee et al. [111] in reinforcing spun fibers of PP as shown in Figure 11.

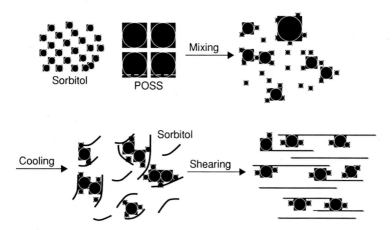

FIGURE 10   Flow-induced self-assembly of POSS by the templates of sorbitol.

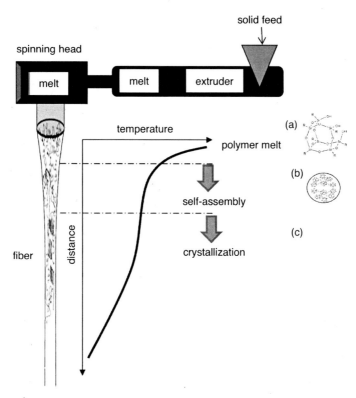

FIGURE 11   Fiber spinning process of PP–POSS nanocomposites. (a) POSS remains in molecularly dispersed state in PP melt. (b) POSS self-assembly into nanoparticles with cooling of fibers. (c) PP crystallizes after POSS self-assembly is complete.

In this case, crystallization of PP was delayed. Also self-assembly of POSS molecules into nanoparticles provided melt strength, so that fibers could be spun with a 50 to 70% reduction in diameter. These fibers then showed 60 to 80% increases in tensile modulus and strength over unfilled PP.

The organic–inorganic hybrid nature of POSS materials offers a number of unique advantages over traditional nanoparticles in the development of polyolefins composites such as layered silicate [112–128], carbon nanotubes [129–136], and nano silica [137–146].

Fu et al. [67] performed crystallization studies of isotactic polypropylene (iPP) blends with octamethyl-POSS under quiescent conditions and under shear. In this work, the DSC traces reveal that octamethyl-POSS crystals acted as nucleating agents in quiescent crystallization of iPP. The addition of POSS also significantly increased the crystallization rate during shear compared with the rate for the neat polymer. So these authors postulated that dispersed POSS molecules behaved as weak cross-linkers in polymer melts and thus increased the relaxation time of the polymer chains after shear. Consequently, polymer chains retained orientation longer, and hence a faster polymer crystallization rate was observed in the presence of POSS. Additionally, Fu et al. [68] studied the rheological behavior of ethylene–propylene (EP) copolymers containing POSS molecules under oscillatory shear and DMA. They reported that EP–POSS nanocomposites exhibited solidlike rheological behavior, apparently inducing physical gelation of EP in the presence of POSS molecules.

Fina et al. [146,147] studied the influence of functional groups of POSS cages on the properties of PP–POSS nanocomposite blends. POSS with different side-chain lengths (e.g., octamethyl-, octaisobutyl-, and octaisooctyl-POSS were considered and melt-mixed with PP). It was observed that octamethyl-POSS acted as a nucleating agent, although poor dispersion was produced. On the other hand, octaisobutyl-POSS did not act as a nucleating agent, but produced good dispersion. However, in the latter case, the POSS aggregates were a few micrometers in size. The thermoxidative properties of the PP matrix improved and the peak weight loss temperature increased in the presence of POSS.

Zheng et al. [88] synthesized PP–POSS copolymer nanocomposites using a metallocene catalyst. These copolymers exhibited a 5 to 15% decrease in melting temperature but a 5 to 15% increase in decomposition temperature (5 to 15%) as measured at a 5% weight loss in air. A decrease in dynamic modulus was detected due to the decrease of crystallinity (e.g., by 10% at 20 wt% POSS loading).

The thermal and tensile properties of POSS-containing polymer composites collected from literature are presented in Table 3. It is seen that the improvement in thermal and tensile properties depends critically on the surface chemistry of the POSS molecules, the polymerization method, the additive content, and the crystallization conditions of host polymers. Previous studies have shown that a self-assembled nanoscale structure of POSS molecules can only be obtained by the POSS–copolymer system. Although PP–POSS and PA–POSS composites showed improved thermal and dynamic mehcnaical properties, there are few successful studies reporting improvement in the mechanical properties of bulk copolymer composites and a nonreactive blend system.

TABLE 3 Summary of Thermal and Mechanical Properties of POSS Nanocomposites

Authors	Polymer/Composite	POSS Type (R = alkyl, X = reactive)	Thermal and Mechanical Properties[a]				
Leu et al. [108]	Polyimide/ copolymer–POSS	R = cyclopentyl X = monofunctional	POSS (%)	T_g (°C)	Mod. (GPa)	E_b (%)	T_b (MPa)
			0	350	1.6	6	51
			18	317	1.58	5	49
			33	308	1.43	4	46
Huang et al. [148]	Polyimide/ copolymer–POSS	X = multifunctional (octaaminophenyl)	POSS (%)	T_g (°C)	Mod. (GPa)	E_b (%)	T_b (MPa)
			0	301	2.5	7.5	82
			2.7	326	2.6	8.0	99
			5.3	340	2.8	7.1	90
Liu et al. [107]	Polyamide 6/copolymer–POSS	R = isobutyl X = monofunctional	POSS (%)	T_g (°C)	Mod. (GPa)	E_b (%)	T_b (MPa)
			0	312	0.7	7.8	37
			1.0	305	1.1	1.8	14
Baldi et al. [106]	Polyamide 6/copolymer–POSS	R = isobutyl X = monofunctional	POSS (%)	x_c (%)	Mod. (MPa)	E_b (%)	
			0	20	490	60	
			5.4	16	290	105	
			10.8	16	310	40	
Zeng et al. [98]	PET (fiber)/ copolymer–POSS/ blend–POSS	R_1 = octaisooctyl R_2 = octyltrisilanol X = monofunctional (co-)propandiolbutyl	POSS (%)	T_g (°C)	Mod. (GPa)	E_b (%)	T_b (GPa)
			0	79	11	36	0.6
			R_1(5)	80	15	17	0.8
			R_2(5)	82	13	27	0.6
			X(2.5)	79	3	27	0.5

(Continued)

TABLE 3 *(Continued)*

Authors	Polymer/Composite	POSS Type (R = alkyl, X = reactive)		Thermal and Mechanical Properties[a]				
Zhao et al. [102]	PC/blend–POSS	R = phenyltrisilanol	POSS (%)	T_g (°C)	Mod. (GPa)	Yield (MPa)	T_b (MPa)	
			0	153	0.67	52	51	
			2.5	153	0.66	51	50	
			5	154	0.66	52	43	
			10	153	0.67	51	42	
Wu et al. [83]	PS/copolymer–POSS	R = isobutyl	POSS (%)	T_g (°C)	M_w*10^3 (g/mol)			
		X = monofunctional	0	98	161			
			6	93	186			
			15	92	195			
Fina et al. [146]	PP/blend–POSS	R_1 = octamethyl	POSS (%)	T_d in air (°C)	T_d in N_2 (°C)			
		R_2 = octabutyl	0	319	460			
			R_1(10)	331	460			
			R_2(10)	349	460			
Fu et al. [68]	PP/blend–POSS	R = octamethyl	POSS (%)	T_g (°C)				
			0	−50				
			10	−45				
Zheng et al. [88]	PP/copolymer–POSS	R = cyclopentyl	POSS (%)	T_m (°C)	T_d in N_2 (°C)			
		X = norbonylene	0	140	382			
			20	134	405			
			58	130	421			

[a]Mod, Young's modulus; E_b, elongation at break; T_b, tensile strength; x_c, crystallinity; T_d, degradation temperature; T_g, glass transition temperature; T_m, melting temperature; M_w, molecular weight.

It is apparent that enhancement of mechanical properties of PP–POSS nanocomposites spun into fibers remains an unexplored area of study. So we postulated that the self-assembled size of POSS molecules would play a key role in providing the desired improvement in mechanical properties in PP–POSS nanocomposites. It was hypothesized that such properties would be controlled by the interactions between properly chosen POSSs and sorbitol-type nucleating agents.

4 GENERAL PRINCIPLES FOR DISPERSION OF NANOFILLERS

The electrostatic attractive force holding the individual nanoscopic filler particles together is large due to the small gap between the particles, which is usually on the order of a few nanometers [149–151]. This is true for layered silicate clay, expanded graphite nanoparticles, and CNTs and CNFs. The electrostatic attractive force is often greater than the shear force produced in typical mixing equipment. Filler particles are often predispersed in low-viscosity resin (e.g., epoxy) or monomers (e.g., caprolactam before curing and polymerization). In these cases, shear forces are often limited. To circumvent this, nanofillers are dispersed in low-viscosity resin or monomer by sonication. Although somewhat successful in breaking particle agglomerates and achieving dispersion, this method often causes breakage of high-aspect-ratio particles and sometimes triggers premature curing or polymerization of the monomers, as achieving good dispersion by sonication usually takes from several hours to an entire day. Some high-density nanoparticles also undergo gravity-induced settling even from well-dispersed states.

Melt compounding in extruders provides large shear forces, due to higher polymer viscosity. Typical twin-screw extruders running at a shear rate of 50 to 100 s^{-1} for a polymer with a 1000 Pa·s shear viscosity exerts 0.05 to 0.1 MPa of shear stress on nanoparticle agglomerates. This works well for compatible polymer–nanofiller systems (e.g., polyamides and layered silicate clays) [152]. Nevertheless, the melt compounding method, although conducive for nanoparticle dispersion, does not always produce desirable results.

In the case of CNFs and CNTs, poor interactions at filler–polymer interfaces lead to insignificant polymer infiltration into nanoparticle bundles. High polymer viscosity, amorphous carbon holding several fibers together, and large aspect ratios of fibers are found to be responsible [153]. In addition, the high shear force encountered in melt compounding causes severe damage to nanofibers, with a consequent reduction in aspect ratio [154] and deterioration of composite properties.

Fiber dispersion can be greatly improved in polar polymers if CNFs and CNTs are functionalized. However, functionalization causes a reduction in nanofiber intrinsic conductivity as well as mechanical strengths. In prior studies on polymer composites of CNTs and CNFs, an array of mixing methods have been used to achieve desired degrees of dispersion: for example, twin-screw extruders [154–157], ball mill [153,158–160], two-roll mill [161], multiple hot pressing [162], internal mixing [155,164–167], and single-screw extruder [164] in addition to in situ polymerization [168–171]. In many of these cases, no information is available on mixing kinematics, rates of shear, and contributions of shear and elongational flows. In addition, the merits of various methods of dispersion were not compared and contrasted.

Recently, a two-dimensional batch chaotic mixer with well-defined rates of shear was used to obtain excellent CNF dispersion and to achieve substantial orientation of CNFs [172]. The composites became electrically conductive at much lower carbon nanofiber contents than when using commercial internal mixers, due to orientation of fibers and much less damage due to low shear conditions. It was noted by several researchers that the nano-confined spaces between layered silicate clay particles offered significantly higher rates of polymerization of caprolactam to produce polyamides [173–177] and curing of epoxy [178–184] than in bulk form. This was used successfully to produce polyamide–nanoclay nanocomposites. Intra gallery caprolactam polymerization occurred at a much faster rate due, to the catalytic effects of alkylammonium ions present in organically treated clay.

The in situ polymerization scheme—successful in explaining exfoliation of clay galleries in polyamide and epoxy systems—cannot be extended readily to thermoplastic polyurethanes, although polyurethane chains can be synthesized in the presence of nanoparticles [185–190]. In this case, linear chains of polyurethane polymer do not exert enough entropic force on clay layers to produce exfoliation. An approach was developed by Pattanayak and Jana [191–194] whereby organically treated clay and polymer were mixed in bulk, and clay–polymer reactions took place during mixing. Pattanayak and Jana [191–194] used commercially available clay, Cloisite 30B (Southern Clay Products), as the reactive clay and exploited the $-CH_2CH_2OH$ groups in the structure of the quaternary ammonium ion, $N^+(CH_2CH_2OH)_2(CH_3)T$, where T represents an alkyl group with approximately 65% $C_{18}H_{37}$, 30% $C_{16}H_{33}$, and 5% $C_{14}H_{29}$ to promote clay–polymer reactions. Some processing-related advantages of bulk polymerization methods are that (1) larger shear forces during mixing can aid clay dispersion; (2) products can be easily finished into useful articles by injection molding, film casting, or other techniques; and (3) conventional mixers, such as single- and twin-screw extruders, can be used to produce such nanocomposites. Nevertheless, bulk polymerization methods have some limitations. For example, diffusional constraints hinder the rates of polymerization as well as clay–polymer tethering reactions. In addition, reaction conditions are seldom isothermal. A concomitant rise in temperature may trigger many side reactions (e.g., formation of biurets and allophanates).

5 SUMMARY

In this chapter we presented a review of how polymer nanocomposites research evolved in the last 20 years. We also discussed several issues that researchers should consider and evaluate critically while delving into research projects on the development of new polymer nanocomposite products. We described relevant up-to-date literature on polymer nanocomposite products of POSS and carbon nanofiber and nanotubes with the general understanding that it also applies to other nanoparticles. Some predictions on mechanical properties are presented as guiding principles. Some guidelines for achieving good nanoparticle dispersion are provided. It is understood that many heuristics developed in the past 20 years of research apply to a number of specific systems and that further work is necessary to develop broader understanding.

REFERENCES

1. Li, X., Chen, H., Dang, Y., Lin, Y., Larson, C. A., and Roco, M. C., A longtitudinal analysis of nanotechnology literature: 1976–2004, *J. Nanopart. Res.* 2008;10:3–22.

2. Li, X., Hu, D., Dang, Y., Chen, H., Roco, M. C., Larson, C. A., and Chan, J., Nano mapper: an Internet knowledge mapping system for nanotechnology development, *J. Nanopart. Res.* 2009;11:529–552.

3. Giordano, G., and Inman, H., Thinking small pays big, *Plast. Eng.* 2010;66(1):6–10.

4. Jancar, J., Douglas, J. F., Starr, F. W., Kumar, S. K., Cassagnau, P., Lesser, A. J., and Sternstein, S. S., Current issues in research on structure–property relationships in polymer nanocomposites, *Polymer* 2010;51:3321–3343.

5. Rocco, M. C., Nanoparticle and nanotechnology research in the USA, *J. Aerosol Sci.* 1998;29:749–760.

6. Du, M., Guo, B., and Jia, D., Thermal stability and flame retardant effects of halloysite nanotubes on polypropylene, *Eur. Polym. J.* 2006;42:1362–1369.

7. Ye, Y., Chen, H., Wu, J., and Ye, L. High impact strength epoxy nanocomposites with natural nanotubes, *Polymer* 2007;48:6426–6433.

8. Marney, D. C. O., Russell, L. J., Wu, D. Y., Nguyen, T., Cramm, D., Rigopoulos, N., Wright, N., and Greaves, M., The suitability of halloysite nanotubes as a fire retardant for nylon, *Polym. Degrad. Stabil.* 2008;93:1971–1978.

9. Hedicke-Hoechstoetter, K., Lim, G. T., and Altstaedt, V., Novel polyamide nanocomposites based on silicate nanotubes of the mineral halloysite, *Compos. Sci. Technol.* 2009;69:330–334.

10. Liu, M., Guo, B., Du, M., Chen, F., and Jia, D., Halloysite nanotubes as a novel β-nucleating agent for isotactic polypropylene. *Polymer* 2009;50:3022–3030.

11. Guo, B., Chen, F., Lei, Y., and Jia, D., Tubular clay composites with high strength and transparency, *J. Macromol. Sci. B*, 2010;49:111–121.

12. Kholmanov, I. N., Kharlamov, A., Barborini, E., Lenardi, C., Li Bassi, A., Bottani, C. E., Ducati, C., Maffi, S., Kirillova, N. V., and Milani, P., A simple method for the synthesis of silicon carbide nanorods, *J. Nanosci. Nanotechnol.* 2002;2:453–456.

13. Raman, V., Bhatia, G., Sengupta, P. R., Srivastava, A. K., and Sood, K. N., Synthesis of silicon carbide nanorods from mixture of polymer and sol–gel silica, *J. Mater. Sci.* 2007;42(14):5891–5895.

14. Wu, R. B., Yang, G. Y., Pan, Y., and Chen, J. J., Synthesis of silicon carbide nanorods without defects by direct heating method, *J. Mater. Sci.* 2007;42(11):3800–3804.

15. Jia, Q., Wu, Y., Ping, X., Xin, Y., Wang, Y., and Zhang, L., Combined effect of nano-clay and nano-carbon black on properties of NR nanocomposites, *Polym. Polym. Compos.* 2005;13:709–719.

16. Zhang, H., Guo, L., Shao, H., and Hu, X., Nano-carbon black filled Lyocell fiber as a precursor for carbon fiber. *J. Appl. Polym. Sci.* 2006;99:65–74.

17. Kiliaris, P., and Papaspyrides, C. D., Polymer /layered silicate (clay) nanocomposites: an overview of flame retardancy, *Prog. Polym. Sci.* 2010;35(7):902–958.

18. Esfandiari, A., Nazokdast, H., Rashidi, A.-S., and Yazdanshenas, M.-E., Review of polymer–organoclay nanocomposites, *J. Appl. Sci.* 2008;8(3):545–561.

19. Goettler, L. A., Lee, K. Y., and Thakkar, H., Layered silicate-reinforced polymer nanocomposites: development and applications, *Polym. Rev.* 2007;47(2):291–317.

20. Ray, S. S., and Okamoto, M., Polymer/layered silicate nanocomposites: a review from preparation to processing, *Prog. Polym. Sci.* 2003;28(11):1539–1641.

21. LeBaron, P. C., Wang, Z., and Pinnavaia, T. J., Polymer-layered silicate nanocomposites: an overview, *Appl. Clay Sci.* 1999;15(1–2):11–29.

22. Pinnavaia, T. J., and Beall, G. W. (Eds.), *Polymer–Clay Nanocomposites*, Wiley, New York, 2000.

23. Utracki, L. A., *Clay-Containing Polymeric Nanocomposites*, Rapra Technology, *Ltd.*, Shropshire, UK, 2004.

24. Krishnamoorti, R., Vaia, R. A., Polymer Nanocomposites: Synthesis, Characterization, and Modeling, ACS Symposium Series 804, American Chemical Society, Washington, DC, 2002.

25. Hussain, F., Hojjati, M., Okamoto, M., and Gorga, R. E., Polymer–matrix nanocomposites, processing, manufacturing, and application: an overview, *J. Compos. Mater.* 2006;40(17):1511–1575.

26. Chen, G., and Zhao, W., Polymer/graphite nanocomposites, *Nano-Biocompos.* 2010; 79–106.

27. Tjong, S. C., Polymer-graphite nanocomposites, in Thomas, S., Zaikov, G. E., and Valsaraj, S. V. (Eds.), *Recent Advances in Polymer Nanocomposites*, 2009, pp. 19–45.

28. Wakabayashi, K., Pierre, C., Dikin, D. A., Ruoff, R. S., Ramanathan, T., Brinson, L. C., and Torkelson, J. M., Polymer–graphite nanocomposites: effective dispersion and major property enhancement via solid-state shear pulverization, *Macromolecules* 2008;41(6):1905–1908.

29. Du, X., Yu, Z.-Z., Dasari, A., Ma, J., Mo, M., Meng, Y., and Mai, Y.-W., New method to prepare graphite nanocomposites, *Chem. Mater.* 2008;20(6):2066–2068.

30. Fukushima, H., Drzal, L. T., Rook, B. P., Rich, M. J., Thermal conductivity of exfoliated graphite nanocomposites, *J. Therm. Anal. Calorimetry* 2006;85(1):235–238.

31. Meng, Y., Polymer/graphite nanocomposites, in Mai, Y.-W., Yu, Z.-Z., (Eds.), (*Polymer Nanocomposites*, 2006, pp.510–539.

32. Dresselhaus, M. S., Dresselhaus, G., Eklund, P. C., Saito, R., and Endo, M. Introduction to carbon materials, in: *Carbon Nanotubes:Preparation and Properties*, Ebbesen, T. W., Ed., CRC Press, Boca Raton, FL, 1997.

33. Mordkovich, V. Z., Carbon nanofibers: a new ultrahigh-strength material for chemical technology, *Theor. Found. Chem. Eng.* 2003;37:429–438.

34. Endo, M., Saito, R., Dresselhaus, M. S., and Dresselhaus, G., From carbon fibers to carbon nanotubes, in Ebbesen, T. W. (Ed.), *Carbon Nanotubes: Preparation and Properties,* CRC Press, Boca Raton, FL, 1997.

35. Lake, M. L., and Ting, J.-M., Vapor grown fiber composites, in: Burchell, T. D., (Ed.), *Carbon Materials for Advanced Technologies*, Pergamon Press, Elmsford, the Netherlands, 1999.

36. Issi, J.-P., Thermal and electrical properties of carbons: relationship to structure, in Yardim, M. F., (Ed.), *Design and Control of Structure of Advanced Carbon Materials for Enhanced Performance*, Kluwer Academic, Dordrecht, The Netherlands, 2001.

37. Ehrburger, P., and Vix-Guterl, C., Surface properties of carbons for advanced carbon-based composites, in: Yardim, M. F. (Ed.), *Design and Control of Structure of Advanced Carbon Materials for Enhanced Performance,* Kluwer Academic, Dordrecht, The Netherlands, 2001.

38. Ros, T. G., van Dillen, A. J., Geus, J. W., and Koningsberger, D. C., Surface oxidation of carbon nanofibres, *Chemi. Eur. J.* 2002;8:1151–1162.

39. Haiber, S., Ai, X. T., Bubert, H., Heintze, M., Bruser, V., Brandl, W., and Marginean, G., Analysis of functional groups on the surface of plasma-treated carbon nanofibers, *Anal. Bioanal. Chem.* 2003;375:875–883.

40. Bubert, H., Ai, X., Haiber, S., Heintze, M., Bruser, V., Pasch, E., Brandl, W., and Marginean, G., Basic analytical investigation of plasma-chemically modified carbon fibers, *Spectrochim. Acta B* 2002;57:1601–1610.

41. Yue, Z. R., Jiang, W., Wang, L., Gardner, S. D., and Pittman, C. U., Surface characterization of electrochemically oxidized carbon fibers, *Carbon* 1999;37:1785–1796.

42. Boehm, H. P., Surface oxides on carbon and their analysis: a critical assessment, *Carbon* 2002;40:145–149.

43. Peebles, L. H., *Carbon Fibers: Formation, Structure, and Properties*, CRC Press, Boca Raton, FL, 1995.

44. Biniak, S., Szymanski, G., Siedlewski, J., and Swiatkowski, A., The characterization of activated carbons with oxygen and nitrogen surface groups, *Carbon* 1997;35:1799–1810.

45. Moreno-Castilla, C., Lopez-Ramon, M. V., and Carrasco-Marin, F., Changes in surface chemistry of activated carbons by wet oxidation, *Carbon* 2000;38:1995–2001.

46. Velasco-Santos, C., A.L, M.-H., and Castaño, V. M., Chemical functionalization on carbon nanotubes: principles and applications, in Dirote, E. V. (Ed.), *Trends in Nanotechnology Research*, Nova Science Publishers, Hauppauge, NY, 2004.

47. Rakov, E. G., Chemistry of carbon nanotubes, in: Gogotsi, Y. (Ed.), *Nanomaterials Handbook,* CRC/Taylor & Francis, Boca Raton, FL, 2006.

48. Liu, P., Modifications of carbon nanotubes with polymers, *Eur. Polym. J.* 2005;41:2693–2703.

49. Jin, Z. X., Sun, X., Xu, G. Q., Goh, S. H., and Ji, W., Nonlinear optical properties of some polymer/multi-walled carbon nanotube composites, *Chem. Phys. Lett.* 2000;318:505–510.

50. Lin, Y., Rao, A. M., Sadanadan, B., Kenik, E. A., and Sun, Y. P., Functionalizing multiple-walled carbon nanotubes with aminopolymers, *J. Phys. Chem. B* 2002;106:1294–1298.

51. Gao, C., Jin, Y. Z., Kong, H., Whitby, R. L. D., Acquah, S. F. A., Chen, G. Y., Qian, H. H., Hartschuh, A., Silva, S. R. P., Henley, S., Fearon, P., Kroto, H. W., and Walton, D. R. M., Polyurea-functionalized multiwalled carbon nanotubes: synthesis, morphology, and Raman spectroscopy, *J. Phys. Chemi. B* 2005;109:11925–11932.

52. Wei, G., Saitoh, S., Saitoh, H., Fujiki, K., Yamauchi, T., and Tsubokawa, N., Grafting of polymers onto vapor grown carbon fiber surface by ligand-exchange reaction of ferrocene moieties of polymer with polycondensed aromatic rings of the wall-surface, *Polymer* 2004;45:8723–8730.

53. Li, G., Wang, L., Ni, H., and Pittman C. U., Jr., Polyhedral oligomeric silsesquioxane (POSS) polymers and copolymers: a review, *J. Inorg. Organomet. Polym.* 2002;11(3):123–154.

54. Scott, D. W., Thermal rearrangement of branched-chain methylpolysiloxanes, *J. Am. Chem. Soc.* 1946;68:356–358.

55. Barry, A. J., Daudt, W. H., Domicone, J. J., and Gilkey, J. W., Crystalline organosilsesquioxanes, *J. Am. Chemi. Soci.* 1955;77:4248–4252.

56. Brown, J. F., Jr. and Vogt, L. H., Jr., The polycondensation of cyclohexylsilanetriol, *J. Am. Chemi. Soc.* 1965;87(19):4313–4317.

57. Feher, F. J., Budzichowski, T. A., Blanski, R. L., Weller, K. J., and Ziller, J. W., Facile syntheses of new incompletely condensed polyhedral oligosilsesquioxanes: [(c-C5H9)7Si7O9(OH)3], [(c-C7H13)7Si7O9(OH)3], *and* [(c-C7H13)6Si6O7(OH)4], *Organometallics* 1991;10(7):2526–2528.

58. Feher, F. J., Phillips, S. H., and Ziller, J. W., Facile and remarkably selective substitution reactions involving framework silicon atoms in silsesquioxane frameworks, *J. Am. Chem. Soci.* 1997;119(14):3397–3398.

59. Hanssen, R. W. J. M., van Santen, R. A., and Abbenhuis, H. C. L., The dynamic status quo of polyhedral silsesquioxane coordination chemistry, *Eur. J. Inorg. Chem.* 2004(4):675–683.

60. Lichtenhan, J. D., Feher, G. J. W., and Frank J. Process for preparation of polyhedral oligomeric silsesquioxanes and systhesis of polymers containing polyhedral oligomeric silsesqioxane group segments, U.S. Patent 5,484,867, University of Dayton, Washington, DC, 1996.

61. Phillips S. H., Haddad, T. S., and Tomczak, S. J., Developments in nanoscience: polyhedral oligomeric silsesquioxane (POSS)-polymers, *Curr. Opin. Solid State Mater. Sci.* 2004;8(1):21–29.

62. Feher, F. J., Newman, D. A., and Walzer, J. F., Silsesquioxanes as models for silica surfaces, *J. Am. Chem. Soci.* 1989;111(5):1741–1748.

63. Feher, F. J., and Newman, D. A., Enhanced silylation reactivity of a model for silica surfaces, *J. Am. Chem. Soc.* 1990;112(5):1931–1936.

64. Zhang, C. and Laine, R. M., Hydrosilylation of allyl alcohol with [HSiMe2OSiO1.5]8: octa(3-hydroxypropyldimethylsiloxy)octasilsesquioxane and its octamethacrylate derivative as potential precursors to hybrid nanocomposites, *J. Am. Chem. Soc.* 2000;122(29):6979–6988.

65. Sellinger, A., and Laine, R. M., Silsesquioxanes as synthetic platforms. thermally curable and photocurable inorganic/organic hybrids, *Macromolecules* 1996;29(6):2327–2330.

66. Fasce, D. P., Williams, R. J. J., Mechin, F., Pascault, J. P., Llauro, M. F., and Petiaud, R., Synthesis and characterization of polyhedral silsesquioxanes bearing bulky functionalized substituents, *Macromolecules* 1999;32(15):4757–4763.

67. Fu, B. X., Yang, L., Somani, R. H., Zong, S. X., Hsiao, B. S., Phillips, S., Blanski, R., and Ruth, P., Crystallization studies of isotactic polypropylene containing nanostructured polyhedral oligomeric silsesquioxane molecules under quiescent and shear conditions, *J. Polym. Sci. B* 2001;39(22):2727–2739.

68. Fu, B. X., Gelfer, M. Y., Hsiao, B. S., Phillips, S., Viers, B., Blanski, R., and Ruth, P., Physical gelation in ethylene-propylene copolymer melts induced by polyhedral oligomeric silsesquioxane (POSS) molecules, *Polymer* 2003;44(5):1499–1506.

69. Hong, B., Thoms, T. P. S., Murfee, H. J., and Lebrun, M. J., Highly branched dendritic macromolecules with core polyhedral silsesquioxane functionalities, *Inorg. Chemi.* 1997;36(27):6146–6147.

70. Murfee, H. J., Thoms, T. P. S., Greaves, J., and Hong B., New metallodendrimers containing an octakis(diphenylphosphino)-functionalized silsesquioxane core and ruthenium(II)-based chromophores, *Inorg. Chem.* 2000;39(23):5209–5217.

71. Byrne, M. T., and Gun'ko, Y., K., Recent advances in research on carbon nanotube–polymer composites, *Adv. Mater.* (Weinheim, Germany) 2010;22(15):1672–1688.

72. Martinez-Hernandez, A. L., Velasco-Santos, C., and Castano, V. M., Carbon nanotubes composites: processing, grafting and mechanical and thermal properties, *Curr. Nanosci.* 2010;6(1):12–39.

73. Pfaendner, R., Nanocomposites: industrial opportunity or challenge? *Polym. Degrad. Stabil.* 2010;95(3):369–373.

74. Auhofer, W., and Kovacs, J. Z., A review and analysis of electrical percolation in carbon nanotube polymer composites, *Compos. Sci. Technol.* 2009;69(10):1486–1498.

75. Al-Saleh, M. H., and Sundararaj, U., A review of vapor grown carbon nanofiber/polymer conductive composites, *Carbon* 2009;47(1):2–22.

76. Nielsen, L., and Landel, R., *Mechanical Properties of Polymers and Composites*, 2nd ed., Marcel Dekker, New York, 1994.

77. Cox, H. L., The elasticity and strength of paper and other fibrous materials, *Br. J. Appl. Phys.* 1952;3:72–79.

78. Baxter, W. J., The strength of metal matrix composites reinforced with randomly oriented discontinuous fibers, *Metall. Trans. A* 1992;23:3045–3053.

79. Clingerman, M. L., Weber, E. H., King, J. A., and Schulz, K. H., Development of an additive equation for predicting the electrical conductivity of carbon-filled composites, *J. Appl. Polym. Sci.* 2003;88:2280–2299.

80. Mather, P. T., Jeon, H. G., Romo-Uribe, A., Haddad, T. S., and Lichtenhan, J. D., Mechanical relaxation and microstructure of poly(norbornyl-POSS) copolymers, *Macromolecules* 1999;32(4):1194–1203.

81. Jeon, H. G., Mather, P. T., and Haddad, T. S., Shape memory and nanostructure in poly(norbornyl-POSS) copolymers, *Polym. Int.* 2000;49(5):453–457.

82. Pyun, J., Matyjaszewski, K., Wu, J., Kim, G.-M., Chun, S. B., and Mather, P. T., ABA triblock copolymers containing polyhedral oligomeric silsesquioxane pendant groups: synthesis and unique properties, *Polymer* 2003;44(9):2739–2750.

83. Wu J., Haddad, T. S., Kim, G.-M., and Mather, P. T., Rheological behavior of entangled polystyrene–polyhedral oligosilsesquioxane (POSS) copolymers, *Macromolecules* 2007;40(3):544–554.

84. Zhang, W., Fu, B. X., Seo, Y., Schrag, E., Hsiao, B., Mather, P., Yang, N.-L., Xu, D., Ade, H., Rafailovich, M., and Sokolov, J., Effect of methyl methacrylate/polyhedral oligomeric silsesquioxane random copolymers in compatibilization of polystyrene and poly(methyl methacrylate) blends, *Macromolecules* 2002;35(21):8029–8038.

85. Kim, G. M., Qin, H., Fang, X., Sun, F. C., and Mather, P. T., Hybrid epoxy-based thermosets based on polyhedral oligosilsesquioxane: cure behavior and toughening mechanisms, *J. Polym. Sci., B* 2003;41(24):3299–3313.

86. Kim, B.-S., and Mather, P. T., Morphology, microstructure, and rheology of amphiphilic telechelics incorporating polyhedral oligosilsesquioxane, *Macromolecules* 2006;39(26):9253–9260.

87. Kim, B.-S., and Mather, P. T., Amphiphilic telechelics with polyhedral oligosilsesquioxane (POSS) end-groups: dilute solution viscometry, *Polymer* 2006;47(17):6202–6207.

88. Zheng, L., Farris, R. J., and Coughlin, E. B., Novel polyolefin nanocomposites: synthesis and characterizations of metallocene-catalyzed polyolefin polyhedral oligomeric silsesquioxane copolymers, *Macromolecules* 2001;34(23):8034–8039.

89. Zheng, L., Hong, S., Cardoen, G., Burgaz, E., Gido, S. P., and Coughlin, E. B., Polymer nanocomposites through controlled self-assembly of cubic silsesquioxane scaffolds, *Macromolecules* 2004;37(23):8606–8611.

90. Zheng, L., Waddon, A. J., Farris, R. J., and Coughlin, E. B., X-ray characterizations of polyethylene polyhedral oligomeric silsesquioxane copolymers, *Macromolecules* 2002;35(6):2375–2379.

91. Waddon, A. J., Zheng, L., Farris, R. J., and Coughlin, E. B., Nanostructured polyethylene-POSS copolymers: control of crystallization and aggregation, *Nano Lett.* 2002;2(10):1149–1155.

92. Iyer, S., and Schiraldi, D. Synthesis and properties of copolymers of polyesters and polyamides with polyhedral oligomeric silsesquioxanes (POSS); comparison with blended materials, *PMSE Preprints* 2005;92:326–327.

93. Cardoen, G., and Coughlin, E. B., Hemi-telechelic polystyrene-POSS copolymers as model systems for the study of well-defined inorganic/organic hybrid materials, *Macromolecules* 2004;37(13):5123–5126.

94. Lee, A., and Lichtenhan, J. D., Viscoelastic responses of polyhedral oligosilsesquioxane reinforced epoxy systems, *Macromolecules* 1998;31(15):4970–4974.

95. Fu, B. X., Namani, M., and Lee, A., Influence of phenyltrisilanol polyhedral silsesquioxane on properties of epoxy network glasses, *Polymer* 2003;44(25):7739–7747.

96. Fu, B. X., Lee, A., and Haddad, T. S., Styrene–butadiene–styrene triblock copolymers modified with polyhedral oligomeric silsesquioxanes, *Macromolecules* 2004;37(14):5211–5218.

97. Drazkowski, D. B., Lee, A., Haddad, T. S., and Cookson, D. J., Chemical substituent effects on morphological transitions in styrene–butadiene–styrene triblock copolymer grafted with polyhedral oligomeric silsesquioxanes, *Macromolecules* 2006;39(5):1854–1863.

98. Zeng, J., Kumar, S., Iyer, S., Schiraldi, D. A., and Gonzalez, R. I., Reinforcement of poly(ethylene terephthalate) fibers with polyhedral oligomeric silsesquioxanes (POSS), *High Performance Polym.* 2005;17(3):403–424.

99. Yoon, K. H., Polk, M. B., Park, J. H., Min, B. G., and Schiraldi, D. A., Properties of poly(ethylene terephthalate) containing epoxy-functionalized polyhedral oligomeric silsesquioxane, *Polym. Int.* 2005;54(1):47–53.

100. Tamaki, R., Tanaka, Y., Asuncion, M. Z., Choi, J., and Laine, R. M., Octa(aminophenyl)silsesquioxane as a nanoconstruction site, *J. Am. Chem. Soc.* 2001;123(49):12416–12417.

101. Zeng, J., Bennett, C., Jarrett, W. L., Iyer, S., Kumar, S., Mathias, L. J., and Schiraldi, D. A., Structural changes in trisilanol POSS during nanocomposite melt processing, *Compos. Interfaces* 2005;11(8–9):673–685.

102. Zhao, Y., and Schiraldi, D. A., Thermal and mechanical properties of polyhedral oligomeric silsesquioxane (POSS)/polycarbonate composites, *Polymer* 2005;46(25): 11640–11647.

103. Capaldi, F. M., Rutledge, G. C., and Boyce, M. C., Structure and dynamics of blends of polyhedral oligomeric silsesquioxanes and polyethylene by atomistic simulation, *Macromolecules* 2005;38(15):6700–6709.

104. Capaldi, F. M., Boyce, M. C., and Rutledge, G. C., The mechanical properties of crystalline cyclopentyl polyhedral oligomeric silsesquioxane, *J. Chem. Phys.* 2006;124(21):214709/214701–214709/214704.

105. Ricco, L., Russo, S., Monticelli, O., Bordo, A., and Bellucci F. ε-Caprolactam polymerization in presence of polyhedral oligomeric silsesquioxanes (POSS), *Polymer* 2005;46(18):6810–6819.

106. Baldi, F., Bignotti, F., Ricco, L., Monticelli, O., and Ricco, T., Mechanical and structural characterization of POSS-modified polyamide 6, *J. Appl. Polym. Sci.* 2006;100(4):3409–3414.

107. Liu, Y.-L. and Lee, H.-C., Preparation and properties of polyhedral oligosilsequioxane tethered aromatic polyamide nanocomposites through michael addition between maleimide-containing polyamides and an amino-functionalized polyhedral oligosilsesquioxane, *J. Polym. Sci., A* 2006;44(15):4632–4643.

108. Leu, C.-M., Chang, Y.-T., and Wei, K.-H., Synthesis and dielectric properties of polyimide-tethered polyhedral oligomeric silsesquioxane (POSS) nanocomposites via POSS-diamine, *Macromolecules* 2003;36(24):9122–9127.

109. Lee, Y.-J., Kuo, S.-W., Huang, W.-J., Lee, H.-Y., and Chang, F.-C., Miscibility, specific interactions, and self-assembly behavior of phenolic/polyhedral oligomeric silsesquioxane hybrids, *J. Polym. Sci., B* 2004;42(6):1127–1136.

110. Perilla, J. E., Lee, B.-J., and Jana, S. C., Rheological investigation of interactions between sorbitol and polyhedral oligomeric silsesquioxane in development of nanocomposites of isotactic polypropylene. *J. Rheol.*, 2010;54(4):761–779.

111. Lee, B.-J., Roy, S., and Jana, S. C., POSS/PP nanocomposites: characterization and properties, presented at the 67th Annual Technical Conference, Society of Plastics Engineers, 2009, pp. 126–130.

112. Ristolainen, N., Hippi, U., Seppala, J., Nykanen, A., and Ruokolainen J., Properties of polypropylene/aluminum trihydroxide composites containing nanosized organoclay, *Polym. Eng. and Sci.* 2005;45(12):1568–1575.

113. Manias, E., Touny, A., Wu L., Strawhecker, K., Lu B, and Chung, T. C., Polypropylene/montmorillonite nanocomposites: review of the synthetic routes and materials properties, *Chem. Mater.* 2001;13(10):3516–3523.

114. Laske, S., Kracalik, M., Gschweitl, M., Feuchter, M., Maier, G., Pinter, G., Thomann, R., Friesenbichler, W., and Langecker, G. R., Estimation of reinforcement in compatibilized polypropylene nanocomposites by extensional rheology, *J. Appl. Polym. Sci.* 2009;111(5):2253–2259.

115. Preschilla, N., Sivalingam, G., Abdul Rasheed, A. S., Tyagi, S., Biswas, A., and Bellare, J. R., Quantification of organoclay dispersion and lamellar morphology in polypropylene–clay nanocomposites with small angle x-ray scattering, *Polymer* 2008;49(19):4285–4297.

116. Nguyen, Q. T. and Baird, D. G., Dispersion of nanoclay into polypropylene with carbon dioxide in the presence of maleated polypropylene, *J. Appl. Polym. Sci.* 2008;109(2):1048–1056.

117. Kandola, B. K., Smart, G., Horrocks, A. R., Joseph, P., Zhang, S., Hull, T. R., Ebdon, J., Hunt, B., and Cook A., Effect of different compatibilisers on nanoclay dispersion, thermal stability, and burning behavior of polypropylene-nanoclay blends, *J. Appl. Polym.* 2008;108(2):816–824.

118. Nguyen, Q. T. and Baird, D. G., An improved technique for exfoliating and dispersing nanoclay particles into polymer matrices using supercritical carbon dioxide, *Polymer* 2007;48(23):6923–6933.

119. Hatzikiriakos, S. G., Rathod, N., and Muliawan, E. B., The effect of nanoclays on the processibility of polyolefins, *Polym. Eng. Sci.* 2005;45(8):1098–1107.

120. Chrissopoulou, K., Altintzi, I., Andrianaki, I., Shemesh, R., Retsos, H., Giannelis, E. P., and Anastasiadis, S. H., Understanding and controlling the structure of polypropylene/layered silicate nanocomposites, *J. Polym. Sci. B* 2008;46(24):2683–2695.

121. Mittal, V., Polypropylene-layered silicate nanocomposites: filler matrix interactions and mechanical properties, *J. Thermoplast. Compos. Mater.* 2007;20(6):575–599.

122. Liu, H., Lim, H. T., Ahn, K. H., and Lee, S. J., Effect of ionomer on clay dispersions in polypropylene-layered silicate nanocomposites, *J. Appl. Polym. Sci.* 2007;104(6):4024–4034.

123. Koo, C. M., Kim, S. O., and Chung, I. J., Study on morphology evolution, orientational behavior, and anisotropic phase formation of highly filled polymer-layered silicate nanocomposites, *Macromolecules* 2003;36(8):2748–2757.

124. Bertini, F., Canetti, M., Audisio, G., Costa, G., and Falqui, L. Characterization and thermal degradation of polypropylene–montmorillonite nanocomposites, *Polym. Degrad. Stabil.* 2006;91(3):600–605.

125. Perrin-Sarazin, F., Ton-That, M. T., Bureau, M. N., and Denault, J., Micro- and nanostructure in polypropylene/clay nanocomposites, *Polymer* 2005;46(25):11624–11634.

126. Wu, J.-Y., Wu, T.-M., Chen, W.-Y., Tsai, S.-J., Kuo, W.-F., and Chang, G.-Y., Preparation and characterization of PP/clay nanocomposites based on modified polypropylene and clay, *J. Polym. Sci., B* 2005;43(22):3242–3254.

127. Tjong, S. C., Bao, S. P., and Liang, G. D., Polypropylene/montmorillonite nanocomposites toughened with SEBS-g-MA: structure–property relationship, *J. Polym. Sci., B* 2005;43(21):3112–3126.

128. Lu, K., Grossiord, N., Koning, C. E., Miltner, H. E., van Mele, B., and Loos J., Carbon Nanotube/isotactic polypropylene composites prepared by latex technology: morphology analysis of CNT-induced nucleation, *Macromolecules* (Washington, DC) 2008;41(21):8081–8085.

129. Masuda, J. and Torkelson, J. M., Dispersion and major property enhancements in polymer/multiwall carbon nanotube nanocomposites via solid-state shear pulverization followed by melt mixing, *Macromolecules* (Washington, DC) 2008;41(16):5974–5977.

130. Hou, Z., Wang, K., Zhao, P., Zhang, Q., Yang, C., Chen, D., Du, R., and Fu, Q., Structural orientation and tensile behavior in the extrusion-stretched sheets of polypropylene/multi-walled carbon nanotubes' composite, *Polymer* 2008;49(16):3582–3589.

131. Miltner, H. E., Grossiord, N., Lu, K., Loos, J., Koning, C. E., and Van Mele, B., Isotactic polypropylene/carbon nanotube composites prepared by latex

technology. thermal analysis of carbon nanotube-induced nucleation, *Macromolecules* (Washington, DC) 2008;41(15):5753–5762.

132. Koval'chuk, A. A., Shchegolikhin, A. N., Shevchenko, V. G., Nedorezova, P. M., Klyamkina, A. N., and Aladyshev, A. M., Synthesis and properties of polypropylene/multiwall carbon nanotube composites, *Macromolecules* (Washington, DC) 2008;41(9):3149–3156.

133. Jose, M. V., Dean, D., Tyner, J., Price, G., and Nyairo, E. Polypropylene/carbon nanotube nanocomposite fibers: process–morphology–property relationships, *J. Appl. Polym. Sci.* 2007;103(6):3844–3850.

134. Dondero, W. E. and Gorga, R. E., Morphological and mechanical properties of carbon nanotube/polymer composites via melt compounding, *J. Polym. Sci., B* 2006;44(5):864–878.

135. Leelapornpisit, W., Ton-That, M.-T., Perrin-Sarazin, F., Cole, K. C., Denault, J., and Simard, B., Effect of carbon nanotubes on the crystallization and properties of polypropylene, *J. Polym. Sci., B* 2005;43(18):2445–2453.

136. Gao, X., Meng, X., Wang, H., Wen, B., Ding, Y., Zhang, S., and Yang, M., Antioxidant behaviour of a nanosilica-immobilized antioxidant in polypropylene, *Polym. Degrad. Stab.* 2008;93(8):1467–1471.

137. Leng, P. B., Md Akil, H., and Lin, O. H., Thermal properties of microsilica and nanosilica filled polypropylene composite with epoxy as dispersing aid, *J. Reinf. Plast. Compos.* 2007;26(8):761–770.

138. Zhou, H. J., Rong, M. Z., Zhang, M. Q., Ruan, W. H., and Friedrich, K., Role of reactive compatibilization in preparation of nanosilica/polypropylene composites, *Polym. Eng. Sci.* 2007;47(4):499–509.

139. Reddy, C. S., and Das, C. K., Polypropylene-nanosilica-filled composites: effects of epoxy-resin-grafted nanosilica on the structural, thermal, and dynamic mechanical properties, *J. Appl. Polym. Sci.* 2006;102(3):2117–2124.

140. Liu, Y., and Kontopoulou, M., The structure and physical properties of polypropylene and thermoplastic olefin nanocomposites containing nanosilica, *Polymer* 2006;47(22):7731–7739.

141. Huang, L., Zhan, R., and Lu, Y., Mechanical properties and crystallization behavior of polypropylene/nano-SiO$_2$ composites, *J. Reinf. Plast. Compos.* 2006;25(9):1001–1012.

142. Sreekala, M. S., Lehmann, B., Friedrich, K., and Rong, M. Z., Nanosilica-reinforced polypropylene composites: microstructural analysis and crystallization behavior, *Int. J. Polym. Mater.* 2006;55(8):577–594.

143. Ristolainen, N., Vainio, U., Paavola, S., Torkkeli, M., Serimaa, R., and Seppaelae, J., Polypropylene/organoclay nanocomposites compatibilized with hydroxyl-functional polypropylenes, *J. Polym. Sci., B* 2005;43(14):1892–1903.

144. Papageorgiou, G. Z., Achilias, D. S., Bikiaris, D. N., and Karayannidis, G. P., Crystallization kinetics and nucleation activity of filler in polypropylene/surface-treated SiO$_2$ nanocomposites, *Thermochim. Acta* 2005;427(1–2):117–128.

145. Wang, D., and Gao, J., Nonisothermal crystallization kinetics and mechanical properties of PP/NANO-SiO$_2$ composites, *Int. J. Polym. Mater.* 2004;53(12):1085–1100.

146. Fina, A., Tabuani, D., Frache, A., and Camino, G. Polypropylene–polyhedral oligomeric silsesquioxanes (POSS) nanocomposites, *Polymer* 2005;46(19):7855–7866.

147. Pracella, M., Chionna, D., Fina, A., Tabuani, D., Frache, A., and Camino, G., Polypropylene–POSS nanocomposites: morphology and crystallization behaviour,

Macromol. Symp. 2006;234(Trends and Perspectives in Polymer Science and Technology):59–67.

148. Huang, J.-C., He, C.-B., Xiao, Y., Mya, K. Y., Dai, J., and Siow, Y. P., Polyimide/POSS nanocomposites: interfacial interaction, thermal properties and mechanical properties, *Polymer* 2003;44(16):4491–4499.

149. Yoshioka, H., *Physica, B.*, 2000;284–288 1756.

150. Kim, B., Park, H., and Sigmund, W. M., *Langmuir* 2003;19:2525.

151. Pailett, M., Poncharal, P., and Zahab, A., *Phys. Rev. Lett.*, 2005;94:186801.

152. Dennis, H. R., Hunter, D. L., Chang, D., Kim, S., White, J. L., Cho, J. W., and Paul, D. R., *Polymer* 2001;42:9513.

153. Tibbetts, G. G., McHugh, J. J., Mechanical properties of vapor-grown carbon fiber composites with thermoplastic matrices. *J Mater. Res.* 1999;14:2871–2880

154. Carneiro, O. S., Covas, J. A., Bernardo, C. A., Caldeira, G., Van Hattum, F. W. J., Ting, J.-M., Alig, R. L., and Lake, M. L., Production and assessment of polycarbonate composites reinforced with vapour-grown carbon fibers, *Compos. Sci. Tech.* 1998;58:401–407.

155. Ma, H., Zeng, J., Realff, M. L., Kumar, S., and Schiraldi, D. A., Processing, structure, and properties of fibers from polyester/carbon nanofiber composites, *Comp. Sci. Tech.* 2003;63:1617–1628.

156. Zeng, J., Saltysiak, B., Johnson, W. S., Schiraldi, D. A., Kumar, S., Processing and properties of poly(methyl methacrylate)/carbon nanofiber composites, *Composites B* 2004;35:173–178.

157. Gorga, R. E., and Cohen, R. E., Toughness enhancements in poly(methyl methacrylate) by addition of oriented multiwall carbon nanotubes, *J. Polym. Sci. B* 2004;42:2690–2702.

158. Finegan, I. C., Tibbetts, G. G., Gibson, R. F., Modeling and characterization of damping in carbon nanofiber/polypropylene composites, *Compos. Sci. Tech.* 2003;63:1629–1635.

159. Tibbetts, G. G., Finegan, I. C., and Kwag, C., Mechanical and electrical properties of vapor-grown carbon fiber thermoplastic composites, *Mol. Cryst. Liq. Cryst.* 2002;387:129–133.

160. Cooper, C. A., Ravich, D., Lips, D., Mayer, J., and Wagner, H. D., Distribution and alignment of carbon nanotubes and nanofibrils in a polymer matrix, *Compos. Sci. Tech.* 2002;62:1105–1112.

161. Wu, G., Asai, S., and Sumita, M., A self-assembled electric conductive network in short carbon fiber filled poly(methyl methacrylate) composites with selective adsorption of polyethylene, *Macromolecules* 1999;32:3534–3536.

162. Haggenmueller, R., Gommans, H. H., Rinzler, A. G., Fischer, J. E., and Winey, K. I., Aligned single-wall carbon nanotubes in composites by melt processing methods, Chem. *Phys. Lett.* 2000;330:219–225.

163. Lozano, K., Bonilla-Rios, J., and Barrera, E. V., A study on nanofiber-reinforced thermoplastic composites: II Investigation of the mixing rheology and conduction properties, *J. Appl. Polym. Sci.* 2001;80:1162–1172.

164. Lozano, K., and Barrera, E. V., Nanofiber-reinforced thermoplastic composites: I. Thermoanalytical and mechanical analyses, *J. Appl. Polym. Sci.* 2001;79:125–133.

165. Shofner, M. L., Lozano, K., Rodriguez-Macias, F. J., Barrera, E. V., Nanofiber-reinforced polymers prepared by fused deposition modeling. *J. Appl. Polym. Sci.* 2003;89:3081–3090.

166. Xu, J., Donohoe, J. P., and Pittman, C. U., Jr., Preparation, electrical and mechanical properties of vapor grown carbon fiber (VGCF)/vinyl ester composites, *Comp A* 2004;35:693–701.

167. Jin, Z., Pramoda, K. P., Xu, G., Goh, S. H., Dynamic mechanical behavior of melt-processed multi-walled carbon nanotube/poly(methyl methacrylate) composites, *Chem. Phys. Lett.* 2001;337:43–47.

168. Higgins, B., and Brittain, W. J., Polycarbonate carbon nanofiber composites, *Eur. Polym. J.* 2005;41:889–893.

169. Xu, Y., Higgins, B., and Brittain, W. J., Bottom-up synthesis of PS-CNF nanocomposites, *Polymer* 2005;46:799–810.

170. Jia, Z., Wang, Z., Xu, C, Liang, J., Wei, B., Wu, D, Zhu, S. Study on poly(methyl methacrylate)/carbon nanotube composites. *Mater. Sci. Eng.* 1999;**A271**;395–400.

171. Sung, J. H., Kim, H. S., Jin, H. J., Choi, H. J., Chin, I. J., Nanofibrous membranes prepared by multiwalled carbon nanotube/poly(methyl methacrylate) composites, *Macromol* 2004;37:9899–9902.

172. Jimenez, G., and Jana, S. C., Electrically conductive polymer nanocomposites of polymethylmethacrylate and carbon nanofibers prepared by chaotic mixing, *Composites A* 2007;38:983–993.

173. Fukushima, Y., Okada, A., Kawasumi, M., Kurauchi, T., and Kamigaito, O., *Clay Miner.* 1988;23:27.

174. Usuki, A., Kojima, Y., Kawasumi, M., Okada, A., Fukushima, Y., Kurauchi, T., and Kamigaito, O., *J. Mater. Res.* 1993;8:1179.

175. Kojima, Y., Usuki, A., Kawasumi, M., Okada, A., Kurauchi, T., and Kamigaito, O., *J. Polym. Sci. A* 1993;31:983.

176. Yano, K., Usuki, A., Okada, A., Kurauchi, and T., Kamigaito, O., *J. Polym. Sci. A* 1993;31:2493.

177. Okada, A., and Usuki, A., *Mater. Sci. Eng. C*, 1995;**C3**:109.

178. Lan, T., Pinnavaia, T. J., *Chem. Mater.* 1994;6:2216

179. Lan, T., Kaviratna, P. D., and Pinnavaia, T. J., *Chem. Mater.* 1994;6:573.

180. Wang, M. S., and Pinnavaia, T. J., *Chem. Mater.* 1994;6:468.

181. Lan, T., Kaviratna, P. D., and Pinnavaia, T. J., *Chem. Mater.* 1995;7:2144.

182. Lan, T., Kaviratna, P. D., and Pinnavaia, T. J., *J. Phys. Chem. Solids* 1996;57:1005.

183. Shi, H., Lan, T., Pinnavaia, T. J., *Chem. Mater.*, 1996;8:1584.

184. Park, J. H., and Jana, S. C., Mechanism of exfoliation of nanoclay particles in epoxy-clay nanocomposites, *Macromolecules*, 2003;36(8):2758–2768.

185. Chen, X., Wu, L., Zhou, S., and You, B., *Polym. Int.* 2003;52:993–998.

186. Rhoney, I., Brown, S., Hudson, N. E., and Pethrik, R. A., *J. Appl. Polym. Sci.*, 2003;91:1335–1343.

187. Osman, M. A., Mittal, V., Morbidelli, M., and Suter, U. W., *Macromolecules* 2003;36:9851–9858.

188. Dai, X., Xu, J., Guo, X., Lu, Y., Shen, D., and Zhao, N., *Macromolecules* 2004;37:5615.

189. Plummer, C. J. G., Rodlert, M., Bucaille, J.-L., Grunbauer, H. J. M., and Manson, J.-A. E., *Polymer*, 2005;46:6543.

190. Zilli, D., Chiliotte, C., Escobar, M. M., Bekeris, V., Rubiolo, G. R., Cukierman, A. L., and Goyanes, S., *Polymer* 2005;46:6090.

191. Pattanayak, A., and Jana, S. C., Thermoplastic polyurethane nanocomposites of reactive silicate clays: effects of soft segments on properties, *Polymer* 2005;46(14):5183–5193.

192. Pattanayak, A., and Jana, S. C., High strength and low stiffness composites of nanoclay-filled thermoplastic polyurethanes, *Polym. Eng. Sci.*, 2005;45(11):1532–1539.

193. Pattanayak, A., and Jana, S. C., Properties of bulk-polymerized thermoplastic polyurethane nanocomposites, *Polymer* 2005;46(10):3394–3406.

194. Pattanayak, A., and Jana, S. C., Synthesis of thermoplastic polyurethane nanocomposites of reactive clay by bulk polymerization methods, *Polymer* 2005;46(10):3275–3288.

5

LASER-ASSISTED FABRICATION TECHNIQUES

P. T. MURRAY

University of Dayton Research Institute, Dayton, Ohio

1   INTRODUCTION

There are numerous laser-based methods of fabricating nanomaterials, all of which exploit the ability of a laser beam to deliver a controlled amount of energy in a selected region at a selected time. These fabrication techniques entail irradiating a target with laser radiation, thereby ablating atomic and molecular species from the target and forming nanomaterials by controlling the conditions (e.g., pressure and temperature) under which the ablated species recombine. The nanomaterials produced by this process can be in the form of nanoparticles or nanotubes.

The use of a laser beam offers a number of parameters that can be adjusted to alter the nature of the fabricated nanomaterial; these parameters include laser wavelength, pulse duration, pulse repetition rate, and laser energy density at the target. By far the majority of studies reported have entailed the use of pulsed lasers, with pulse durations

Nanoscale Multifunctional Materials: Science and Applications, First Edition.
Edited by Sharmila M. Mukhopadhyay.
© 2012 John Wiley & Sons, Inc. Published 2012 by John Wiley & Sons, Inc.

on the order of a few nanoseconds. Such lasers are excellent sources for research studies but are of doubtful utility, due to their very small duty cycle, for eventual use in large-scale nanomaterial manufacturing. Considerably fewer studies have reported the use of continuous lasers, but such sources show more promise for scale-up.

Presented in this chapter are examples of nanomaterial fabrication techniques that are laser assisted. The techniques covered are (1) pulsed laser ablation, in both the gas and liquid phases; (2) through thin-film ablation; and (3) continuous laser heating. The synthesis of carbon nanotubes by pulsed laser ablation is not covered here since that topic is discussed in Chapter 3.

2   PULSED LASER TECHNIQUES

2.1   Pulsed Laser Ablation in the Gas Phase

Nanoparticle synthesis by pulsed laser ablation (PLA) is conceptually a very simple process that is a variation of the thin-film fabrication technique of pulsed laser deposition (PLD). The process is illustrated in Figure 1. In this process, a target of the desired nanoparticle material is placed in a vacuum chamber and is irradiated

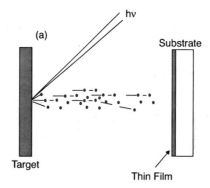

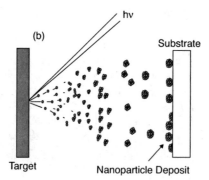

FIGURE 1   Pulsed laser ablation process and its use in (a) thin-film deposition and (b) nanoparticle synthesis.

with the output of a high-powered pulsed laser. The laser–target interaction results in the ejection of atomic and molecular species from the target. If this process is carried out in high vacuum, the mean free paths of the ejected species are sufficiently long that their trajectories result in collisions with a macroscopic object such as a vacuum chamber wall, or with an appropriately placed substrate, as depicted in Figure 1(a). In this case, the process is designated PLD, which has been used for over 25 years and is now a widely used technique for thin-film deposition [1].

Numerous studies [2–7] have been published on the dynamics of the PLA process; these studies have shown that the species ejected from the target have kinetic energies that depend on a number of parameters, including the laser wavelength, laser pulse duration, and laser beam energy density at the target. It has generally been observed that the species ejected have kinetic energies on the order of 1 to 50 eV. The ejected species can recombine in the gas phase, to form nanoparticles, if their relative kinetic energy is decreased to a value that is less than the energy of their mutual attraction, which is typically on the order of a few electron volts. This is most easily accomplished by ablating the target in the presence of a background gas. Under these conditions, the ejected species collide with the background gas and undergo energy loss with each collision. After several such collisions, the kinetic energy of the ablated species is sufficiently small that the ejected species begin to recombine in the gas phase to form nanoparticles of the target material. This is illustrated in Figure 1(b). The nanoparticles can be collected by an appropriately placed substrate for subsequent analysis or use.

Shown in Figure 2 is a time-lapse photograph taken in the author's laboratory during PLA of a graphite target by 248-nm radiation. The target can be seen on the left side of the image. The path of the laser beam is collinear with the visible emission. In this experiment, the ablation chamber was filled with argon to a static pressure of 665 Pa, resulting in the formation of nanoparticles which filled the ablation chamber. That the laser beam is visible to the unaided eye is due to the nanoparticles absorbing the ultraviolet (UV) laser light and reemitting visible radiation. Indeed, this reemitted light has been used as a real-time indicator that nanoparticles were being formed in the chamber. It is interesting to note that there is structure present in the visible emission, especially in the upper right portion of the image near the vacuum chamber wall. This is caused by local variations in the nanoparticle density. It can be seen that there are oscillations in the nanoparticle density, which is caused by the pulsed nature of the fabrication process; each oscillation is caused by a single laser pulse. The pile-up near the chamber wall is probably caused by a slowing down of the nanoparticles are they approach the chamber wall. It would be very useful to use time-gated UV laser illumination during PLA to characterize the hydrodynamics occurring during this process.

Presented in Figure 3 are transmission electron (TEM) micrographs of titanium nanoparticles formed by PLA. The two TEM micrographs were acquired from the same nanomaterial but at different magnifications. In this work, the nanoparticles were collected on a TEM grid that was situated parallel to the target and in a position along the target normal that would allow the highest density of nanoparticles to be collected [as depicted in Figure 1(a)]. It can be seen clearly that the nanoparticles

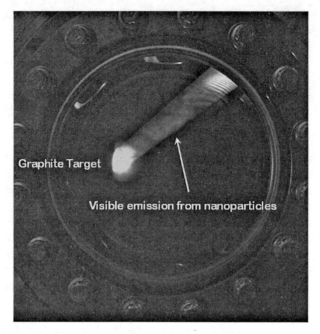

FIGURE 2   Time-lapse photograph recorded during nanoparticle formation by pulsed laser ablation. The target is situated on the left of the image. The UV laser beam path is collinear with the visible emission, which is caused by nanoparticle photoabsorption and reemission in the visible region of the electromagnetic spectrum.

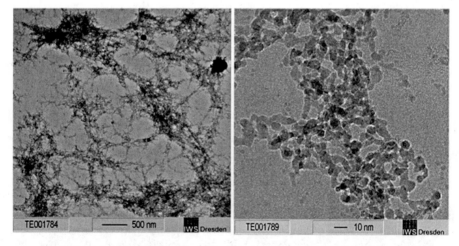

FIGURE 3   TEM micrographs of Ti nanoparticles formed by PLA. (From [8].)

formed by this fabrication technique are highly agglomerated, forming a network of material. Similar results have been observed [8] with PLA of other materials, including Si, Fe, and C. The average nanoparticle size of the deposit presented in Figure 3 is 7 nm, although it can be seen that there are several larger (hundreds of nanometers) nanoparticles present in the micrograph. These larger particles are observed routinely in deposits formed by PLD and are caused by a process called *splashing*, which is a thermal phenomenon caused by superheating of an irradiated spot on the target. It has been observed that the splashed particles are ejected primarily along the target normal.

Sasaki and co-workers [9] reported results of the formation of iron-complex oxide nanoparticles by PLA. Their experimental setup is presented in Figure 4, where it can be seen that their substrate was placed in a position away from the surface normal that would minimize deposition of larger splashed particles. An AFM micrograph of their results is presented in Figure 5. Although the field of view of Figure 5 precludes observing larger particles, these workers concluded that placing a substrate away from the surface normal did indeed minimize the collection of larger nanoparticles. Their iron-complex oxide nanoparticle deposit was also agglomerated. Presented in Figure 6 are the size distributions extracted from their AFM images. It is interesting to note that the size distribution depended on the nature of the background gas used during PLA.

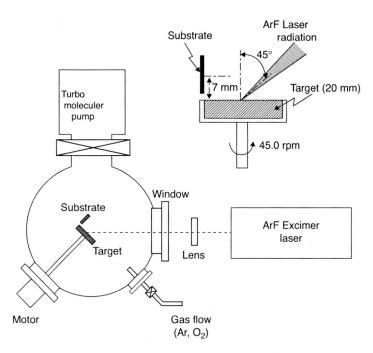

FIGURE 4   Experimental setup of Sasaki et al. for synthesizing Fe complex oxide nanoparticles by PLA. (Reprinted from [9] with permission of Elsevier © 1998.)

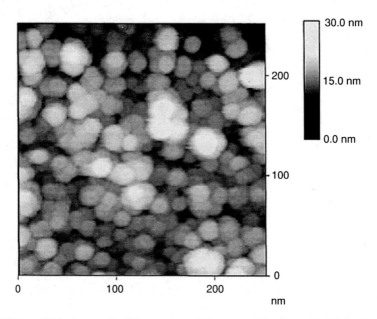

FIGURE 5 AFM micrograph of Fe complex oxide nanoparticles formed by PLA. (Reprinted from [9] with permission of Elsevier © 1998.)

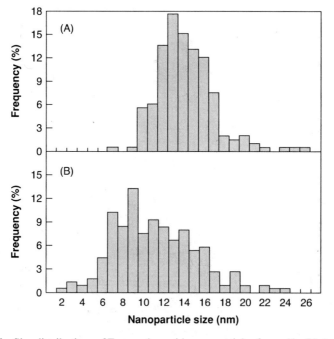

FIGURE 6 Size distributions of Fe complex oxide nanoparticles formed by PLA. (Reprinted from [9] with permission of Elsevier © 1998.)

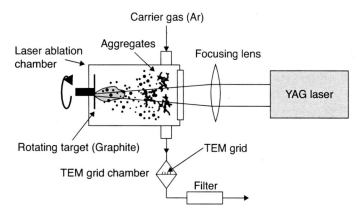

FIGURE 7   Experimental setup to form carbon nanoparticle aggregates by PLA. (Reprinted from [10] with permission of the American Chemical Society © 2006.)

Rong and co-workers [10] formed carbon nanoparticle aggregates by PLA and reported the mechanical properties of the deposit. These workers have cleverly taken advantage of the agglomerated nature of the deposit, have noted that such aggregates occur in a number of applications (including that of carbon black in rubber), and have denoted them as network chain aggregates. The authors have also proposed that these networks might be used in deformable electronic surfaces. A schematic of their experimental setup is presented in Figure 7. A TEM micrograph of their carbon nanoparticle aggregate is presented in Figure 8. The average length of their aggregates was 2 mm.

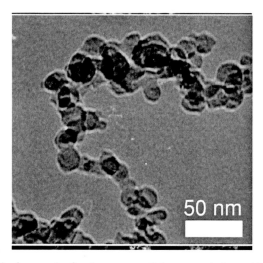

FIGURE 8   TEM micrograph of carbon nanoparticle aggregate formed by PLA. (Reprinted from [10] with permission of the American Chemical Society © 2006.)

These workers carried out PLA in a continuous flow of argon, which resulted in the nanoparticles being entrained in the flow and swept out of the ablation chamber into a trap, where the particles were later collected. This approach is clearly more efficient in particle collection and is more amenable to scale-up than is using a substrate that collects a fraction of the deposit.

Other workers have synthesized numerous materials in nanoparticle form by PLA. These include nanoparticles of Si [11–20], W [21], Au [22–25], Pd [26], Ag [27,28], diamond [29], TiO_2 [30,31], Al_2O_3 [31], SiO_2 [32], and $YBa_2Cu_3O_7$ [33]. Overall, PLA within the gas phase is a very versatile technique that allows the formation of nanoparticles of a wide variety of materials. The chief limitations of the technique are limited throughput, the formation of larger nanoparticles by splashing, and the formation of nanoparticle agglomerates, which may, or may not, be a problem, depending on the application.

2.2   Pulsed Laser Ablation in a Liquid

PLA within a liquid environment is an intriguing concept that may offer a number of advantages over the more conventional gas-phase approach:

1. Increased efficiency for collecting the product
2. A potential for forming, collecting, and storing nanoparticles that may be air sensitive
3. A potential for separating large splashed nanoparticles by normal settling in solution
4. A potential for exploiting solution-based chemistry to fabricate nanoparticles with more complex stoichiometry
5. Less expensive fabrication costs since vacuum hardware is not required

A typical experimental setup is presented in Figure 9. The procedure entails immersing a target within a liquid, laser-ablating the target, and collecting the product. Golightly and Castleman [34] reported the formation of titanium nanoparticles by this process (ablation was carried out in water, ethanol, 2-propanol, and *n*-hexane); a TEM micrograph of their results is presented in Figure 10 (for the process carried out in 2-propanol). The nanoparticles are spherical in shape, and their size distribution is shown in Figure 11. The most probable nanoparticle size is on the order of 15 nm.

Shown in Figure 12 are TEM micrographs of Ti nanoparticles formed in *n*-hexane. The image at the top of Figure 12 was acquired from a sample produced with a laser energy of 50 mJ/pulse, while the image at the bottom of Figure 12 was acquired with a laser energy of 100 mJ/pulse. The corresponding size distributions are presented in Figure 13. The result is somewhat counterintuitive in that increasing laser energy results in a broader size distribution. The authors also reported that Ti nanoparticles formed in water had oxygen incorporation, and those formed in alcohol and *n*-hexane had carbon incorporation.

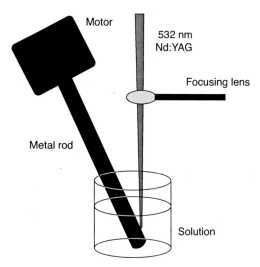

FIGURE 9   Experimental setup for forming Ti nanoparticles by PLA in liquid. (Reprinted from [34] with permission of the American Chemical Society © 2006.)

Shown in Figure 14 is the experimental setup used by Crouse et al. [35], who formed nanoparticles of aluminum in tetrahydrofuran with the addition of a small amount of oleic acid, whose purpose was to inhibit agglomeration. These authors reported that the solution changed color within the first few minutes of ablation, as shown in Figure 15. Presented in Figure 16(a) to (c) are TEM images of their

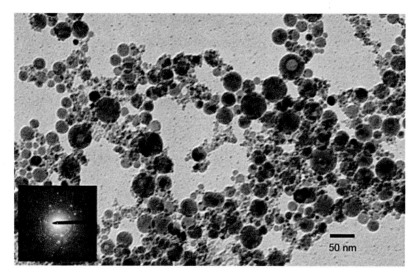

FIGURE 10   TEM micrograph of Ti nanoparticles formed by PLA in 2-propanol. (Reprinted from [34] with permission of the American Chemical Society © 2006.)

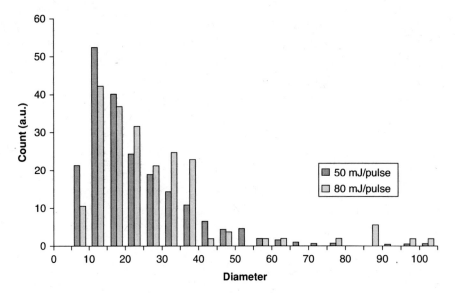

FIGURE 11 Size distributions of Ti nanoparticles formed by PLA in 2-propanol. (Reprinted from [34] with permission of the American Chemical Society © 2006.)

nanoparticle product. Figure 16(a) shows that their aluminum nanoparticles were not agglomerated when ablation was carried out in THF containing oleic acid. The size distribution of the aluminum nanoparticles is presented in Figure 16(d). However, the authors noted that ablating the aluminum target without oleic acid formed the product shown in Figure 16(c). In this case, the product was a mass of agglomerated aluminum nanoparticles. The authors concluded that the oleic acid formed a thin shell around the aluminum nanoparticles, which prevented their agglomeration. The authors also analyzed their product by energy-dispersive spectroscopy (shown in Figure 17). The results indicated that their aluminum nanoparticles had less oxygen than did a sample of Al_2O_3 nanoparticles. The authors concluded that the interior of the nanoparticles contained essentially pure aluminum, and that oxygen was present only as a shell. The chief result of their work, however, was the production of nonagglomerated nanoparticles in liquid.

Other workers have reported [36,37] PLA in a liquid environment, and the technique appears to be a promising method of fabricating nanoparticles. The chief disadvantage would appear to be the incorporation of contaminant species within the nanoparticles. However, this may be used as an advantage in some applications.

2.3 Through Thin-Film Ablation

There are several applications that require a thin layer of nanoparticles that are spread uniformly and without agglomeration over a surface. Among these applications are carbon nanotube growth, fuel cells, and advanced sensors. There is a need for

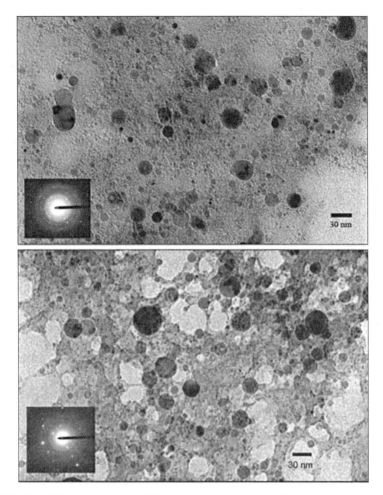

FIGURE 12   TEM micrographs of Ti nanoparticles formed by PLA in n-hexane. (Reprinted from [34] with permission of the American Chemical Society © 2006.)

improved processes to form these nanoparticles. Among the numerous methods that have been developed, PLA has proved to be especially effective because of the potential for congruent ablation, the ability to produce nanoparticles of high purity, the ability to deposit nanoparticles on room-temperature substrates, and the relative simplicity of the process. One problem with the conventional PLA process is the observation that the material deposited, in addition to containing nanoparticles, also contains large, micrometer-sized particles. These are formed through splashing, which occurs when a transient liquid layer is formed in the irradiated volume of the target; liquid droplets can be ejected from the target by the recoil pressure of the expanding gas on the transient liquid layer. Traditionally, splashing has been minimized (but not eliminated) by ablating very smooth targets or by using low

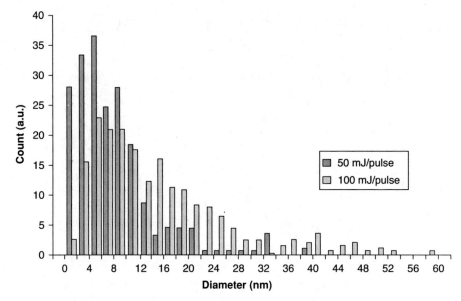

FIGURE 13 Size distributions of Ti nanoparticles formed by PLA in n-hexane. (Reprinted from [34] with permission of the American Chemical Society © 2006.)

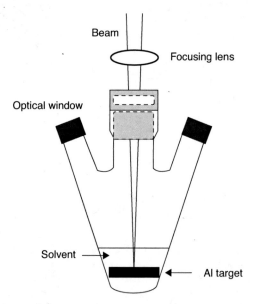

FIGURE 14 Experimental setup for forming Al nanoparticles by PLA in tetrahydrofuran and oleic acid. (Reprinted from [35] with permission of Elsevier © 2010.)

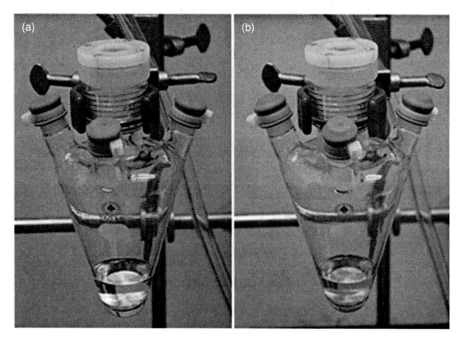

FIGURE 15 Experimental setup for forming Al nanoparticles by PLA in tetrahydrofuran and oleic acid (a) before PLA and (b) after PLA. (Reprinted from [35] with permission of Elsevier © 2010.)

laser energy densities. Target splashing is an issue in nanoparticle synthesis since the deposition of large particles within a field of nanoparticles makes it more difficult to exploit the unique properties of isolated, nonagglomerated nanoparticles. Target splashing is also an issue because it represents material waste.

The technique of through thin-film ablation (TTFA) was developed [38] as a solution to this problem. The technique, which is illustrated in Figure 18, entails using a very thin film as a target. The target film is deposited onto a support that is transparent to the ablating laser; this target film is irradiated from the back side. The use of a very thin (tens of nanometers) film as a target allows one to carry out laser ablation and to form nanoparticles with essentially no contribution from larger nanoparticles. The important difference from conventional PLA lies in the different target geometry.

Conventional laser ablation entails irradiating a bulk target from the front. With this configuration, the region of highest transient temperature rise (as a result of laser impact) lies at the surface of the target; the temperature rise will decrease as a function of depth into the target. At some critical distance from the surface, the temperature rise will be insufficient to cause explosive ejection of atomic species but will be sufficient to cause target melting. At this point and deeper, the transient liquid layer of molten target will be formed, and this will be the source of large particles that are splashed from the target.

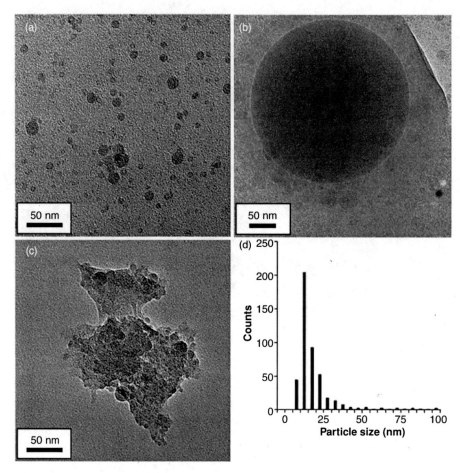

FIGURE 16 (a,b) TEM micrographs of Al nanoparticles formed by PLA in tetrahydrofuran and oleic acid, (c) same as previous but without oleic acid, (d) size distribution of Al nanoparticles. (Reprinted from [35] with permission of Elsevier © 2010.)

In contrast, TTFA entails irradiating the target film from the back side. In this geometry, the region of highest transient temperature rise will lie at the interface between the target support and the target film. Target material at this interface will be predominantly atomized, and regions away from the interface (toward the target surface) will experience smaller temperature rises. At some critical distance from the interface, the temperature rise will be insufficient to cause explosive ejection but will be sufficient to melt the target. By choosing a target thickness that is less than this critical distance, one can ablate a target with minimal melting and therefore with a minimal amount of splashed particles.

Shown in Figure 19 is a time-lapse photograph taken in the author's laboratory during TTFA of an iron film with 248-nm laser radiation. The target and substrate are situated on the left and right sides of the figure, respectively. The laser beam struck

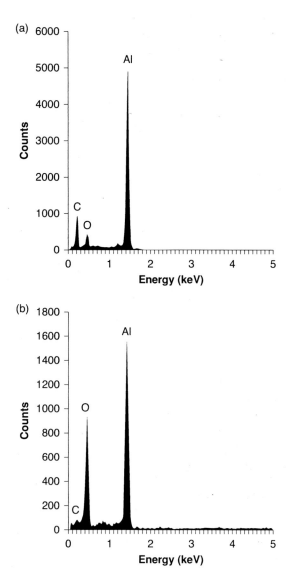

FIGURE 17 Energy dispersive spectrum of (a) Al nanoparticles and (b) commercial Al_2O_3 nanoparticles. (Reprinted from [35] with permission of Elsevier © 2010.)

the target from behind at an angle of 45°. The bright object seen in the figure is due to visible emission from the hot species ejected from the target as a result of laser impact. This feature is very different from the typical appearance of a PLA plume; the path of the abated species is remarkably linear and long. The target–substrate distance in the figure was 75 mm. The collimated nature of the ablated material suggests that the technique may be used in some direct-write applications.

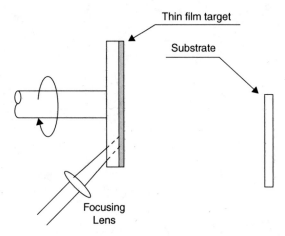

FIGURE 18 Experimental setup for Through Thin Film Ablation (TTFA). The thin film target is ablated from the backside, and nanoparticles are deposited on a substrate. (Reprinted from [38] with permission of Elsevier © 2008.)

Presented in Figure 20 are TEM micrographs [38] of silver nanoparticles formed by TTFA. These deposits were each made with a single laser shot, and the images shown in Figure 20 were acquired from deposits formed by TTFA in vacuum ($\sim 10^{-4}$ Pa) and in 665 Pa of argon, respectively. The *silver nanoparticles are clearly not agglomerated*. Shown in Figure 21 are the corresponding nanoparticle size

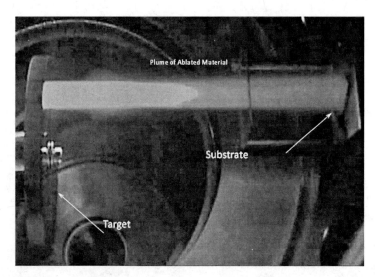

FIGURE 19 Time-lapse photograph taken in the author's laboratory during TTFA of an Fe film with 248-nm laser radiation. The target and substrate are situated on the left and right sides, respectively, of the figure. The laser beam struck the target from behind at an angle of 45°.

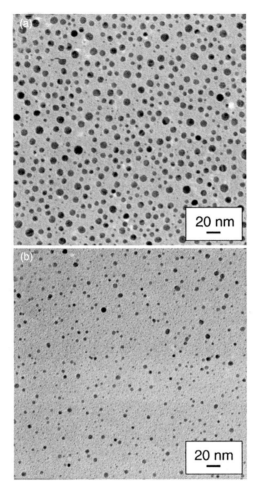

FIGURE 20   TEM micrographs of Ag nanoparticles formed by TTFA. These deposits were each made with a single laser shot. The deposits were formed by TTFA (a) in vacuum ($\sim 10^{-4}$ Pa) and (b) in 665 Pa of Ar. (Reprinted from [38] with permission of Elsevier © 2008.)

distributions. Both distributions are bimodal, and the distribution presented in Figure 21(b) has a most probable size of 1 nm.

Nanoparticles of other materials, including Fe and Pt, have been fabricated in the author's laboratory, and in all cases, the nanoparticles have been nonagglomerated. TTFA therefore appears to be applicable to a wide range of materials. Iron nanoparticles formed by TTFA have been used to catalyze the growth of carbon nanotubes, and Pt nanoparticles formed by the same process have been used in fuel cells and in nanosensor applications. The ability to form a layer of nonagglomerated nanoparticles offers the potential for more fully exploiting unique properties of isolated nanoparticles.

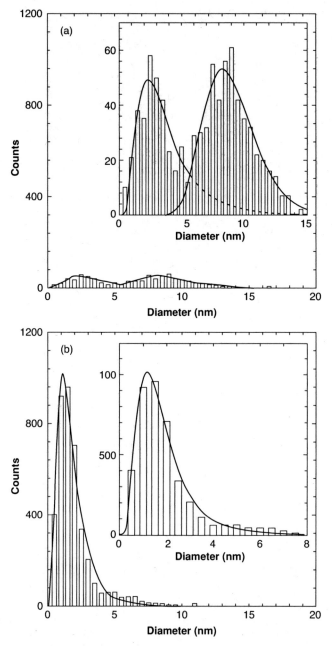

FIGURE 21   Size distributions of Ag nanoparticles formed by TTFA in (a) vacuum ($\sim 10^{-4}$ Pa) and (b) in 665 Pa of Ar. (Reprinted from [38] with permission of Elsevier © 2008.)

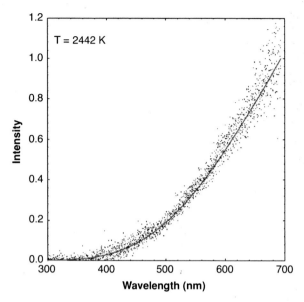

FIGURE 22   Time-averaged optical emission spectrum from TTFA of Fe. The dots represent the experimental data points, and the solid curve is a fit to the using the Planck blackbody radiation formula. (From [39].)

Considerable effort has been devoted to characterizing the dynamics of the TTFA process in order to better understand and better control it. This work has entailed using various probes, including optical emission spectroscopy (OES), time-of-flight (TOF) measurements, and time-gated spectroscopy as well as fast photography. The results for iron are presented here. Shown in Figure 22 are the time-averaged OES results from TTFA of iron. The dots represent the experimental data points, which were acquired by dispersing the radiation collected 12 mm in front of the target surface. It can be seen that the data are a smoothly increasing function of the emission wavelength. The solid red curve in the figure is a fit to the data using the Planck blackbody radiation formula. The fit is excellent. The effective temperature extracted from these data is 2442 K, which, interestingly, is above the melting temperature but below the boiling point of bulk iron. These data suggest that the visible emission detected from the TTFA plume is caused by emission from nanodroplets of liquid Fe. The larger component in the bimodal size distribution seen in TTFA deposits arises from these nandroplets, which are ejected directly from the target.

Shown in Figure 23 are images of the ejected material that were acquired at the indicated delay times after the laser pulse. There is a fast component that can be clearly seen at 2 and 5 μs after the laser pulse. This component has a speed on the order of 5 km/s. In addition, there is a slower component in the ablated flux that is more clearly seen at delay times greater than 20 μs. The corresponding speed for this component is on the order of 300 m/s.

These data are corroborated by the TOF results presented in Figure 24. The solid line represents the intensity of visible light detected by a photomultiplier tube that

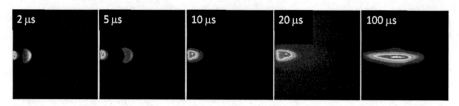

| 2 μs | 5 μs | 10 μs | 20 μs | 100 μs |

FIGURE 23   High-speed camera images of the ejected material that were acquired at the indicated delay times after the laser pulse. The target is situated on the left of each image. (From [39].)

was situated 25 mm from the target. There is a sharp, fast component that reaches a maximum near 5 μs (with a corresponding speed of 5 km/s). In addition, there is a slower, broader peak that reaches a maximum near 80 μs (with a corresponding speed of about 300 m/s). This is the slower component that is also visible at longer delay times in Figure 23.

To identify the nature of these two components, time-gated spectroscopy was carried out by imaging the visible light emitted at a distance of 25 mm from the target at selected delay times after the laser pulse. The spectrum acquired at 5 μs is presented by the inset as Figure 24(a). A number of sharp lines are shown in this spectrum, all of which can be assigned to transitions in electronically excited, atomic iron. This demonstrates that the fast component is due to hot, fast Fe atoms that are

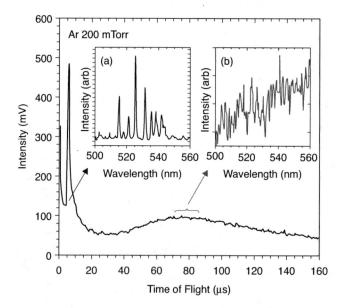

FIGURE 24   Time-of-flight distribution of visible emission detected by a photomultiplier tube that was situated 25 mm from the target. Gated emission spectra acquired (a) 5 μs and (b) between 75 and 85 μs after the laser pulse. (From [39].)

ejected from the TTFA target. Figure 24(b) shows the spectrum acquired between 75 and 85 μs after the laser pulse. This spectrum is fairly featureless and can be described by the blackbody formula. This validates the assertion that the slower component in the ablated flux is due to slow, hot nanoparticles.

Together, the data indicate that TTFA results in the ejection of hot fast atoms, followed by hot slower nanoparticles that are ejected directly from the target. Carrying out TTFA in a background gas has little effect on the latter, due to their very large mass. However, doing TTFA in a background gas causes the ablated atoms to lose energy by collisions with the background gas. These collisions give rise to the formation of nanoparticles. The bimodal size distribution can therefore be understood as arising from the ablated atoms (producing small nanoparticles) and the directed ejeted nanoparticles (comprising the large nanoparticles). This is further validated by the observation that the size distribution of the small mass component is independent of target thickness, while the size distribution of the larger particles scales with the target thickness.

TTFA is a relatively new technique for forming nanoparticles that will undoubtedly find its niche in some specialized uses. The technique allows one to form a layer of nonagglomerated nanoparticles. The technique is, however, limited to making very small batches of nanomaterial, and the prospects of its being scaled up are questionable.

3   CONTINUOUS LASER HEATING

Nanomaterial fabrication with continuous lasers is less common but has more potential for eventual use in manufacturing, due to the unity duty cycle. One of the most promising applications for this type of fabrication is the synthesis of boron nitride (BN) nanotubes, as reported by Laude et al. [40] and Lee et al. [41]. The experimental procedure entailed irradiating, with the output of a high-powered continuous CO_2 laser ($\lambda = 10.6 \, \mu m$), a BN target within a chamber that had a continuous supply of N_2. This caused melting and evaporation of the BN target. The evaporated species recombined in the gas phase to form nanotubes, and these were entrained in the N_2 flow and were trapped by a filter from which they were extracted and characterized. One of the very interesting aspects of this process is that it *allows the formation of BN nanotubes without the use of a catalyst*. The tubes thus formed need no further purification to remove metal catalyst, unlike the situation with carbon nanotubes.

Shown in Figure 25 is a TEM micrograph of the BN nanotubes formed by Laude and co-workers. One of the interesting observations from their work is that some nanotubes appeared to have originated from boron nanoparticles, suggesting that the nanoparticles served as the catalyst for nanotube growth.

More recently, Smith et al. [42] reported the synthesis of large quantities of BN nanotubes by a process called the pressurized vapor/condenser method. One of the attractive features of this process is its flexibility in terms of laser as well as boron source. The authors reported BN nanotube synthesis using both a 1-kW free electron laser (operating at 1.6 μm) as well as a kW-class" welding laser operating

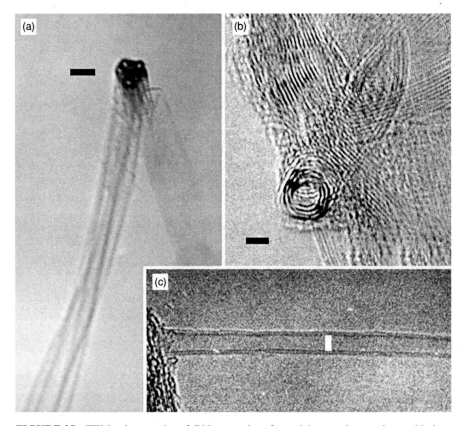

FIGURE 25 TEM micrographs of BN nanotubes formed by continuous laser ablation. (Reprinted from [40] with permission of the American Institute of Physics © 2000.)

at 10.6 μm. They also reported using hot-pressed BN, cold-pressed BN, amorphous boron powders, and cast boron as laser targets to produce BN nanotubes.

The experimental procedure for this process entailed irradiating a boron source with the output of one of these continuous lasers; target irradiation was carried out in several atmospheres (2 to 20) of N_2. One of the unique features of their process is the insertion of a cooled metal wire into the plume of evaporated boron. This serves to cool the hot boron gas locally to such an extent that the gas condenses to form seed particles, from which BN nanotubes grow. The metal wire is translated within the evaporant plume in order to initiate growth. Presented in Figure 26 is an image that shows the unique product of a production run using this process. The BN nanotube product has the appearance of combed cotton and is oriented along the growth axis. The length of the product is approximately 15 cm.

Shown in Figure 27(a) to (c) are scanning electron micrographs of the product. The growth direction is indicated by the arrow in Figure 27(a), and it can be seen that

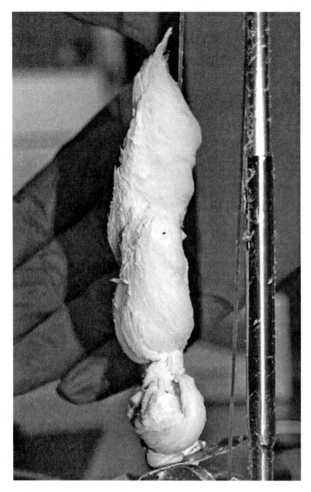

FIGURE 26 BN nanotube product formed by continuous laser ablation. (Reprinted from [42] with permission of IOP © 2009.)

the fibers are aligned parallel to the growth direction. It is interesting to note that a rounded, solidified boron droplet is marked by the arrow in Figure 27(c). Presented in Figure 27(d) are TEM micrographs of the BN nanotube product; nanotubes consisting one of, three, and five walls can be seen in this figure.

This new process would appear to have excellent potential for the manufacture of BN nanotubes for numerous applications. It is also important to note that this process requires no catalyst, thus obviating the need for extensive postproduction cleanup. It will be exciting to see whether this process can be extended to the manufacture of other nanomaterials. As of this writing, this process is considered the state of the art for large-scale production of BN nanotubes.

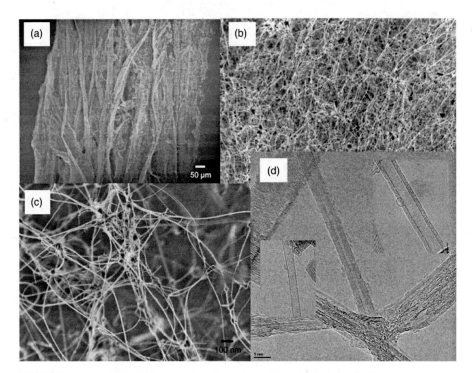

FIGURE 27   (a–c) SEM micrographs; (d) TEM micrograph of BN nanotube product formed by continuous laser ablation. (Reprinted from [42] with permission of IOP © 2009.)

4   SUMMARY

Laser-assisted fabrication techniques have played a key role in the development of new nanomaterials. Pulsed laser ablation techniques are especially attractive because they are inherently simple and because they can be used easily and quickly to carry out feasibility studies, but they suffer from small throughput. Continuous laser heating processes show excellent potential for larger-scale manufacturing and can be carried out with existing industrial lasers. It will be very interesting to watch future developments as this field matures.

REFERENCES

1. D. B. Chrisey and G. K. Hubler (Eds.), *Pulsed Laser Deposition of Thin Films*, Wiley, New York, 1994.
2. D. L. Pappas, K. L. Saenger, J. J. Cuomo, and R. W. Dreyfus, Characterization of laser vaporization plasmas generated for the deposition of diamond-like carbon, *J. Appl. Phys.* **72**(9), 3966–3970 (1992).

3. P. T. Murray and D. T. Peeler, Dynamics of graphite photoablation: kinetic energy of the precursors to diamond-like carbon, *Appl. Surf. Sci.* **69**(1–4), 225–230 (1993).

4. R. Braun and P. Hess, Time of flight investigation of infrared laser-induced multilayer desorption of benzene, *J. Chem. Phys.* **99**(10), 8330–8340 (1993).

5. P. T. Murray and D. T. Peeler, Pulsed laser deposition of carbon films: dependence of film properties on laser wavelength, *J. Electron. Mater.* **23**(9), 855–859 (1994).

6. M. A. Capano, Time of flight analysis of the plume dynamics of laser ablated 6H silicon carbide, *J. Appl. Phys.* **78**(7), 4790–4792 (1995).

7. A. V. Bulgakov and N. M. Bulkova, *Sov. J. Phys. D* **28**(8), 1710–1718 (1995).

8. P. T. Murray, B. Koehler, J. Kaspar, J. Schreiber, and S. Lipfert, Nanomaterials produced by laser ablation techniques: synthesis and passivation of nanoparticles, *SPIE Proc.* **6175**, 61750D1–D8 (2006).

9. T. Sasaki, S. Terauchi, N. Koshizaki, and H. Umehara, The preparation of iron complex oxide nanoparticles by pulsed-laser ablation, *Appl. Surf. Sci.* **127–129**, 398–402 (1998).

10. W. Rong, W. Ding, L. Madler, R. S. Ruoff, and S. K. Friedlander, Mechanical properties of chain aggregates by combined AFM and SEM: isolated aggregates and networks, *Nano Lett.* **6**(12), 2646–2655 (2006).

11. T. A. Burr, A. A. Seraphin, E. Werwa, and K. D. Kolenbrander, Carrier transport in thin films of silicon nanoparticles, *Phys. Rev. B* **56**, 4818–4824 (1997).

12. D. B. Geohegan, A. A. Puretzky, G. Duscher, and S. J. Pennycook, Time-resolved imaging of gas phase nanoparticle synthesis by laser ablation, *Appl. Phys. Lett.* **72**, 2987–2989 (1998).

13. K. Hata, M. Fujita, S. Yoshida, S. Yasuda, T. Makimura, K. Murakami, H. Shigekawa, W. Mizutani, and H. Tokumoto, Selective adsorption and patterning of Si nanoparticles fabricated by laser ablation on functionalized self-assembled monolayer, *Appl. Phys. Lett.* **79**, 692–694 (2001).

14. D. H. Lowndes, C. M. Rouleau, T. G. Thundat, G. Duscher, E. A. Kenik, and S. J. Pennycook, Silicon and zinc telluride nanoparticles synthesized by pulsed laser ablation: size distributions and nanoscale structure, *Appl. Surf. Sci.* **129**, 355–361 (1998).

15. D. H. Lowndes, C. M. Rouleau, T. G. Thundat, G. Duscher, E. A. Kenik, and S. J. Pennycook, Silicon and zinc telluride nanoparticles synthesized by low energy density pulsed laser ablation into ambient gases, *J. Mater. Res.* **14**, 359–370 (1999).

16. T. Makimura, Y. Kunii, N. Ono, and K. Murakami, Silicon nanoparticles embedded in SiO_2 films with visible photoluminescence, *Appl. Surf. Sci.* **129**, 388–392 (1998).

17. T. Makino, N. Suzuki, Y. Yamada, T. Oshida, T. Seto, and N. Aya, Size classification of Si nanoparticles formed by pulsed laser ablation in helium background gas, *Appl. Phys. A* **69**, S243–S247 (1999).

18. A. A. Seraphin, E. Werwa, and K. D. Kolenbrander, Influence of nanostructure size on the luminescence behaviour of silicon nanoparticle thin films, *J. Mater. Res.* **12**, 3386–3392 (1997).

19. Y. H. Tang, X. H. Sun, F. C. K. Au, L. S. Liao, H. Y. Peng, C. S. Lee, S. T. Lee, and T. Sham, Microstructure and field-emission characteristics of boron-doped Si nanoparticle chains, *Appl. Phys. Lett.* **79**, 1673–1675 (2001).

20. J. Muramoto, Y. Nakata, T. Okada, and M. Maeda, Visualization and control of Si nanoparticle behavior in laser-ablation plume, *Conference on Lasers and Electro-Optics*, Conference Edition, 1998 Technical Digest Series, Vol. 6, San Francisco, 1998.

21. E. Ozawa, Y. Kawakami, and T. Seto, Formation and size control of tungsten nanoparticles produced by Nd : YAG laser irradiation, *Scripta Mater.* **44**, 2279–2283 (2001).

22. M. F. Becker, J. R. Brock, H. Cai, D. E. Henneke, J. W. Keto, J. Y. Lee, W. T. Nichols, and H. D. Glicksman, Metal nanoparticles generated by laser ablation, *Nanostruct. Mater.* **10**, 853–863 (1998).

23. J. Bekesi, R. Vajtai, P. Simon, and L. B. Kiss, Subpicosecond excimer laser ablation of thick gold films of ultra-fine particles generated by a gas deposition technique, *Appl. Phys. A* **69**, S385–S387 (1999).

24. S. Link and M. A. El-Sayed, Shape and size dependence of radiative, nonradiative and photothermal properties of gold nanocrystals, *Int. Rev. Phys. Chem.* **19**, 409–453 (2000).

25. F. Mafune, J. Y. Kohno, Y. Takeda, T. Kondow, and H. Sawabe, Formation of gold nanopaticles by laser ablation in aqueous solution of surfactant, *J. Phys. Chem. B* **105**, 5114–5120 (2001).

26. C. B. Hwang, Y. S. Fu, Y. L. Lu, S. W. Jang, P. T. Chou, C. R. C. Wang, and S. J. Wu, Synthesis, characterization, and highly efficient catalytic reactivity of suspended palladium nanoparticles, *J. Catal.* **195**, 336–341 (2000).

27. M. F. Becker, J. R. Brock, H. Cai, D. E. Henneke, J. W. Keto, J. Y. Lee, W. T. Nichols, and H. D. Glicksman, Metal nanoparticles generated by laser ablation, *Nanost. Mater.* **10**, 853–863 (1998).

28. F. Mafune, J. Y. Kohno, Y. Takeda, T. Kondow, and H. Sawabe, Structure and stability of silver nanoparticles in aqueous solution produced by laser ablation, *J. Phys. Chem. B* **104**, 8333–8337 (2000).

29. M. Hajra, N. N. Chubun, A. G. Chakhovskoi, C. E. Hunt, K. Liu, A. Murali, S. H. Risbud, T. Tyler, and V. Zhirnov, Field emission characteristics of silicon tip arrays coated with GaN. and diamond nanoparticle cluster, in *Proceedings of the 14th International Vacuum Microelectronics Conference* (Cat. No. 01TH8586), 2001, pp. 121–122.

30. S. K. Friedlander, K. Ogawa, and M. Ullman, Elasticity of nanoparticle chain aggregates: implications for polymer fillers and surface coatings, *Powder Technol.* **118**, 90–96 (2001).

31. K. Ogawa, T. Vogt, M. Ullman, S. Johnson, and S. K. Friedlander, Elasticity of nanoparticle chain aggregates of TiO_2, Al_2O_3, and Fe_2O_3 generated by laser ablation, *J. Appl. Phys.* **87**, 63–73 (2000).

32. M. S. El Shall, S. Li, T. Turkki, D. Graiver, U. C. Pernis, and M. I. Baraton, Synthesis and photoluminescence of weblike agglomeration of silica nanoparticles, *J. Phys. Chem.* **99**, 17805–17809 (1995).

33. P. Barnes, P. T. Murray, T. Haugan, R. Rogow, and G. P. Perram, Nanoparticle formation from $YBa_2Cu_3O_{7-x}$ for potential flux pinning mechanism, *Physica C* **377**(4), 578–582 (2002).

34. J. S. Golightly and A. W. Castleman, Jr., Analysis of titanium nanoparticles created by laser irradiation under liquid environments, *J. Phys. Chem. B* **110**, 19979–19984 (2006).

35. C. A. Crouse, E. Shin, P. T. Murray, and J. E. Spowart, Solution assisted laser ablation synthesis of discrete aluminum nanoparticles, *Mater. Lett.* **64**, 271–274 (2010).

36. I. Dolgaev, A. V. Simakin, V. V. Voronov, G. A. Shafeev, and F. Bozon-Verduraz, Nanoparticles produced by laser ablation of solids in liquid environments, *Appl. Surf. Sci.*, **186**(1–4), 546–551 (2002).

37. P. Sylvestre, A. V. Kabashin, E. Sacher, and M. Meunier, Femtosecond laser ablation of gold in water: influence of the laser-produced plasma on the nanoparticle size distribution, *Appl. Phys. A* **80**(4), 753–758 (2005).

38. P. T. Murray and E. Shin, Formation of silver nanoparticles by through thin film ablation, *Mater. Lett.* **62**, 4336–4338 (2008).

39. P. T. Murray and E. Shin, Thin film, nanoparticle, and nanocomposite fabrication by through thin film ablation, *Proc. SPIE* **7404**, 74040F (2009).

40. T. Laude, Y. Matsui, A. Marraud and B. Jouffrey, Long ropes of boron nitride nanotubes grown by a continuous laser heating, *Appl. Phys. Lett.* **76**(22), 3239–3241 (2000).

41. R. S. Lee, J. Gavillet, M. Lamy de la Chapelle, and A. Loiseau, J.-L. Cochon, D. Pigache, J. Thibault, F. Willaime, Catalyst-free synthesis of boron nitride single-all nanotubes with a preferred zig-zag configuration, *Phys. Rev. B* **64**, 121405 (2001).

42. M. W. Smith, K. C. Jordan, C. Park, J.-W. Kim, P. T. Lillehei, R.Crooks, and J. S. Harrison, Very long single- and few-walled boron nitride nanotubes via the pressurized vapor/condenser method, *Nanotechnology* **20**, 505604 (2009).

6

EXPERIMENTAL CHARACTERIZATION OF NANOMATERIALS

A. G. Jackson

Center for Nanoscale Multifunctional Materials, Wright State University, Dayton, Ohio

1   INTRODUCTION

As a result of the breakthroughs in technology made in the past 10 to 20 years, experimenters are able to characterize properties of materials with probes capable of nanometer resolution. For large data sets (multiple spectra, multiple images, and large images), advances in computer power have also provided means for rapid processing that yield quantitative values for properties not obtainable on a wide scale a few

Nanoscale Multifunctional Materials: Science and Applications, First Edition.
Edited by Sharmila M. Mukhopadhyay.
© 2012 John Wiley & Sons, Inc. Published 2012 by John Wiley & Sons, Inc.

years ago. *Ab initio* calculations and modeling of nanoscale materials now provide the experimenter with visualizations of behaviors that can be compared directly with data obtained experimentally, allowing each approach to refine parameters and experimental techniques.

In this chapter, overviews of only a few of the many experimental methods in use are presented, each of which has resulted in advances in understanding that have created openings to areas not possible before. Of the many methods available, two are presented: microscopies and x-rays. An overview of each is provided that will allow an interested researcher to obtain a sense of the method and its use. Sources have also been included to enable a researcher to find detailed explanations and literature on applications of the method. This approach was chosen for several reasons. First, the number of methods is very large. Some are extensions of existing methods. Second, detailed explanations of each method, although desirable, are not practical here because it would require a very large book to present each method competently. Searching the Internet will produce millions of hits for nanomaterials or methods. My advice here is to select very specific materials or methods and then select university sites or journal sites to find detailed references.

Methods presented here cover microscopies and two applications of x-ray diffraction that are relatively recent for crystalline or amorphous solid materials. Various microscopies, electron and optical, have been refined considerably over the last decade and can now be used to provide images with fractional nanometer resolutions and dynamic images at the nanosecond scale, a very exciting development that represents an important change in electron microscopy. Computer control of electron microscope lens systems has enabled improvement in resolution by factors of 2 or more, allowing less than a tenth of a nanometer for transmission and less than 1 nm for scanning electron microscopes. Optical microscopy, an almost forgotten tool, has been revived with the scanning near-field system, which allows resolutions considerably below the diffraction limit, thereby opening this venerable technique to imaging nanoscale features.

Another venerable technique, x-ray diffraction, is being used to obtain quantitative data on the structure of amorphous solids and on nanoparticles. Interestingly, the application of pair distribution function (PDF) analysis of the usual diffractogram reveals nanoscale data not thought obtainable on coordination shells in locally ordered materials, because of the poor Bragg peak shape and size and the very significant diffuse background present. Application of PDF to liquids has been carried out for a long time, and extension of this analysis method to diffractograms of solids is an example of the effects of available technology to accomplish analyses not practical before.

For characterization one needs a probe or probes, suitably prepared material, and a detection device(s) with processing electronics. The nanoscale domain requires probes capable of interaction diameters on the order of nanometers or smaller to achieve the spatial resolutions desired. Probes consist of electrons, photons (e.g., x-ray, ultraviolet, optical, infrared), and/or particles (neutrons, ions). Probe–signal combinations are indicated in Table 1, and some examples of the methods are given

TABLE 1   Probes Used and Signal Generated or the Interaction That Occurs

Signal Probe	X-Ray Photon	Electron	Optical Photon	Particle
X-ray	Diffraction fluorescence	Characteristic excitation/emission	Emissions related to electronic structure of material	Desorption, molecular fragments, ions
Electron	Species-specific emission	Diffraction, electron states, excitations	Species-specific emissions	Desorption, ions
Optical photon	Not known	Photoconductivity	Emissions related to electronic structure	Stimulated desorption or dissociation
Particle (ion, neutron)	Characteristic energy	Charge current	State excitations	Sputtering, ions, electrons, x-rays

TABLE 2   Methods and Probe Types

Signal Probe	X-Ray Photon	Electron	Optical Photon	Particle
X-ray	Characteristic x-rays, diffraction	XPS	Fluorescence	Ionization spectroscopy
Electron	EDA, EPMA	AES, diffraction, TEM, STEM, SEM, HRTEM, HRSTEM, SPM, EELS	Luminescence	Electron desorption spectroscopies, mass spectroscopy
Non-x-ray photon	[?]	Photoelectric effect	Absorption, emission, phosphorescence	Optical desorption spectroscopies
Particle (ion, neutron)	[?]	FIB, secondary emission	Stress-induced fluorescence	SIMS, ion beam etching

in Table 2. In addition, multiple probe combinations are mentioned, such as photons to generate photoelectrons, which are then used to interact with the sample.

2 MICROSCOPY

Advances in microscopy have been very large in the last 10 years or so, because of advances in electron optics coupled with computer controls, availability of control at the nanoscale of mechanical probes, and refinements of optical imaging using near-field phenomena not possible without such nanoscale control of the probe and rapid processing of the signals to produce meaningful images.

An emerging area in electron microscopy utilizes the ability to generate ultrashort laser pulses for exploration in the time domain of crystalline temporal behavior. Such studies have opened the possibilities for refining understanding of phase transitions and metastable states in materials.

X-ray instrumentation is now capable of micrometer-sized beams, allowing spatial examination of microvolumes as well as element analyses from these volumes. Advances in computing power and availability of intense beam lines produced a re-examination of the simple x-ray diffractogram that has yielded coordination analyses from such simple data.

Software for processing very large data sets (>10 to 1000 gigabytes) and for modeling atomic-scale systems is available as open-source or commercial packages, providing means for rapidly comparing experimental data and images against models, resulting in refinement of models and enabling advances in detailed understanding of behavior at the nanoscale.

For characterization of nanomaterials, various microscopies provide detailed information on the structure via diffraction, morphology via imaging of the material (density, spatial arrangement of components), and composition via beam–specimen interactions (beam scattering, phase changes, contrast differences). Several spectroscopies are intimately associated with electron microscopy in particular.

2.1 Electron Microscopy

Electron microscopy is usually divided into transmission and scanning approaches. Merging of the two approaches into scanning transmission microscopies began some years ago and has now reached the point where a single instrument can image nanoscale features at low voltage (about 1 to 30 kV) for surface topography, atomic number, phase distribution, and crystal orientation. Separately, instruments designed for high voltages (100 to 400 kV) have achieved breakthroughs in resolution to the picometer scale not thought achievable a decade ago.

Imaging of features that correspond to arrangement of the atoms and molecules in the material are essentially routine, and comparisons to models can be done to verify models of such observed arrangements. Simulations of transmission electron microscope (TEM) images have progressed to the point that simulated images faithfully reproduce features observed in real images, such as defects and atomic arrangements. With continued improvements in computer capability, the speed of generating simulated images to compare with observed features certainly will reach real-time computation in the next few years. Some examples of simulated images of defects in copper, shown in Figure 1, illustrate the fidelity with which models of image formation can reproduce features accurately.

Atomic Resolution Advances in TEM/STEM   Haider et al. [2] point out that although Scherzer knew in the late 1940s that the two principal axial aberrations in TEM and STEM (scanning TEM) could be corrected with suitable electrostatic or

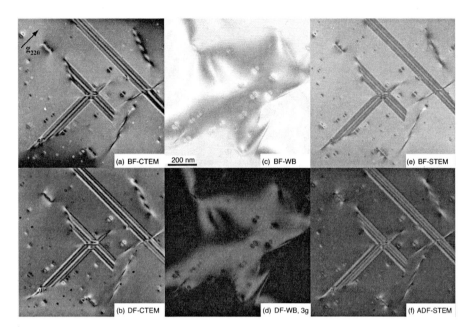

FIGURE 1   Simulation of images from Cu. (From [1].)

magnetic correctors, the technology was not at the point where it was feasible to accomplish fabrication of the needed parts. By 1998 this had changed dramatically, and these authors reported on development and application of hexapole correctors to reduce the chromatic aberration to very small values at 200 kV. Axial chromatic aberration and off-axis aberrations were reduced using this approach. The improvement in resolution reported was notable: 0.24 to 0.14 nm for a 200-kV beam. Kuwabara et al. [3] describe results of combining a scanning tunneling microscope (STM) with a TEM in reflection mode to demonstrate the feasibility of such a combination.

Batson et al. [4] report on computer-controlled aberration correction in a STEM to achieve resolution of less than 0.1 nm at 120 kV. The material used was gold deposited onto carbon film. As a result of the improved resolution and the aberration control, observation of small "rafts" of gold were reported that could be observed only for resolution better than 0.2 nm. Video sequences (10 frames/s) are also reported, and these show the movement and interaction of gold atoms. Additionally, a $Ge_{30}Si_{70}$ sample was imaged to show the dramatic enhancement of resolution, suggesting the achievement of 0.1 nm spatial resolution and possibly even better:

> This successful aberration correction is linked to recent technical advances: (1) computation of electron optical parameters is now possible for nonrotationally symmetric systems, allowing practical designs to be simulated with high accuracy; (2) mechanical fabrication tolerances have advanced materially in the past 15 years; (3) high

stability electronic components have become available in the past 10 years, allowing the packaging of many, very high stability, computer-controlled power supplies in a small space; and (4) high-speed small computers are now available for real-time processing of the shadow map Ronchigram data to obtain aberration parameters [4].

Batson et al. represent the potential jump in capability to explore nanoscale structures upon application of technological advances to refinement of characterization methods.

Expansion of the use of special imaging techniques has contributed to improvements in resolution. Krivanek et al. [5] describe work on chemically identifying individual atoms in a monolayer hexagonal boron nitride (BN) film by utilizing computer-controlled aberration correction and STEM imaging of the film via an annular dark field for a low voltage of 60 kV.

Annular dark-field (ADF) STEM requires a small probe diameter, so that scanning the beam across the specimen produces spatial resolution of signals from each step in the scan. Although ADF scattering occurs by a Rutherford mechanism that yields nonlinear contrast proportional to the atomic number ($Z^{1.7}$), the signal from low-Z atoms is weak. Fortunately, the controlled aberration correction allows sufficient signal strength to produce useful image contrast and chemical identification of the species from which the electrons scattered.

Contrast from boron and nitrogen could be clearly differentiated to allow identification of each type of atom in a BN monolayer. Some image processing was required to remove beam overlap by the tail of the beam distribution, smoothing to average out noise, and normalizing intensity to the boron signal.

A valuable method application that is described briefly is the use of intensity histograms of features in the image that clearly show frequency distributions that can be associated with boron and nitrogen. Data for carbon and oxygen are rather sparse, as the authors note. This approach is described in detail in the supplementary information available from the *Nature* web site for the paper. Application of this statistical approach allowed identification of atom species in the image, and their Figure 3 is really remarkable for the atomic-level detail it exhibits for spatial and chemical information.

Alem et al. [6] examined thin BN plates using high-resolution STEM (HRSTEM) at 80 kV. The presence of unusual defects was noted (triangular shapes and edges that are zigzag or "armchair"). Identification of species by comparing intensities in *Multislice* (the preferred software package/algorithm for simulating images) computed images is reported, and cautions are mentioned related to potential issues with beam alignment that produce images that are easily misinterpreted.

Meyer et al. [7] report on a study of BN thin films characterized via high-resolution TEM (HRTEM) for 80-kV exposure. Results are similar to those reported by Alem et al. [6].

Low-voltage HRSTEM is highly desirable because of the reduction in radiation damage to specimens, which produces defects or destruction of the material. Suenaga et al. [8] present results for an erbium peapod [(Er-C82)n-SWNT; a *peapod* is a single walled nanotube (SWNT) that contains atoms lined up like peas in a pod] which show that characterization of individual atoms can be accomplished by using electron energy loss spectroscopy (EELS)/STEM instrumentation. Their instrument utilizes fifth-order aberration corrections together with correction of sixfold astigmatism and spherical aberration. Additional results included in this paper describe the characterization of calcium peapods. In each case the EELS counts are sufficient for a 0.2-nm beam to generate meaningful spectra that allow an image of the elemental distribution in the nanoparticle to be viewed.

An interesting and useful summary of the major advance in attaining atomic resolution and chemical identification of single atoms in a sample in TEM appears in the June 2010 issue of *Physics Today*. The importance is not in the resolution improvement, but that light elements can be resolved and identified using EELS at 60 keV, keeping the radiation damage to a minimum.

Ultrafast Electron Microscopy Zuo et al. [9] have reported on a coherent nanoarea electron diffraction approach in which an electron beam is focused on the focal plane of the preobjective lens in a TEM (see Figures 2 to 4). This generates a column of electrons parallel to the optic axis that is roughly 10 nm in diameter. With this probe size and method devised by Zuo et al. for reconstructing phases of the pattern, details of structures are obtainable via the diffraction pattern.

Zewail and his group at Cal Tech have developed ultrafast electron microscopy methods that allow observations of events on the picosecond time scale. Thermal effects and mechanical responses of the lattice to energy inputs offer the potential for understanding the dynamic behavior of crystal lattices that will allow refinement of models that can be used to design materials with specific responses. His group is also extending this technique to organic and biological materials [10–12].

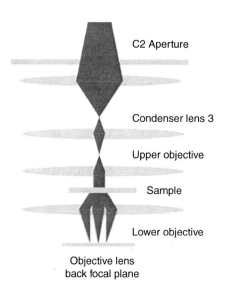

FIGURE 2   Nanodiffraction mode in TEM. (From [9].)

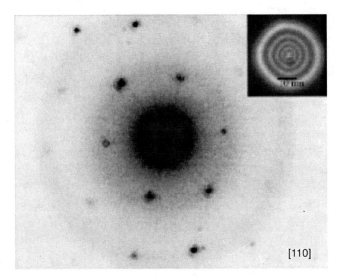

FIGURE 3   Diffraction pattern from the 110 zone of a nanoparticle. (From [9].)

From this group, Park et al. [11] have reported electron microscopy of time-dependent nanoscale phenomena resulting from using a combination of ultrafast laser pulses (femtosecond) and selected area diffraction (about a 1- to -10μm diameter area). Heat stress effects on an 11-nm-thick gold film were seen after a 40-mJ/cm^2 · ns pulse. Forbidden spots appeared 50 ns after the pulse, indicating that the absorption of the pulse energy had distorted the structure by producing parallel to the electron beam

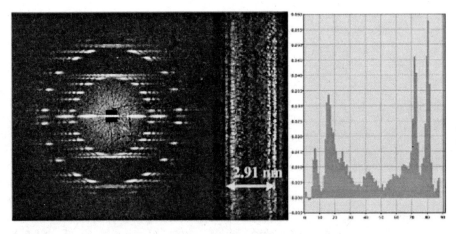

FIGURE 4   Diffraction pattern from a single double-walled carbon nanotube (left); (middle) reconstructed image of a nanotube based on a phase reconstruction algorithm; (right) profile of the reconstructed potential across the middle of the image.

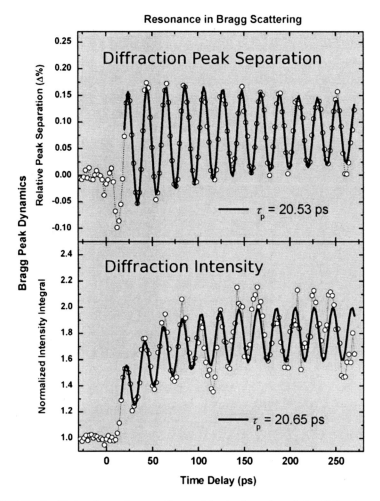

FIGURE 5 Oscillation in peak separation and in the intensity of the diffraction pattern. (From [10] and [11].)

a temporary "bulge" that relaxed after a few microseconds to the original structure (Figure 5). Park et al. calculated the thermal response expected and concluded that the bulge arose from the local heating that was dissipated via transport to the rest of the film and the holder.

The mechanical resonance of graphite was observed after a femtosecond laser pulse with 7 mJ/cm^2. By measurement of the change in separation of diffraction peaks and the intensity present in each frame for a series of patterns taken over a 300-ps span, the picosecond "ringing" was determined to have a period of 20.6 ps and was interpreted to be due to cavity resonance in the film. By utilizing energy loss data from the film, Park et al. deduced Young's modulus values for the graphite, which are comparable to values reported in the literature. Oscillations in the structure were also seen in the images collected.

Dynamic electron microscopy on this scale suggests a wide range of possibilities that are only now being considered for measuring the thermal and mechanical properties of nanomaterials. Park et al. [11] report on martensitic transformation based on nanosecond and microsecond electron diffraction that provides data unattainable any other way on the kinetics of the transformation from body-centered cubic to face-centered cubic. Flannigan et al. [12] report on cantilever movements in a nanosecond time frame that allows the measurement of the frequency of modes of motion of nanometer displacements.

Nanoscale Tomography Jarausch et al. [13] report on a different microscopy advance in which the tomography of nanoscale features is accomplished, providing information on the chemical state of the material (Figure 6). Utilizing focused ion beam (FIB) etching of a multiphase specimen to produce a nanosized cylinder containing layers of phases [14–16], High-angle annular dark field (HAADF) imaging in STEM was used together with EELS to generate data for a set of rotational orientations of

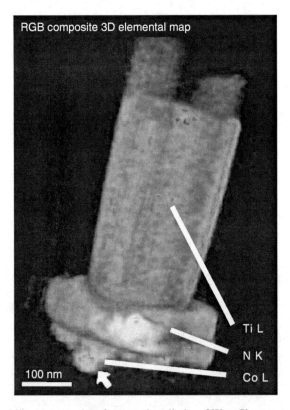

FIGURE 6   Microtomography of nanoscale cylinder of W-to-Si contact. (From [13].)

the cylinder-shaped specimen. From these, a tomography image of EELS element concentrations was constructed that shows the details of the object in terms of the element distribution, bonding, and phase. The Ti–N–Co distribution could be clearly distinguished at the nanometer scale, ranging from about a 10-nm section containing Co, a 100-nm nitrogen-rich phase, and the large Ti-rich phase.

Reconstruction of the microstructure from FIB–OIM data has been reported by Lee et al. [17]. Grain structure is made visible using this approach, but more interesting is the ability to remove phases from the image selectively to reveal small phases that are present that otherwise would have been missed by using the polishing approach only.

Scanning electron microscopy (SEM) has made significant advances in resolution and low-voltage operation that allow imaging of nanomaterials both conductive and nonconductive with resolutions of 1 nm or better. Buhr et al. [18] discuss the problem of dimensional characterization using SEM and STEM/SEM and conclude that acceptable nanoscale measurements are possible, but that a number of issues remain to be studied, including charging of nanoparticles, calibration of scan systems in horizontal and vertical scans, optimization of measurement conditions, size standards at the nanoscale, statistical analysis of sufficient particles, and manufacturer-specific instrument effects on measurements.

2.2   Optical Microscopy

Optical microscopy has until recently been the neglected microscopy for nanoscale, because of the diffraction limit arising from the usual optical wavelengths. With the advent of practical near-field optical microscopy, however, this barrier has been removed, and probing nanomaterials with optical photons has taken on a new life. (A well-written description of NSOM is available at the Olympus web site in the Specialized Techniques tutorials tab.)

Far-Field Basics   As a result of the nature of image formation, ordinary optical microscopy has a fundamental limitation in spatial resolution that is fixed by the wavelength of the light source. Quantitative estimate of the resolution in terms of the distance between two intensity maxima is based on consideration of two point sources and their images. A point source produces an image that consists of a central maximum together with several considerably weaker secondary maxima. This is the *Airy disk* of the point source, and it represents the effects of the imaging optics on the waves that pass through the lenses. Because the diameter of the main maximum in the Airy disk depends on the wavelength of the photons and on the ability of the lenses to transfer a faithful reproduction of the signal, the imaged point source is broadened with respect to the source diameter.

When two point sources are moved closer together, they overlap the main maximum of each until the two maxima appear to be a broader single maximum. The closest distance at which the two maxima are discernible occurs when the maximum of the second point source is centered on the first minimum of the first point source (Rayleigh criterion). At this separation the intensity at the minimum between peaks is about 5% lower in value than the peaks.

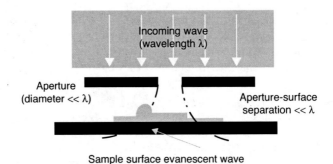

FIGURE 7 Basic SNOM configuration.

In terms of the aperture diameter, wavelength, and distance to the surface, the minimum angular separation is given as

$$\theta = \frac{2\lambda}{a} \tag{1}$$

where θ is the angular separation, λ the wavelength, and a the aperture diameter. If the angle is small, the *sine* and *tangent* are about equal. Hence, the distance from the center of the first maximum in the Airy disk to the first minimum is written as

$$r = L\theta \tag{2}$$

where r is the spot radius, L the distance from the aperture to the surface, and θ the angular separation. This radius is the smallest distance that can be resolved in the image.

These considerations are for the case where the separation between the aperture and the surface is large (more than 1 mm or so) and aperture diameters are similarly a few millimeters. This is the far-field situation commonly used to image objects.

Near-Field Basics If, on the other hand, the separation between aperture and object is less than 1 μm, and the aperture size is a few hundred nanometers, the situation is quite different. This is the near-field situation that for optical wavelengths effectively was not achievable in practice until the technology for submicron position control had been developed [19]. With the development of such control and of apertures of a few hundred nanometers in diameter made from glass fibers, advantage could be taken of the near-field wave. By using illumination from a laser source passed through a metal-coated fiber drawn to a few hundred nanometers in diameter, imaging of the nanoscale topography could be done. Although the wavelength was larger than the diameter of the fiber near the surface, the evanescent wave passing through the aperture was scattered from the surface and detected, generating an image produced by scanning the sample.

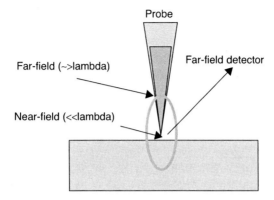

FIGURE 8   Near-field optical microscopy arrangement.

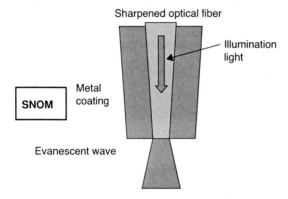

FIGURE 9   Scanning near-field configuration.

In Figure 7 is shown a schematic of the basic scanning near-field optical microscope (SNOM) configuration [20,21]. Variations of this approach have been reported (e.g., see Ref. 22), as indicated in Figures 8 to 10. Apertureless SNOM (ASNOM) [23] offers some advantages over the simple SNOM setup. Aperture and apertureless methods provide nanoscale spatial resolution determined by the aperture/tip diameter.

2.3   Scanning Tunneling Microscopy

Scanning tunneling microscopy (STM) has numerous manifestations that take advantage of the interactions between an atomically sharp probe placed within a few nanometers of the surface and features on the surface [24–26]. Force, current, and magnetic field can be used individually to image the interaction, mapping it with the resolution of fractions of nanometers. Variations of scanning near-field microscopy also use a tip in conjunction with an illumination source to image the surface [27].

STM is based on the quantum phenomenon of electron tunneling through an energy barrier. Because the electron behaves as a wave and is described by a wavefunction, as with photons of visible light, there is an evanescent electron wave that penetrates

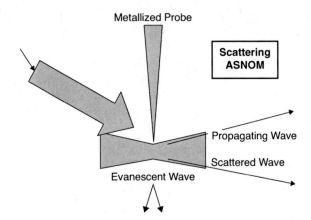

FIGURE 10   Apertureless scanning near-field configuration.

the barrier. The amplitude of this evanescent wave decays exponentially with distance into the barrier. In principle, if the distance between surfaces of two materials is small enough for overlap of the evanescent waves, there is a finite probability that electrons can move from one material to the other (see Figures 11 to 13). This "tunneling" can be enhanced by applying a voltage to one of the materials to produce a difference in the Fermi levels, which reduces the barrier height between the materials. Realizing these conditions experimentally became possible when piezoelectric control systems were built that provided the means for reproducible nanometer movements of an atomically sharp probe relative to the surface and/or movement of the surface laterally.

By applying a voltage to an atomically sharp tip placed a few nanometers from a surface, the energy barrier difference between the tip material and the surface material is lowered enough for electrons to "spill" out of the tip (negative bias on the tip) or be extracted from the surface (positive bias on the tip). By measuring the current

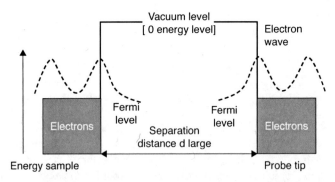

FIGURE 11   Tunneling as a function of the separation of probe and sample. Electron wavefunctions do not overlap when the separation is a few nanometers or more.

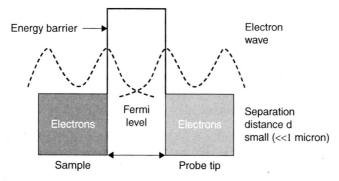

FIGURE 12 Tunneling as a function of the separation of probe and sample. Electron wave-functions overlap when the separation is a few nanometers or less.

to the tip while scanning the sample surface, an image is produced that represents the variation of electron density on the surface. Thus, the image displays intensity variations that mimic the atom locations.

STM has two basic modes for collecting data: constant current mode and constant height mode. By use of piezoelectric control of the vertical movement of the probe, the current can be maintained at a constant value (constant current mode). Alternatively, the probe position can be fixed and the variation in current measured (constant height mode). Each mode is useful, and application of either depends on the sample surface. Having the ability to control a probe at the nanoscale has given rise to a number of variations on the basic modes. (For a list, see Mayer [28].)

Current resulting from the overlap of electron wave functions and an applied voltage is approximated by

$$I \sim V\rho_d(0, E_F)e^{-2kd} \sim V\rho_d(0, E_F)e^{-1.025\sqrt{\phi}} \tag{3}$$

where E_F is the Fermi energy, ρ_d the density of electron states, d the probe–sample separation, and ϕ the difference in work functions of the materials [26].

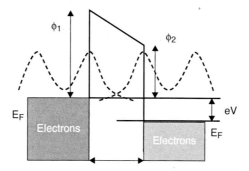

FIGURE 13 When electron wavefunctions overlap, the difference in work functions and applied potential determine the direction of electron flow.

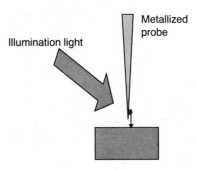

FIGURE 14   Configuration for light-illuminating STM.

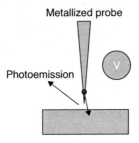

FIGURE 15   Arrangement for photon-emitting operation of STM.

Measurement of van der Waals forces at the surface, electron density of states, spin polarization states, magnetic and electric field variations across a surface, and structural geometry are a few of the properties that can be investigated using STM systems and their variations.

In Figures 14 and 15 are shown configurations for photoexcitation and photon-emission from surfaces. These can be used with SNOM configuration, thereby utilizing the advantages of this technique relative to a sample environment. Light-illuminating STM configuration measures the tunneling current arising from photoexcited carriers in the sample. Spin states can be detected using polarized illumination. In photon-emission configuration, photons are measured that are created by inelastic scattering of electrons in the gap between the tip and the surface. The spectra contain information about the local density of states [29].

3   X-RAY DIFFRACTION

3.1   Diffractometer XRD

In use for decades, x-ray diffraction (XRD) is a tool that has provided enormous amounts of data that have allowed an understanding of crystalline structures and their behavior to be refined to a routine practiced by anyone with a simple (or complex) instrument. Recently, an additional aspect [30,31] of the intensity–2 θ data set has emerged as a significant player in the characterization of crystalline and amorphous nanomaterials.

In its simplest form, x-ray diffraction depends on the wavelength λ of the probe radiation, on the angle θ at which the scattered radiation is collected, and on the average crystal geometry of the sample (d is the interplanar separation of equivalent planes), stated succinctly in the Bragg equation as

$$\lambda = 2d \sin \theta \tag{4}$$

A plot of intensity versus 2θ yields a spectrum that contains maxima corresponding to planes that have met the Bragg condition. By use of the Bragg equation, the angular locations in the spectrum are then converted to d-values, from which Miller indices are identified for the crystal structure. This works adequately when the sample size is a few tens of micrometers or larger (e.g., as powder), but as the particle size decreases, other effects become dominant, such as changes in crystal structure due to relaxation of surface, diffuse scattering relative to the peak size increases, decrease in the peak size because of the smaller number of scattering atoms, and increase in the peak width. All of these render simple XRD unusable when the particle size drops below a few tenths of a micrometer.

3.2 Pair Distribution Function from XRD Diffractograms

A different approach that still utilizes an intensity–angle plot to collect data has been used successfully to characterize the nature of small particles [31–34]. Originally used for liquids, the pair distribution function (PDF) approach takes the intensity–angle data and applies a Fourier transformation to produce a direct space representation of the distances from a reference center (atom) outward to each distance along a radius at which the next atom set lies (i.e., it generates a radial distribution function that identifies the coordination sphere for the material). Short- or long-range order are not crucial to this method, and the broad background plays a major role rather than an undesirable effect. That's the good news. The bad news is that the reciprocal lengths needed are three to four times larger than are usually encountered. This arises because the transform equations from reciprocal to direct space involve the wavevector Q:

$$Q = \frac{4\pi \sin \theta}{\lambda} \tag{5}$$

where θ is the half-angle of scattering and λ is the wavelength of the x-rays. Petkov [31] presents a brief discussion of the method in which he defines the total scattering structure function $S(Q)$ associated with the coherent scattering portion of the intensity as

$$S(Q) = 1 + \frac{\left[I^{\text{coh}}(Q) - \sum c_i |f_i(Q)|^2\right]}{|\sum c_i |f_i(Q)||^2} \tag{6}$$

where c_i is the concentration of element i and f_i is the x-ray scattering factor of element i. Taking the Fourier transform of $S(Q)$ yields the PDF:

$$G(r) = 4\pi r[\rho(r) - \rho_0(r)] \tag{7}$$

in which $\rho(r)$ is the local atomic number density and ρ_0 is the average atomic number density. In terms of the scattering structure function, the radial distribution function is

$$G(r) = \frac{2}{\pi} \int_{Q=0}^{Q_{max}} Q[S(Q) - 1] \sin(Qr)\,dQ \tag{8}$$

Calculation of the coordination radii via the PDF can be accomplished if the angle spectrum can be obtained using a high-intensity beam for the x-rays (e.g., as in a synchrotron). Large reciprocal distances of up to 300 nm^{-1} (30 Å^{-1}) are needed [35] to allow the Fourier transformation of the spectrum to produce a direct lattice plot of the pair distribution function. The radial distribution function for the system $G(r)$ locates the radii (up to about 1 nm or 10 Å) of positions of atoms in the nanomaterial from which scattering took place [36]. Fortunately, free software is available for accomplishing the conversion from angle space to reciprocal length space. The necessary preprocessing of the raw data is handled by the software as well (the web site http://www.ccp14.ac.uk/solution/high q pdf/index.html contains a list of software).

This development is attractive experimentally because the measurements are only slightly extended over the usual diffractogram measurements, yet from the data one can extract detailed information about coordination present in the local structure.

3.3   Microbeam X-Ray Probes

In addition to the PDF approach, in the last five to 10 years there have been important advances in obtaining submicron-sized x-ray beams with diameters in the range 10 to 20 nm [37]. Use of synchrotron sources is common for this approach, but alternative sources such as laser-excited liquid droplets suggest that a commercial laboratory system is near realization [38]. Fabrication of submillimeter Fresnel plates has been accomplished for use as an x-ray lens to allow focusing of x-rays to spot sizes of 10 nm or so [39], a truly amazing development made possible by advances in technology of submicron etching, as in FIB processing of precision cuts in materials [40,41].

Fresnel zone plate consists of concentric rings that alternate transparent and opaque. Its utility lies in its ability to focus monochromatic radiation to a spot much smaller in diameter than the source. For visible-light wavelengths, construction of a Fresnel lens is well established and widely used. For x-rays, whose wavelengths are about three orders of magnitude smaller, construction of a Fresnel zone plate has only recently become practical. Although the concept of an x-ray plate was discussed in the late 1970s and in the 1980s, the resolutions possible were about 50 to 100 nm, still two orders of magnitude larger than x-ray photon wavelengths.

Recently, Vila-Comamala et al. [42] report on a method to produce Fresnel plates with 12-nm resolution via an atomic layer deposition approach that yields a high-refractive-index coating on the Fresnel plate that effectively doubles the resolution from 20 to 25 nm to about 10 to 15 nm.

With nanometer beam sizes, the opportunities for exploring nanomaterials are open to determining atomic behavior at resolutions previously possible only with electron sources. The photon flux [photons/(mm^2·s·$mrad^2$·0.1% bandwidth (BW)] from a synchrotron is on the order of 10^{15} photons/(mm^2·s·$mrad^2$·0.1% BW), or about 10^3 photons/(nm^2·s·$mrad^2$·0.1% BW) [37,43]. If a 20-nm-diameter beam is obtained, the beam area (314.2 nm^2) provides about 30,000 to 40,000 photons/(s·$mrad^2$·0.1% BW) in the beam that are delivered to the specimen, which yields a small number of x-rays. In terms of the flux needed, the focused beam needs about 10^6 photons/(nm^2·s), which, because of the small x-ray yield from the material, means times of seconds or minutes are needed to collect a statistically meaningful signal [44].

To avoid radiation damage, the exposure must be below the damage value [45]. Because of the low contrast, processing of the image data is especially important here. This is not a new aspect. In transmission electron microscopy, the effects of beam interactions with the specimen are well known, and methods of image collection via rastering (e.g., scanning TEM) and adding multiple images allow such sensitive materials to be viewed with minimal damage, and refinement of similar techniques to accommodate the x-ray case may be possible.

4 SUMMARY

Nanomaterials require characterization methods that push the frontier of probe technologies, a process that over the last decade has generated a sea change in achievable spatial resolutions, advanced temporal methods at the nanoscale that are breathtaking, and extended data and image collection and processing approaches to reveal atomic-scale data that was unapproachable because of limitations in computing power and collection technologies. For a recent more detailed overview of the characterization of nanomaterials, limited to microscopies, see Wang's book [46].

Microscopies in particular illustrate this change. Optical microscopy has been revived as a powerful approach to nanoscale characterization through near-field instrumentation and software. Electron microscopy has advanced to the tens of picometer spatial resolution only dreamed of just a decade ago. Scanning tunneling microscopy now has so many variants that just about any property can be examined at the nanometer scale. The power of x-ray diffraction to generate detailed data on the structure of nanomaterials has been advanced with the application of pair distribution function analyses, and the achievement of submillimeter Fresnel plates opens the door to studying nanomaterials using x-rays with spatial resolution rivaling that of scanning electron microscopy.

Acknowledgments

The author gratefully acknowledges helpful discussions with I. Barney, Wright State University, and M. DeGraef, Carnegie Mellon University, and thanks S. Mukhopadhyay, Editor, for patience during preparation of the manuscript.

REFERENCES

1. DeGraef, M., personal communication, Simulated copper TEM images, 2010.
2. Haider, M., et al., Electron microscopy image enhanced, *Nature*, 1998;392:768–769.
3. Kuwabara, M., La, W., and Spence, J. C. H., Reflection electron microscope imaging of an operating scanning tunneling microscope, *J. Vac. Sci. Technol. A*, 1989;7:2745–2751.
4. Batson, P. E., Dellby, N., and Krivanek, O. L., Sub-Angstrom resolution using aberration corrected electron optics, *Nature* 2002;418:617–620.
5. Krivanek, O. L., et al., Atom-by-atom structural and chemical analysis by annular dark-field electron microscopy, *Nature*, 2010;464:571–574.
6. Alem, N., et al., Atomically thin hexagonal boron nitride probed by ultrahigh-resolution transmission electron microscopy, *Phys. Rev. B*, 2009;80:155425-1-7.
7. Meyer, J. C., et al., Selective sputtering and atomic resolution imaging of atomically thin boron nitride membranes, *Nano Lett.*, 2009;9:2683–2689.
8. Suenaga, K., et al., Electron energy-loss single atom spectroscopy at 60 kV, *Nat. Chem.*, 2009;1:415.
9. Zuo, J. M., et al., Coherent nano-area electron diffraction, *Microsc. Res. Tech.*, Vol. 2004;64:347–355.
10. Park, H. S., et al., 4D ultrafast electron microscopy: imaging of atomic motions, acoustic resonances, and moire fringe dynamics, *Ultramicroscopy*, 2009; 7–19.
11. Park, H. S., et al., Direct observation of Martensitic phase-transformation dynamics in iron by 4D single-pulse electron microscopy, *Nano Lett.*, 2009;9(11): 3954–3962.
12. Flannigan, D. J., et al., Nanomechanical motions of cantilevers: direct imaging in real space and time with 4D electron microscopy, *Nano Lett.*, 2009;9:875–881.
13. Jarausch, K., et al., Four-dimensional STEM-EELS: enabling nano-scale chemical tomography, *Ultramicroscopy*, 2009;109:326–337.
14. Giannuzzi, L. A., and Stevens, F. A., *Introduction to Focused Ion Beams: Instrumentation, Theory, Techniques and Practice*, Springer-Verlag, New York, 2004.
15. Orloff, J., Utlaut, M. and Swanson, L., *High Resolution Focused Ion Beams: FIB and Its Applications*, Springer-Verlag, New York, 2004.
16. Orloff, J., Swanson, L. W., and Utlaut, M., Fundamental limits on imaging resolution in focused ion beam systems, *J. Vac. Sci. Technal. B*, 1996;14:3759.
17. Lee, S.-B., Rollett, A. D., and Rohrer, G. S., Three-dimensional microstructure reconstruction using FIB-OIM, *Materi. Sci. Forum*, 2007;558–559:915–920.
18. Buhr, E., et al., Characterization of nanoparticles by scanning electron microscopy in transmission mode, *Meas. Sci. Technol.*, 2009;20.
19. Pohl, D. W., Denk, W., and Lanz, M., Optical stethoscopy: image recording with resolution, *Appl. Phys. Lett.*, 1984;44:651–653.

20. Bachelot, R., et al., Probing photonic and optoelectronic structures by apertureless scanning near-field optical microscopy, *Microsc. Res. Tech.*, 2004;64:441–452.

21. Hsu, J. W. P., Near-field scanning optical microscopy of electronic and photonic materials and devices, *Mater. Sci. Eng.*, 2001;33:1–50.

22. Oshikane, Y., Kataoka, T. and Okuda, M., Observation of nanostructure by scanning near-field optical microscope with small sphere probe, *Sci. Technol. Adv. Mater.*, 2007;8:181–185.

23. Zayats, A., and Richards, D., (Eds.), *Nano-optics and Near-field Optical Microscopy*, Artech House, Norwood, MA, 2009.

24. Schneider, W.-D., Scanning tunneling microscopy and spectroscopy of nanostructures, *Surf. Sci.*, 2002;514:74–83.

25. Xu, H., et al., Nanoscale characterization by scanning tunneling microscopy, *Cosmos*, 2007;3:23–50.

26. Vitali, L., Lecture notes: rastertunnelmikroskopie (in English), Max-Planck-Institute for Solid State Research, Nanoscale Science Department, http://www.fkf.mpg.de/kern/index.html, 2007.

27. Greene, M. E., et al., Application of scanning probe microscopy to the characterization and fabrication of hybrid nanomaterials, *Microsc. Res. Tech.*, 2004;64: 415–434.

28. Mayer, E., Hug, H. J., and Bennewitz, R., *Scanning Probe Microscopy: the Lab on a Tip*, Springer-Verlag, New York, 2004.

29. Fujita, D., Onishi, K., and Niori, N., Light emission induced by tunneling electrons from surface nanostructures observed by novel conductive and transparent probes, *Microsc. Res. Tech.*, 2004;64:403–414.

30. Proffen, Th., et al., Structural analysis of complex materials using the atomic pair distribution function: a practical guide, *Z. Kristallogr.*, 2003;218:132–143.

31. Petkov, V., Structure of nanocrystalline materials by the atomic pair distribution function technique, *Adv. X-Ray Anal.*, 2003;46:31–36.

32. Nijenhuis, J. te, Gateshki, M., and Fransen, M. J., Possibilities and limitations of X-ray diffraction using high-energy X-rays on a laboratory system, *Z. Kristallogr* (Suppl. 30) (2009); 163–169.

33. Michel, F. M., et al., The structure of ferrihydrite: a nanocrystalline material, *Science*, 2007;316:1726–1729.

34. Michel, F. M., et al., Short to medium range atomic order and crystalline size of the initial FeS precipitate from PDF analysis, *Chem. Mater.*, 2005;17:6246.

35. Chupas, P. J., et al., Rapid-acquisition pair distribution function (RA-PDF) analysis, *J. Appl. Crystallogr.*, 2003;36:1342–1347.

36. Qiu, X., et al., Reciprocal-space instrumental effects on the real-space neutron atomic pair distribution function, *J. Appl. Crystallogr.*, 2004;37:110–116.

37. Kirz, J., and Jacobsen, C., The history and future of x-ray microscopy, in *9th International Conference on X-Ray Microscopy*, IOP Publishing, London, 2009; Vol. 186.

38. Hertz, H. M., et al. Laboratory x-ray micro imaging: sources, optics, systems and applications, *in 9th International Conference on X-Ray Microscopy*, IOP Publishing London, 2009,Vol. 186.

39. Chao, W., et al., Soft x-ray microscopy at a spatial resolution better than 15 nm, *Nature*, 2005;435:1211–1213.

40. Schroer, C., Focusing hard x rays to nanometer dimensions using Fresnel zone plates, *Phys. Rev. B*, 2006;74:033405-1 to 033405-4.

41. Kang, H. C., et al., Focusing of hard x-rays to 16 nanometers with a multilayer Laue lens, *Appl. Phys. Lett.*, 2008;92:22114-1 to 3.

42. Vila-Comamala, J., et al., *Advanced X-Ray Diffractive Optics*, IOP Publishing, London, 2009;186:012078.

43. Kim, K.-J., X-Ray data booklet, Section 2.1, Characteristics of synchrotron radiation, http://xdb.lbl.gov, 2009.

44. Schropp, A., and Schroer, C. G., Dose requirements for resolving a given feature in an object by coherent x-ray diffraction imaging, *New J. Phys.* 2010;12:035016.

45. Li, Wenjie, et al., Image enhancement of x-ray microscope using frequency spectrum analysis, in *9th International Conference on X-Ray Microscopy*, IOP Publishing, London, 2009, Vol. 186.

46. Wang, Z. L., *Characterization of Nanophase Materials*, Wiley-VCH, Hoboken, NJ, 2000.

7

MODELING AND SIMULATION OF NANOSCALE MATERIALS

SOUMYA S. PATNAIK

Propulsion Directorate, Air Force Research Laboratory, Wright-Patterson Air Force Base, Dayton, Ohio

MESFIN TSIGE

Department of Polymer Science, University of Akron, Akron, Ohio

Nanoscale Multifunctional Materials: Science and Applications, First Edition.
Edited by Sharmila M. Mukhopadhyay.
© 2012 John Wiley & Sons, Inc. Published 2012 by John Wiley & Sons, Inc.

1 INTRODUCTION

With the ever-increasing presence of nanoscale materials in current industrial products, there is a tremendous need for a greater understanding of the properties of matter
at the nanometer scale. It is well known that matter behaves in complex ways and
exhibits exotic properties at nanometer length scales. However, understanding the
behavior of matter at such length scales using experimental methods has in general
been very difficult. Computer simulations have proven very useful in predicting the
properties of novel materials yet to be synthesized, as well as predicting difficult-
to-measure or poorly understood properties of existing materials. Contrary to what
is seen in experiments, nanometer-sized systems are very convenient (even uniquely
suited) for computer simulations since the fundamental laws of quantum mechanics
can be applied directly. Modeling and simulation approaches for investigating molecular properties of nanoscale materials can be divided into two broad groups: electronic
methods (approaches dealing explicitly with electrons, such as density functional theory) and atomistic methods (e.g., molecular dynamics and Monte Carlo). The time
and length scales involved in these two categories of simulation methods are shown
in Figure 1. The main emphasis of this chapter is on introducing and explaining the
basic concepts required to understand these methods. Since modeling and simulation of nanomaterials is interdisciplinary in nature, the chapter focuses more on the
basic principles of each method than on the application to a particular discipline.
A few illustrative examples are also discussed for each computational method. We
start with a quantum description of materials, followed by an atomistic description.
A variety of first principle (or ab initio) models and theories that describe atomic
interactions in the materials (i.e., classical force fields) are introduced and discussed
in detail.

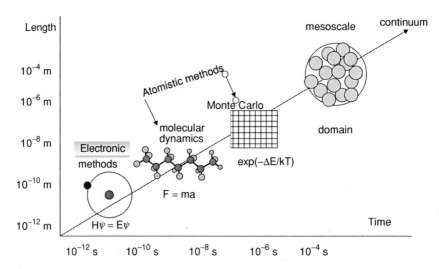

FIGURE 1 Time and length scales involved in electronic and atomistic simulations.

2   AB INITIO METHODS

In principle, quantum mechanics, which is to date the best mathematical description of the behavior of electrons, can exactly predict any property of an atom or a molecule. In practice, however, quantum mechanical equations cannot be solved exactly for any atom or molecule that has more than one electron (i.e., any atom more complicated than a hydrogen atom). Luckily, very powerful approximation methods, formulated in the twentieth century and discussed briefly below, can be used to solve quantum mechanical equations to almost perfect accuracy.

To understand the complexity of this problem, it is enough to consider the simplified, nonrelativistic quantum mechanical Hamiltonian of a system of N particles given by

$$\hat{H} = -\sum_{i=1}^{N} \frac{-\hbar^2}{2m_i} \nabla_i^2 + \sum_{\substack{i,j=1 \\ i<j}}^{N} \frac{q_i q_j}{r_{ij}}$$

The first term represents the kinetic energy of the particles, and the second term represents the potential energy due to coulombic interactions between particles. External potentials are not included in the Hamiltonian, which is often the case for nanosystems. To describe the characteristics of the system, the corresponding many-particle Schrödinger equation,

$$\hat{H}\psi = E\psi$$

must be solved, where ψ is the many-particle wavefunction and E is the total energy of the system. The wavefunction depends on the $3N$ coordinates of the electrons and the positions of the nuclei.

Unfortunately, the N-body Schrödinger equation given above is a complex partial differential equation in $3N$ dimensions. The computational complexity of this equation on *classical computers* grows exponentially with system size (N). This clearly indicates that an exact solution to the many-body Schrödinger equation may probably never be achieved, even using today's petaflop supercomputers. Thus, approximations must be introduced to reduce the dimensionality of the configuration space.

One of the most commonly used approximations is the Born–Oppenheimer [1] or adiabatic approximation, which takes advantage of the fact that electrons are much lighter than nuclei, and thus there will be a strong separation in the time scale between the motion of electrons and nuclei. This allows the calculation of the electron wavefunction and energy in a potential field of initially fixed nuclei. As the nuclei are moved to new coordinates, the electrons' wavefunctions and energies are recalculated. For many systems, the Born–Oppenheimer approximation is found to be an excellent approximation.

Another common approximation method used in simplifying the Schrödinger equation is the central field approximation [2–4]. In this approximation, each electron of an atom is regarded as being in motion in an effective potential that combines the attractive interaction between the electron and the nucleus and the repulsive interaction between all other electrons. This approximation represents the effective potential energy of an electron by a spherically symmetric potential.

Quantum mechanical simulation methods that predict molecular properties based on electronic structure, with no assumptions made in solving the Schrödinger equation (i.e., with no empirical data used other than applying the fewest approximations described above) are referred to as ab initio or *first-principle methods*. Below, we describe briefly the most commonly used ab initio methods, in the context of ab initio molecular dynamics simulations, for obtaining very accurate interatomic forces and energies. The advantages and limitations of ab initio methods for predicting the physical properties of systems are also presented. The power of these methods is highlighted through the use of practical examples extracted from the literature.

2.1 Ab initio Molecular Dynamics

Ab initio molecular dynamics (AIMD) methods are simulation methods that, in principle, can provide exact properties of many-body quantum systems by combining molecular dynamics simulations with high-level ab initio calculations [5,6]. As discussed in the next section, the major problems in classical molecular dynamics simulations are the inability of existing force fields to describe chemical events involving bond formation and breaking, as well as including polarization effects. AIMD overcomes these problems by generating finite-temperature dynamics using forces obtained directly from "on-the-fly" electronic structure calculations. Thus, the heart of an AIMD calculation is the correct representation of the electronic structure of the system.

In AIMD calculations, the following three basic assumptions are usually taken into account [5]:

1. The system under consideration is composed of N nuclei and N_e electrons.
2. The Born–Oppenheimer approximation is valid.
3. The dynamics of the nuclei can be treated classically on the ground-state electronic surface.

The classical dynamics of the nuclei can then be described by the following Newton's equations of motion:

$$M_I \ddot{\mathbf{R}}_I = -\frac{\partial E(\mathbf{R})}{\partial \mathbf{R}_I}$$

Here $\mathbf{R}_I$ denotes the position of the Ith nucleus, dots indicate time derivatives, and $E(\mathbf{R})$ is the position-dependent nuclear interaction energy.

The major difference between classical MD and ab initio MD is in the approach used in calculating the interaction energy $E(\mathbf{R})$ when solving the equation above.

The basic idea behind AIMD simulations is to perform, at each time step, an MD-independent ab initio calculation to determine the interaction energy $E(\mathbf{R})$ of the system at the present nuclear configuration. This allows the execution of MD simulations with forces derived from ab initio methods and has proven to be extremely helpful for a variety of problems in chemistry, physics, and now in biology.

There are many ab initio methods for calculating the required interaction energy of a many-body system in an AIMD simulation. The most commonly used ab initio methods are generally divided into two main groups: Hartree–Fock-based and density functional theory–based methods. The Hartree–Fock (HF) [7,8] scheme is the simplest type of ab initio electronic structure calculation, providing a rigorous one-electron (mean-field) approximation in which the instantaneous coulombic electron–electron repulsion is not taken into account specifically, only its average effect. Unlike in the Hartree–Fock method, in density functional theory (DFT) [9] an interacting system of fermions is described by its density rather than by its many-body wavefunction. This means that for a system containing N electrons, the basic variable of the system depends only on the three spatial coordinates rather than the $3N$ degrees of freedom. An overview of HF and DFT methods is provided below.

2.2 Hartree–Fock Method

The Hartree–Fock method, which early on proved to be fundamental to much of electronic structure theory, is based on the pioneering works of Hartree [7] and Fock [8]. The method is used as an approximate method for the determination of the ground-state energy with the corresponding ground-state wavefunction of a quantum many-body system. In HF calculations, the primary approximation is the central-field approximation, where the explicit coulombic electron–electron repulsion term in the Hamiltonian of a system is replaced by an average effect of the repulsion. Because of this approximation, HF calculations always give energies that are greater than the exact energy [10].

In Hartree–Fock calculations, the exact N-body wavefunction of the system is approximated by a product of one-electron wavefunctions called *orbitals*, which should obey the fermionic nature of electrons: that is, they must be antisymmetric. Each of the orbitals describes the behavior of an electron in the net field of all the other electrons. In modern HF calculations, in the interest of saving computation time, the orbitals are usually expressed as a linear combination of Gaussian-type orbitals [$\exp(-\alpha r^2)$] [10].

Hartree–Fock calculations start by guessing a set of approximate one-electron wavefunctions—orbitals. For each electron, the effect of all other electrons is expressed in a single effective potential, which is used to solve the corresponding Schrödinger equation, resulting in a new set of orbitals. The entire process is then repeated until the change in the energies and corresponding orbitals from one iteration to the next is sufficiently small. Note that there is no absolute guarantee that the calculation will always converge. However, there are various ways of combating this problem that are implemented in most HF calculation programs, but we do not discuss them in this chapter.

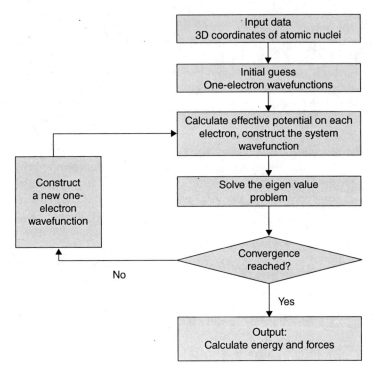

FIGURE 2   Simplified schematic representation of the self-consistent loop for Hartree–Fock calculations.

Hartree–Fock calculations correctly predict bond lengths (within a small percentage) and energy differences between different conformations. However, they are less accurate in predicting cohesive energies and are also not good for systems with high electronic density and delocalized metallic states. This may have to do with the major weakness of the HF method—the effect of electron correlation is neglected. Although electron correlation is neglected, it is important to note that the effect of electron exchange is taken fully into account in HF calculations. For some systems, however, neglecting electron correlation may lead to large differences between HF results and experimental results. A number of new approaches have been devised to include electron correlations in HF calculations and are collectively called *post-Hartree–Fock methods* [11,12]. Post–Hartree–Fock methods usually give more accurate results than HF calculations, but with the price of added computational cost. The flowchart in Figure 2 shows a simplified picture of an HF calculation.

Although HF has generally been the central starting point for most ab initio quantum chemistry methods for molecules, it has a number of difficulties in its application that can make it less attractive [13–17]. For example, an entire set of orbitals is needed to calculate the single-electron Schrödinger equation. Post-Hartree–Fock methods follow a complicated procedure to include the correlation corrections that are crucial in some systems. In addition, for large molecules with complex many-electron systems, HF calculations become too complicated. On the bright side, the density functional theory discussed below has been shown to overcome some of these major difficulties.

2.3   Density Functional Theory

The density functional theory formulates the problem of dealing with a system of N interacting electrons in an external potential $v(r)$ in terms of the electron density $n(r)$ rather than the many-body wavefunction $\psi(r_1, r_2, \ldots, r_N)$. The DFT is based solely on two fundamental theorems formulated by Hohenberg and Kohn in the 1960s [18,19]. The first theorem states that for any system of interacting particles in an external potential $v(r)$, there exists a one-to-one correspondence between $v(r)$ and the ground-state particle density, $n_0(r)$. More explicitly, the theorem states that there is a unique ground-state wavefunction $\psi_0(\{r\})$ for an external potential $v(r)$, resulting in a unique ground-state density $n_0(r)$ as represented by the following map:

$$v(\mathbf{r}) \rightarrow \psi_0(\{r\}) \rightarrow n_0(r)$$

This means that if $n_0(r)$ is known, $v(r)$ and $\psi_0(\{r\})$ are known; in principle, all properties of the system can be determined completely.

The second theorem defines an energy functional, $E[n]$, for the system in terms of the density, $n(r)$. This functional is minimized by the density corresponding to the ground state, and thus the energy functional coincides with the ground-state energy of the system. The proofs for both theorems are simple and may be found in Refs. 5 and 9. Here, we give a brief review of the theorems.

According to Hohenberg and Kohn, the total energy of the system is a functional $E[n]$ of the ground-state density and is written as [18–21]

$$E[n] = \int v(r)\, n(r)\, dr + F_{\text{HK}}[n]$$

where the Hohenberg–Kohn functional $F_{\text{HK}}[n]$ is defined such that $E[n]$ has its unique minimum for the correct ground-state density, $n_0(r)$. The physical meaning of $F_{\text{HK}}[n]$ is

$$F_{\text{HK}}[n] = \langle \psi[n] | (\hat{T} + \hat{V}_{ee}) | \psi[n] = T[n] + V_{ee}[n]$$

where $\psi[n]$ is the ground-state wavefunction associated with the ground-state density $n_0(r)$, and $T[n]$ and $V_{ee}[n]$ are the kinetic energy and Coulomb energy terms, respectively. This means that $F_{\text{HK}}[n]$ contains the many-body kinetic energy of the electrons and Coulomb energy as well as electron exchange and electron correlation effects. The second theorem states that if $\psi[n]$ is not the ground-state wavefunction of the external potential $v(r)$ whereas $\psi[n_0]$ is, then [22]

$$\langle \psi[n] | (\hat{T} + \hat{V}_{ee} + \hat{V}) | \psi[n] \rangle > \langle \psi[n_0] | (\hat{T} + \hat{V}_{ee} + \hat{V}) | \psi[n_0] \rangle = E[n_0]$$

The advantage of the Hohenberg–Kohn formulation is that the multivariable Schrödinger equation is transformed to a variational principle problem involving the density, which is a single variable. Although the Hohenberg–Kohn theorem is exact, the energy functional is unknown, and in order to be used in practical calculations, an explicit expression for it is required. The main obstacle in constructing the energy functional is found to be the kinetic energy term. Thus, to proceed forward, approximations are needed.

In 1965, Kohn and Sham made a significant step forward that helped DFT to become the most widely used electronic structure calculation method today for many practical problems [23]. The key point in the Kohn–Sham method is the introduction of *N fictitious noninteracting electrons* for estimating the kinetic energy of the real interacting many-body system. This method assumes that the density of the fictitious system is equal to the ground-state density of the original interacting system. The density of the auxiliary system can be expressed as [5,20,21,23]

$$n(r) = \sum_{i=1}^{N} |\phi_i(r)|^2$$

where the $\phi_i(r)$ are orthonormal single-particle wavefunctions. In the Kohn–Sham formulation, the energy functional to the full interacting many-body problem is written in terms of $\phi_i(r)$ and is given by

$$E[n] = \int v(r)n(r)\,dr + \frac{1}{2} \int \frac{n(r)n(r')}{|r-r'|}\,dr\,dr' - \frac{1}{2} \sum_{i=1}^{N} \langle \phi_i | \nabla^2 | \phi_i \rangle + E_{XC}[n]$$

In this formulation, the second term is the Hartree term and the third term is the kinetic energy term for the noninteracting electron system. In the Kohn–Sham formulation above, the many-body effects associated with exchange and correlation of electrons are represented by the last term, known as the exchange-correlation functional energy, $E_{XC}[n]$.

If the exchange-correlation functional is known, the Kohn–Sham equations for the noninteracting electron system can be solved and the exact ground-state energy and density of the many-body system can be determined. Unfortunately, the exchange-correlation functional is unknown and must be approximated. The approximation of the exchange-correlation functional is very crucial to any application of density functional theory [24,25].

Luckily, approximating $E_{XC}[n]$ proves to be much easier than approximating the Hohenberg–Kohn functional, $F_{HK}[n]$. For certain classes of systems, even a crude approximation in which $E_{XC}[n]$ is approximated by taking the exchange and correlation energies of a homogeneous electron gas of the same local density $n(r)$ is found to be a sufficient approximation [24]. This approximation, which was initially formulated by Kohn and Sham [23] and is usually considered as the first approximation to $E_{XC}[n]$, is known as the *local density approximation* (LDA) and is given by

$$E_{XC}^{LDA}[n] = \int n(r)\varepsilon_{XC}\,(n(r))\,dr$$

where $\varepsilon_{XC}(n(r))$ is the exchange and correlation energy per particle of a homogeneous electron gas of density $n(r)$. In reality, this is true if the inhomogeneity of $n(r)$ is small, but in practice, LDA is used even if the inhomogeneity is large. This is because of the main approximation of LDA: that any region of space can be considered locally to be a homogeneous electron gas of density $n(r)$. The total exchange-correlation energy is then the sum of the local exchange-correlation energies from every region of space.

LDA, in general, gives good results (often with surprising accuracy) for systems with slowly varying charge densities. It has been found to be very useful and reasonably accurate for metallic and semiconductor solids [26–29]. However, it performs very poorly for strongly correlated systems, where an independent-particle picture breaks down and spatial variations in the electron density vary too rapidly. For example, LDA predicts an incorrect ground state for the titanium atom, gives very poor descriptions of hydrogen bonding, and predicts several high-temperature superconductors to be metallic, among other faults [24,30]. Therefore, great care must be taken when using LDA. LDA has been quite popular in the physics community, but due primarily to its poor description of bonding, it is less popular and has not been fully embraced by the chemistry community.

A significant improvement in the accuracy of the approximation of the exchange-correlation functional has been made by introducing the gradient of the electron density, called the *generalized gradient approximation* (GGA) [31–33]. In this approximation, E_{XC} is a functional not only of the local density as in LDA, but also of the local density gradient, and is expressed as

$$E_{XC}^{GGA}[n(r)] = \int n(r)\varepsilon_{XC}(n(r),\,\nabla n(r))\,dr$$

GGA has extended the applicability of DFT to hydrogen-bonded complexes, surfaces, and solids, to mention a few, and is thus well recognized and accepted by the chemistry community. For van der Waals systems and thermochemistry of molecules and solids, GGA performs better than LDA, but the results depend strongly on the GGA functional chosen. It is important to note that the gradient expansion works only for small gradients in the density. If large density gradients exist, GGA performs very poorly and gives results that are worse than LDA [34]. Therefore, to reduce or eliminate the effect of large gradients in GGA, a cutoff procedure is generally used.

In the last three decades, several GGA, hybrid-GGA, and hybrid meta-GGA functionals have been proposed and tested. The first GGA functionals that have been used extensively for practical calculations are PW86 [35,36], BLYP [37,38], and PW91 [33]. These have largely been replaced by Perdew–Burke–Ernzerh of (PBE) [39] formalism, which until recently has been the most widely used GGA. The advantage of PBE is that it does not contain any parameters that should be adjusted in order to reproduce experimental data or accurate ab initio data, and it has a simpler analytical form. However, there are new GGA functionals (AM05 [40,41], WC [42], PBEsol [43], SOGGA [44]) that, in many cases, perform better than PBE, but there are still classes of solids for which PBE is the best functional choice. For the lattice constant of solids, these new GGA functionals on average give more accurate results than PBE, but most of them are not accurate for the cohesive energy of solids. One has to realize that it is very difficult, if not impossible, to construct a GGA functional that can accurately predict all the properties of a system.

Before closing the discussion on Kohn–Sham DFT, it should be stressed that elucidation of methods to better approximate the exchange and correlation functional is still an area of intense research. Current trends on this issue indicate that several functionals are proposed in a period of one year.

In summary, ab initio simulations have been applied from materials science to biochemistry. Recently, they have also been applied to some geophysics problems [45]. The significant increase in computing power in the past two decades has helped ab initio simulations to become an important tool to complement experimental investigations. These days, thousands of papers dealing with ab initio simulations are being published each year alone, which makes it difficult to provide a comprehensive survey of their applications. Most of the applications may be found in recent reviews [46–48], and it is recommended that interested readers consult these reviews. Below, we present a typical application of ab initio simulations to surface science.

2.4   Example: Structure of Poly(dimethylsiloxane) Bound to a Hydroxylated Silica Surface

Among the most commonly used applications of density functional theory are the determination of binding energy and structure of an adsorbate on a surface. In a recent study, Tsige et al. [49] used quantum and classical approaches to investigate the structure of poly(dimethylsiloxane) (PDMS) oligomers adsorbed on the surface of a thin slab of hydroxylated {100} α-quartz. In the quantum approach, the interaction of monomer and dimer PDMS molecules with a hydroxylated silica surface was investigated using two complementary electronic structure methods, where the silica surface was modeled as either a thin slab of hydroxylated silica oxide or as a molecular cluster (both constructed from the α-quartz crystal surface).

The quantum simulations for the periodic slab approach were conducted using the VASP plane-wave pseudopotential code developed at the Technical University of Vienna [50–52]. VASP treats the valence electrons in the system explicitly by using the generalized gradient approximation (GGA) of Perdew and Wang [35] and represents the inner-core electrons by the Vanderbilt ultrasoft pseudopotential scheme [53] as implemented by Kresse and Hafner [54]. For the molecular cluster approach, the GAUSSIAN 98 suite of programs [55] was implemented. In this case, all electrons

(a)   (b)

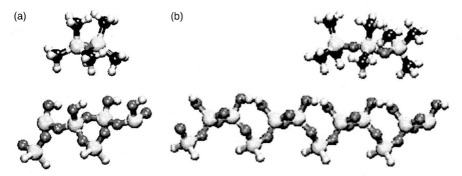

FIGURE 3 Minimum energy structures for (a) monomer and (b) dimer PDMS oligomers adsorbed on top of a thin slab model of hydroxylated {100} α-quartz (two-dimensional periodic boundary conditions are used).

are treated explicitly using the hybrid density functional B3LYP [56] in a localized Gaussian basis set. Here, only results from the periodic slab approach are presented, and interested readers may find the results from both approaches in Ref. 49.

The minimum energy structures for the monomer and dimer oligomers bound to the periodic hydroxylated α-quartz surface using the periodic slab approach are shown in Figure 3. Note that the substrate is doubled in size for the dimer case to avoid interaction of the oligomer with its periodic image. To determine binding energies, minimum energy structures were found for isolated silica and the oligomers as well as for each silica–oligomer complex. The difference between the energy of the complex and the sum of the energies of the isolated components gives the binding energy between the oligomer and the silica surface at its optimum configuration.

The binding energies determined using the periodic slab approach are −1 and −2 kcal/mol for the monomer and dimer, respectively. These values reflect weak binding of the PDMS oligomers on the hydroxylated silica surface. Close examination of the optimum PDMS/silica complex structures shows that instead of forming hydrogen bonds between the oxygen atoms on the oligomer backbone and –OH groups on the surface, the dominant interaction, due to steric hindrance of the methyl groups, is between a hydrogen atom on a single methyl group and an oxygen atom on a surface –OH group. This result agrees very well with other studies on the same type of system. A quantum mechanical study using semiempirical models reported that no hydrogen bonding was observed at the interface [57]. In an experimental adsorption study, low energies of surface interactions were reported and attributed to shielding by methyl groups of the oxygen atoms in the oligomer backbone from the surface hydroxyl groups [58]. In another experimental study [59], a binding energy per monomer of −1 kcal/mol was determined, which agrees well with the results calculated for the monomer and dimer cases. The assumption is that each additional monomer in the PDMS oligomer would contribute an additional weak interaction with the surface with binding energy of −1 kcal/mol per PDMS monomer. To verify this properly, it will be crucial to do electronic structure calculations with larger oligomers. Unfortunately, electronic structure calculations for bigger oligomers are computationally very expensive and go beyond the capability of current DFT calculation methods.

It is clear from the simple example above and also from what we know so far that although ab initio calculations are very effective in predicting physical and chemical properties of a system, they are computationally very expensive and are limited to very few atoms. For a system containing a large numbers of atoms, such as a system containing several long-chain polymers, ab initio calculations are not practical; instead, classical methods, such as molecular dynamics, discussed below, could be appropriate in extracting some of the physical (and *perhaps* chemical) properties.

3   CLASSICAL METHODS

Classical molecular modeling methods represent computational techniques that use Newtonian mechanics to describe the behavior of the molecules. The common feature of all these methods is the atomistic-level description of the molecular systems (i.e., the lowest units are individual atoms or a small group of atoms). This is in contrast to the ab initio methods described earlier, where electrons are considered explicitly. In classical methods, the Born–Oppenhaimer approximation is considered to operate, allowing for the energy of a molecule in the ground state to be a function of the nuclear coordinates only. By reducing the complexity of the molecular system, classical molecular modeling methods allow for a large number of atoms to be considered during simulations, and currently these methods are used widely to study nanoscale structure–property relationships of a variety of materials [60–64]. However, since electronic motion is neglected and the energy is considered to be a function of nuclear positions only, these methods cannot be used to study properties that depend on the electronic distribution of the atoms. Typically, atoms are described as particles with associated mass and charge, and the interaction between atoms is described in terms of bonded and nonbonded interactions. A mathematical description of these molecular interactions is known as the *potential function*, which along with a suitable set of characteristic parameters is defined to be the *force field*. Two of the most commonly used classical molecular modeling techniques are *molecular dynamics* (MD), which models the behavior of the system with the propagation of time, and *Monte Carlo* (MC), which is based on stochastic methods. Although molecular dynamics was described in Section 2.1 in the context of ab initio MD, which combines MD simulations with high-level ab initio calculations, allowing for electronic structure calculations, in this section more details are provided, concentrating primarily on classical MD, which relies on an empirical force field and is used more widely in studying the dynamic properties of materials, including transport coefficients, rheological properties, and time-dependent responses to perturbation. First we describe the force-field and molecular dynamics method, followed by examples of the use of molecular dynamics to study nanoscale properties. This material is introductory in nature and provides an overview of modeling and simulation techniques for analyzing the nanoscale behavior of materials. For a more thorough understanding of molecular modeling techniques, readers are referred to other detailed presentations [65–69].

3.1 Force Field

Most of the widely used force fields rely on a simple model to describe the interactions within a molecular system. Chemical bonds are represented by springlike interactions, such as stretching of bonds, opening and closing of angles, and rotations about bonds. Interactions between nonbonded atoms (defined as atoms in separate molecules or atoms in the same molecule but separated by at least three bonds) are represented using the Coulomb potential for electrostatics and the Lennard-Jones potential for van der Waals interactions. A typical functional form of the force-field-based potential energy of a molecular system is

$$E_{\text{pot}} = \sum_{ij \text{ bonded}} K_{r,ij}(r_{ij} - r_{0,ij})^2 + \sum_{ijk \text{ bonded}} K_{\theta,ijk}(\theta_{ijk} - \theta_{0,ijk})^2$$

$$+ \sum_{ijkl \text{ bonded}} V_{\phi,ijkl}[1 + \cos(n\phi_{ijkl} - \phi_{0,ijkl})] + \frac{1}{4\pi\varepsilon_0\varepsilon_r} \sum_{\substack{ij \text{ nonbonded} \\ (1,2 \text{ and } 1,3 \text{ excl.})}} \frac{q_i q_j}{r_{ij}}$$

$$+ \sum_{\substack{ij \text{ nonbonded} \\ (1,2 \text{ and } 1,3 \text{ excl.})}} \varepsilon_{ij} \left(\left(\frac{r_{0,ij}}{r_{ij}} \right)^{12} - 2 \left(\frac{r_{0,ij}}{r_{ij}} \right)^6 \right) \tag{1}$$

E_{pot} denotes the total potential energy of the molecular system, which is a function of the positions of the total number of atoms and is defined as the difference in energy between a real molecule and an ideal molecule (with bond lengths and angles at reference values). The association of the force-field parameters with specific internal coordinates such as bond lengths, bond angles, and rotation around bonds makes parameterization easier, and energetic penalties are associated with deviation of the bonds and angles from their reference values. In equation (1) the first term represents interactions between all bonded atoms and is modeled by a harmonic potential that increases as the bond length r_{ij} deviates from the reference length $r_{0,ij}$. Similarly, the second term represents the energy contribution due to the valence angles deviating from reference values, and the third term represents the energy changes as the bonds rotate. Although not included in the potential function described in equation (1), additional terms such as improper torsional terms for maintaining stereochemistry and cross-terms representing coupling between internal coordinates are also found in many advanced force fields [70–72]. The electrostatic property of the molecule is commonly modeled by assigning partial charges to the atoms and using Coulomb's law to calculate the electrostatic energy (fourth term). Many force fields also include schemes for including polarization effects. The fifth term represents van der Waals interactions using the *Lennard-Jones 12-6* function, which has been found to be one of the best known potentials for modeling the dispersive and exchange-repulsive forces.

To define a force field, the values of the force constants and the reference bond and angle values need to be defined along with the functional form. These are collectively referred to as the *parameters* of the force field and are specific to *atom-type*. Apart from information on the atomic number, the atom type includes additional information, such as the hybridization state and local environment of the atom. By describing an atom based on atom type, force fields can be more sensitive to the local environment of the specific atom. For example, in most force fields, the force constant and reference angle of sp^3-hybridized carbon atoms is different than those of sp^2-hybridized carbon atoms. The force between bonded atoms is very strong and is reflected in the high magnitude of the force constants for bond stretching (K_r). Since less energy is needed to bend an angle from its equilibrium value, the force constant for angle bending (K_θ) is relatively smaller and torsional force constants (V_ϕ) are even smaller. Due to the stiffness in the force constants for bond stretching and bending, significant energy is needed to cause a deviation from the equilibrium values of bond lengths and angles. A major contribution to any variation in structure and relative energies of molecules is therefore from a torsional energy contribution, along with those from nonbonded interactions.

Transferability of force fields is key to the use of molecular modeling as a predictive tool; therefore, great effort is directed toward parameterizing force fields based on a wide range of molecules so that the same set of parameters can be used for predicting properties of related molecules. Most force fields rely on experimentally derived parameters, equilibrium bond lengths, bond angles, and vibration constants, which can be derived from x-ray, nuclear magnetic resonnce, infrared, microwave, Raman spectroscopy, and ab initio calculations on a given class of small molecules. Partial atomic charge is not an experimentally observable quantity and is generally determined by assigning charges based on electronegativity and then further refining by comparing with quantum mechanics or calculated thermodynamic properties with experiments. The Lennard-Jones 12-6 function contains two adjustable parameters: the van der Waals equilibrium distances $r_{0,ij}$, which can be assigned on the basis of experimentally known crystallographic data, and the well depth, ε_{ij}. Parameterization of well depths is more demanding than parameterization of equilibrium distances, since there is less experimental justification for well depths. Although accurate well-depth values are known for rare gases, additional adjustments are needed to parameterize them for other atoms by fitting to experimental surface tension, solvation, and vaporization energies. For polyatomic systems, mixing rules are generally used to determine the van der Waals parameter between dissimilar atoms.

The empirical nature of force fields dictates that the accuracy with which they can represent molecular properties of interest depends primarily on the level of physical accuracy used in parameterizing them. To address specific properties, apart from the general functional form represented in equation (1), many specialized force fields have been developed. Some of the widely used ones are the Amber [73] and CHARMM [74] force fields for biomolecular simulations and the OPLS [75] force field for organic liquids. To overcome another significant shortcoming of traditional force fields which due to their inherent fixed bonding character are unable to model bond formation/breaking or chemical reactions, several force fields based on bond-order formalism have also been developed. Two such potentials are the Brenner potential [76] which has been used successfully for nanotube simulations, and the ReaxFF potential [77], which was developed primarily to model chemical reactions in hydrocarbons. The force fields described so far are based on a localized bond description which works very well for organic compounds but is unable to model accurately ionic, metallic, and semiconductor materials with highly delocalized bonds. Changing the pairwise potential model to many-body potentials has been found to be useful in cases such as the embedded-atom method [78] for metals and Tersoff potentials for semiconductors [79]; however, development of an universal force field is still a challenge, and no clear solution that can optimize both efficiency and accuracy is yet available.

3.2 Energy Minimization

For most polyatomic molecules the potential energy surface is a multidimensional complicated function of the atomic coordinates, and energy minimization techniques are very useful in finding minimum points on this energy surface. Atoms are assigned coordinates in Cartesian space, and energy minimization attempts to reproduce equilibrium molecular geometries by adjusting bond lengths, bond angles, and torsion angles to equilibrium values. Most commonly used minimization algorithms use the derivative of the energy with respect to the coordinates. *Steepest descent* and *conjugate gradient* are two such first-order derivative methods, and the *Newton–Raphson* method is a widely used second-order derivative method. Depending on the molecular size and the nature of its atomic interactions, a molecule can have a large number of minima on its energy surface. The highest point on the pathway between two energy minima is called the *saddle point*, whereas the minimum with the lowest energy is called the *global energy minimum*. Lower-energy states are more stable and commonly investigated because of their role in chemical and biological processes. Energy minimization generates a static picture of the energy configuration of the system and can be sufficient to determine the thermodynamic properties for specific cases, such as for very small molecules or for gas phases where all energy minimum configurations can be identified. However, for many molecular systems of interest, a full quantification of all the energy minimum configurations is not possible and molecular simulation techniques such as molecular dynamics and Monte Carlo methods are used to generate representative configurations. One of the primary applications of energy minimization in molecular modeling is also to prepare the initial configuration for subsequent MD or MC simulations by relaxing unfavorable interactions.

3.3 Molecular Dynamics

Molecular dynamics is a deterministic simulation method that calculates the real dynamics of a system and is used to determine time-averaged properties. The simulation is performed by numerically integrating Newton's equations of motion over small time steps (usually 1 fs). For any molecular system, the force acting on each atom due to its interaction with neighboring atoms can be determined by differentiating the potential energy. From this force, the acceleration of the atoms can be determined using Newton's second law ($F = ma$). The position, velocity, and acceleration of the atoms with time provide the trajectory of atomic motion, which can subsequently be used to determine the time-averaged values of the properties of the system.

Since the motion of an atom in a molecular system is coupled with that of all the other atoms with which it interacts, this represent a system of coupled differential equations and an analytical solution is difficult. Methods based on finite difference techniques are often used to do step-by-step numerical integration and generate molecular dynamics trajectories. In MD simulations, after initializing by assigning force vector for each atom in the molecule, acceleration of each atom is calculated using the equation $a = F/m$, where m is the mass of the atom and F is the total force on each atom. This force F is taken to be the vector sum of all the forces, due to its interaction with neighboring atoms. Once the position, velocity, and acceleration are determined for a time t, finite difference–based methods that use Taylor series expansions are used to calculate the position and dynamic properties at time $t + \delta t$. The *Verlet algorithm* [80] is one of the most widely used finite difference methods in molecular dynamic calculations. It uses the position and acceleration at time t and position at time $t - \delta t$, to calculate the position at time $t + \delta t$. The choice of time step is important, as too low a value can give rise to very long simulation times, whereas too large a value can lead to instabilities in the simulation. Energy is conserved during molecular dynamics, so an appropriate time step is often chosen to be one that generates an acceptable value in the root-mean-square (rms) variation in the total energy.

These simulated atomic trajectories can be used further to determine the time-averaged values of the properties of the system. Using the *ergodic hypothesis*, which equates time average with ensemble average, the ensemble average of the properties of the system can also be determined. To be computationally manageable, the simulations are carried out on smaller representative models, and *periodic boundary conditions* are used widely to mimic the macroscopic system. *Minimum image convention* is adopted so that each atom interacts with the nearest atom or its image in the periodic array. Although cubic periodic cells in which the simulation box is surrounded by its periodic image cells on all six sides are very commonly used, other cells, such as hexagonal prism, truncated octahedron, and rhombic dodecahedron, are also used to match the inherent geometry of the system. Depending on the molecular system to be modeled, in some cases periodic boundary conditions are used only in selected directions. However, in all cases, it is important to evaluate the effect of the sample size and the specific boundary conditions on the calculated properties of interest. As the number of atoms N in a simulation model increases, unlike the number of bonded energy terms, which increase proportional to N, the number of nonbonded energy terms increase as N^2. Since the Lennard-Jones potential falls rapidly with distance, significant computational effort can thus be saved by limiting the calculation of the nonbonded energies for atoms within a cutoff distance. Molecular systems with strong electrostatic interactions do not, however, adapt very well to low cutoff distances and methods such as Ewald summation [81], and cell multipole methods [82] are used to handle long-range nonbonded forces.

A typical MD simulation starts with an initial configuration of the system, which is usually based on an energy-minimized structure. Initial velocities are assigned to the atoms based on a Maxwell–Boltzmann distribution for the temperature desired. The first stage is the *equilibration phase*, where the system is brought to an equilibration and the thermodynamic properties reach a stable value. In a microcanonical ensemble, the kinetic and potential energies are expected to fluctuate, but the total energy remains a constant. For simulations at constant temperature, velocity scaling is carried out to adjust the temperature. Once equilibration is achieved, the final *simulation phase* starts and the system is allowed to evolve without adjustments. No further velocity scaling is performed, so the temperature is a calculated property of the system. Depending on whether the primary goal is to calculate a single physical, or thermodynamic property, to study a physical process or to analyze the conformational properties, simulation times can range from picoseconds to hundreds of nanoseconds, during which data are collected for further analysis. Whenever practical, the standard practice of varying the simulation time and size of the system to alleviate some of the systematic errors in the simulations is adopted. Although traditional MD simulations are performed in the microcanonical or constant NVE (constant number of particles, volume, and energy) and correspond to an adiabatic process, they can be modified to sample other ensembles. Currently, two other commonly used ensembles are canonical or constant NVT (constant number of particles, volume, and temperature) and isothermal-isobaric or constant NPT. Thermostat methods such as Nose–Hoover [83,84] or Berendsen [85] are used to control the temperature in NVT simulations, while NPT simulations need both thermostats and barostat [85] to control the temperature and pressure.

Apart from classical MD, there has also been development in hybrid methods that incorporate the strengths of other methods within a molecular dynamics framework. One such method is the hybrid MD/MC [86]. Unlike molecular dynamics, where successive configurations are connected in time, in Monte Carlo simulations [87] each configuration depends only on its predecessor and not on previous configurations. Configurations are generated randomly and accepted or not based on a specific set of criteria. For each configuration accepted, the desired properties are calculated and the average value is obtained. Therefore, unlike MD simulations, which can be used to predict the time dependence of molecular properties, MC methods are more appropriate for cases where a rapid exploration of phase space is needed. In Monte Carlo simulations, the total energy has no kinetic energy contribution. Traditionally, Monte Carlo samples the canonical ensemble (constant NVT). The complementary nature of the two methods has led to the development of hybrid MD/MC methods [86] where better sampling is achieved by alternating the simulation between MD and MC.

3.4 Applications

As mentioned earlier, MD simulations can be used for a wide range of properties. The intensive thermodynamic properties such as internal energy (U), heat capacity (C_v), temperature (T), and pressure (P) are time-averaged properties and can be determined from the ensemble average from MD simulations. The relationship between these quantities based on standard statistical mechanics can be given as follows:

$$U = \langle E = E_{\text{pot}} + E_{\text{kinetic}} \rangle = \frac{\sum_i^M E_i}{M} \tag{2}$$

$$C_v = \frac{\langle (E - \langle E \rangle)^2 \rangle}{k_B T^2} \tag{3}$$

$$P = \frac{1}{V} \left(N k_B T - \frac{1}{3} \sum_{i=1}^{i=N} \sum_{j=i+1}^{i=N} r_{ij} f_{ij} \right) \tag{4}$$

$$T = \frac{K_B T}{2} (3N - C) \tag{5}$$

where E is the total energy, M the number of configurations, N the number of particles, V the volume, r_{ij} and f_{ij} the distance and force between atoms i and j, and C the number of constraints in the system. The internal energy U can be obtained from the ensemble average of all the energies [equation (2)]; the heat capacity C_v, which is defined as the partial derivative of the internal energy with respect to temperature, can be expressed by equation (3); pressure P can be calculated from equation (4); and temperature T can be calculated from the kinetic energy of the system [equation (5)]. MD simulations are also used to study the elastic properties of molecules, and constant-stress MD simulations can be carried out to measure the stress–strain behavior of a material subjected to an applied load [88,89]. The simulations consist of increasing the magnitude of an applied tensile stress at a constant rate and monitoring the resulting longitudinal and lateral strains. A quicker method based on the static method is also used to derive the elastic constants for glassy polymers [90,91]. Below the glass transition temperature, the internal energy contribution to the elastic response to deformation is much more significant than the entropic contributions, thereby making it possible to estimate the elastic stiffness coefficients from numerical estimates of the second derivatives of the potential energy with respect to strain.

Apart from the aforementioned thermodynamical properties, MD simulations are ideally suited to calculating transport properties such as the diffusion coefficient and thermal conductivity. They generate configurations that change with time, thus generating time-dependent correlations of dynamic variables. In general, two methods are available to compute transport properties [92–94]. One approach is to calculate the appropriate autocorrelation function from which to derive the transport properties: the diffusion coefficient from the velocity autocorrelation function and the thermal conductivity from the heat-flux autocorrelation function. An alternative and more intuitive method, called the *direct method*, has found wide applications [94–96]. For thermal conductivity, λ, the direct method translates to applying a constant heat flux and calculating the thermal conductivity from the resulting temperature gradient using Fourier's law, which states that under steady-state conditions, the amount of heat flow per unit area in unit time is directly proportional to the temperature gradient at the cross section. λ can thus be defined as

$$\lambda = \frac{Q/A\Delta t}{dT/dz} \tag{6}$$

where Q is heat flow through the cross section, A the cross-sectional area, Δt the time during which heat is flowing, and dT/dz is the steady-state temperature gradient. Similarly, the self-diffusion coefficient D can be determined from the mean-square displacement of the atoms during a simulation of time t using the Einstein relation:

$$D = \lim_\infty \left\langle \frac{|r(t) - r(0)|^2}{6t} \right\rangle \tag{7}$$

One of the most widely used applications of MD simulations is in the study of nanoscale structure and conformation. The availability of efficient parallel molecular dynamics algorithms and increased computational power has led to large-scale molecular dynamics simulations. These simulations have been very successful in providing insight into molecular structure and processes not easily accessible by experimental measurements, and next we present a typical example in surface science.

3.5   Example: Ordering of Liquid Squalane Near a Solid Surface

A molecular dynamics study has been carried out to obtain new insights into the orientation of liquid squalane near a solid surface [97]. Squalane, $C_{30}H_{62}$, is a highly branched alkane molecule consisting of 24 backbone carbon atoms along with six symmetrically placed methyl groups. Earlier studies on the interfacial properties of alkanes have indicated that there can be a varying order of layering on substrate, depending on side-chain branching. In this study, the structure and orientation of squalane molecules adsorbed on amorphous and crystalline hydroxylated SiO_2 surfaces were investigated using canonical MD simulations at varying temperatures. The thicknesses of multilayer squalane films studied were about 40 and 80 Å. The simulations revealed several insights into the molecular ordering of the squalane molecules at the interfaces. The density profiles of the squalane films showed oscillations near the liquid–solid interface, which is a signature of layering in the squalane film. The amplitude of the density oscillation is found to be strongly dependent on the roughness of the substrate (i.e., amorphous vs. crystalline substrate). The first two layers closest to the substrate were found to be tightly bound, while layers farther away from the substrate showed a broadening in the density profile peaks resulting from increased interlayer spacing. This is consistent with experimental observation. No oscillations were induced at the air interface, but the squalane chain segments

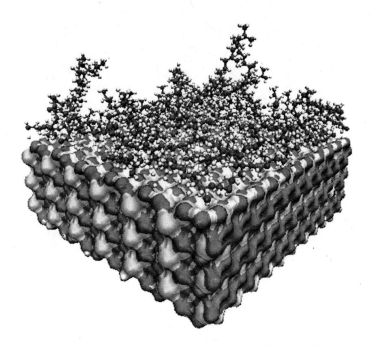

FIGURE 4   Ordering of liquid squalane near a hydroxylated silicon dioxide surface, showing only those squalane molecules that are in the first layer closest to the substrate.

were found to lie preferentially parallel to both the liquid–solid and liquid–vapor interfaces, with the CH_3 groups protruding into the surfaces. The molecular plane of the squalane molecules was found to orient mainly parallel to these interfaces. But, very interestingly, in going farther away from the substrate, the orientation of the molecular plane changes from mainly parallel to perpendicular to the interface.

A representative snapshot showing only those squalane molecules that are in the first layer, closest to the hydroxylated crystalline silica substrate is shown in Figure 4. This shows clearly that most of the squalane molecules adsorbed on the substrate are oriented parallel to the interface but with no in-plane ordering.

In summary, empirically based molecular dynamics has evolved to be a very powerful simulation tool which allows for simulation of large systems at reasonable computational cost. Although more accurate, quantum mechanical methods require computational effort that is currently prohibitive for many applications.

4 SUMMARY

In this chapter we provided an overview, with illustrative examples, of some of the most commonly used computational tools for studying atomic and molecular properties of nanoscale materials. However, the reader should recognize that different time and length scales are usually involved in the broad and diverse array of nanoscale materials. In many cases this requires the application of simulation techniques ranging from molecular scale (those discussed in this chapter) to macroscale (such as finite element). In order to make significant progress in understanding nanomaterials, there is a strong desire to integrate the various modeling techniques over a range of time and length scales. Although very challenging, developing such multiscale methods [61,62,64,98] is emerging as a new research area in computer simulation and modeling and has already had a considerable impact on many scientific and engineering disciplines.

Acknowledgments

The authors acknowledge financial support from the Air Force Office of Scientific Research. M.T. is also grateful to the National Science Foundation for financial support.

REFERENCES

1. M. Born and J. R. Oppenheimer, *Ann. Phys.* **84**, 457 (1927).

2. L. D. Landau and E. M. Lifshitz, *Quantum Mechanics*, Course of Theoretical Physics, Vol. 3, Pergamon Press, London, 1959.

3. L. I. Schiff, *Quantum Mechanics*, McGraw-Hill, New York, 1968.

4. A. Messiah, *Quantum Mechanics*, Vol. 2, Wiley, New York, 1990.

5. R. M. Martin, *Electronic Structure: Basic Theory and Practical Methods*, Cambridge University Press, New York, 2008.

6. D. Marx and J. Hutter, *Ab Initio Molecular Dynamics: Basic Theory and Advanced Methods*, Cambridge University Press, New York, 2009.

7. D. R. Hartree, *Proc. Cambridge Philos. Soc.* **24**(89), 111 (1928); *The Calculation of Atomic Structures*, Wiley-Interscience, New York, 1957.

8. V. A. Fock, *Z. Phys.* **61**, 126 (1930); *Z. Phys.* **62**, 795 (1930).

9. D. Sholl and J. A. Steckel, *Density Functional Theory: A Practical Introduction*, Wiley-Interscience, Hoboken, NJ, 2009.

10. D. Young, *Computational Chemistry: A Practical Guide for Applying to Real World Problems*, Wiley-Interscience, Hoboken, NJ, 2001.

11. J. C. Cramer, *Essentials of Computational Chemistry*, Wiley, Hoboken, NJ, 2002.

12. F. Jensen, *Introduction to Computational Chemistry*, 2nd ed., Wiley, Hobokon, NJ, 1999.

13. J. C. Greer, R. Ashrichs, and I. V. Hertel, *Z. Phys. D* **18**, 413 (1991).

14. S. A. Maluendes and M. Dupuis, *Int. J. Quantum Chem.* **42**, 1327 (1992).

15. B. Hartke and E. A. Carter, *Chem. Phys. Lett.* **189**, 358 (1993).

16. B. Hartke, D. A. Gibson, and E. A. Carter, *Int. J. Quantum Chem.* **45**, 59 (1993).

17. J. Jellinek, V. Bonacic-Koutecky, P. Fantucci, and M. Wiechert, *J. Chem. Phys.* **101**, 10092 (1994).

18. P. Hohenberg and W. Kohn, *Phys. Rev. B* **136**, 864 (1964).

19. W. Kohn and L. J. Sham, *Phys. Rev. A* **140**, 1133 (1965).

20. R. M. Dreizler and E. K. U Gross, *Density Functional Theory*, Springer-Verlag, Berlin, 1990.

21. R. G. Parr and Y. Weitao, *Density-Functional Theory of Atoms and Molecules*, Oxford University Press, New York, 1994.

22. M. Levy, *Proc. Natl. Acad. Sci.* **76**, 6062 (1979).

23. W. Kohn and L. J. Sham, *Phys. Rev.* **140**, A1133 (1965).

24. R. Car, *Quant. Struct. Act. Relat.* **21**, 97 (2002).

25. R. Lftimie, P. Minary, and M. E. Tuckerman, *Proc. Natl. Acad. Sci.* **102**, 6654 (2005).

26. P. Kruger and J. Pollmann, *Phys. Rev. Lett.* **74**, 1155 (1995).

27. D. M. Ceperley and B. J. Alder, *Phys. Rev. Lett.* **45**, 566 (1980).

28. R. Q. Hood, M. Y. Chou, A. J. Williamson, G. Rajagopal, and R. J. Needs, *Phys. Rev. Lett.* **78**, 3350 (1997).

29. R. Q. Hood, M. Y. Chou, A. J. Williamson, G. Rajagopal, and R. J. Needs, *Phys. Rev. B* **57**, 8972 (1998).

30. P. Fulde, *Electron Correlations in Molecules and Solids*, Springer-Verlag, New York, 1995.

31. D. C. Langreth and M. J. Mehl, *Phys. Rev. B* **28**, 1809 (1983).

32. A. D. Becke, *Phys. Rev. A* **38**, 3098 (1988).

33. J. P. Perdew, J. A. Chevary, S. H. Vosko, K. A. Jackson, M. R. Pederson, D. J. Singh, and C. Fiolhais, *Phys. Rev. B* **46**, 6671 (1992); **48**, 4978 (Erratum) (1993).

34. P. Haas, F. Tran, P. Blaha, K. Schwarz, and R. Laskowski, *Phys. Rev. B* **80**, 195109 (2009).

35. J. P. Perdew and Y. Wang, *Phys. Rev. B* **33**, 8800 (1986).

36. J. P. Perdew, *Phys. Rev. B* **33**, 8822 (1986); **34**, 7406 (1986).

37. A. D. Becke, *Phys. Rev. A* **38**, 3098 (1988).

38. C. Lee, W. Yang, and R. G. Parr, *Phys. Rev. B* **37**, 785 (1988).

39. J. P. Perdew, K. Burke, and M. Enzerhof, *Phys. Rev. Lett* **77**, 3865 (1996); **78**, 1396 (1997).

40. R. Armiento and A. E. Mattsson, *Phys. Rev. B* **72**, 085108 (2008).

41. A. E. Mattsson and R. Armiento, *Phys. Rev. B* **79**, 155101 (2009).

42. Z. Wu and R. E. Cohen, *Phys. Rev. B* **73**, 235116 (2006).

43. J. P. Perdew, A. Ruzsinszky, G. I. Csonka, O. A. Vydrov, G. E. Scuseria, L. A. Constantin, X. Zhou, and K. Burke, *Phys. Rev. Lett.* **100**, 136406 (2008).

44. Y. Zhao and D. G. Truhlar, *J. Chem. Phys.* **128**, 184109 (2008).

45. B. B. Karki, L. Stixrude, and R. M. Wentzcovitch, *Rev. Geophys.* **39**, 507 (2001).

46. D. Marx and J. Hutter, in *Modern Methods and Algorithms of Quantum Chemsitry*, J. Grotendorst (Ed.), Forschungszentrum Julich http://www.fz-juelich.de/wsqc/proceedings, 2000.

47. J. S. Tse, *Annu. Rev. Phys. Chem.* **53**, 249 (2002).

48. F. Gygi and G. Galli, *Mater. Today* **8**, 26 (2005).

49. M. Tsige, T. Soddemann, S. B. Rempe, G. S. Grest, J. D. Kress, M. O. Robbins, S. W. Sides, M. J. Stevens, and E. Webb III, *J. Chem. Phys.* **118**, 5132 (2003).

50. G. Kresse and J. Hafner, Phys. Rev. *B* **47**, 558 (1993).

51. G. Kresse and J. Furthmuller, *Comput. Mater. Sci.* **6**, 15 (1996).

52. G. Kresse and J. Furthmuller, *Phys. Rev. B* **54**, 1169 (1996).

53. D. Vanderbilt, *Phys. Rev. B* **41**, 7892 (1990).

54. G. Kresse and J. Hafner, *J. Phys. Condens. Matter* **6**, 8245 (1994).

55. M. J. Frisch et al., *Gaussian 98*, Revision A.2, Gaussian, Inc., Pittsburgh, PA, 1998.

56. A. D. Becke, *J. Chem. Phys.* **98**, 5648 (1993).

57. E. Nikitina, V. Khavryutchenko, E. Sheka, H. Barthel, and J. Weis, *Compos. Interfaces* **6**, 3 (1999).

58. Y. V. Kazakevich and A. Y. Fadeev, *Langmuir* **18**, 3117 (2002).

59. S. Peutz, E. J. Kramer, J. Baney, C.-Y. Hui, and C. Cohen, *J. Polym. Sci. B* **36**, 2129 (1998).

60. A. Redondo and R. LeSar, Modeling and simulation of biomaterials, *Annu. Rev. Mater. Res.* **34**, 279 (2004).

61. M. Fermeglia and S. Pricl, Multiscale modeling for polymer systems of industrial interest, *Prog. Org. Coat.* **58**, 187 (2007).

62. G. Scocchi, P. Posocco, A. Danani, S. Pricl, and M. Fermegli, To the nanoscale, and beyond! Multiscale molecular modeling of polymer–clay nanocomposites, *Fluid Phase Equilibria* **261**, 366 (2007).

63. G. Scocchi, P. Posocco, M. Fermegli, and S. Pricl, Polymer–clay nanocomposites: a multiscale molecular modeling approach, *J. Phys. Chem. B* **111**, 2143 (2007).

64. Q. H. Zeng, A. B. Yu, and G. Q. Lu, Multiscale modeling and simulation of polymer nanocomposites, *Prog. in Polym. Sci.* **33**, 191 (2008).

65. M. P. Allen and D. J. Tidesley, *Computer Simulation of Liquids*, Clarendon Press, Oxford, UK, 1987.

66. J. M. Haile, *Molecular Dynamics Simulations*, Wiley, Hoboken, NJ, 1992.

67. A. R. Leach, *Molecular Modeling: Principles and Applications*, Pearson Prentice Hall, Upper Saddle River, NJ, 2001.

68. D. Frenkel and B. Smit, *Understanding Molecular Simulations: From Algorithms to Applications*, Academic Press, San Diego, CA, 2002.

69. A. Hinchliffe, *Molecular Modeling for Beginners*, Wiley, Hoboken, NJ, 2003.

70. J. A. Maple, M.-J. Hwang, T. P. Stockfisch, U. Dinur, M. Waldman, C. S. Ewig, and A. T. Hagler, Derivation of class II force fields: I. Methodology and quantum force field for the alkyl functional group and alkane molecules, *J. Comput. Chem.* **15**, 162 (1994).

71. H. Sun and D. Rigby, Polysiloxanes: ab initio force field and structural conformational and thermophysical properties, *Spectrochim. Acta A* **53**, 1301 (1997).

72. H. Sun, Compass: an ab initio force-field optimized for condensed-phase applications—overview with details on alkane and benzene compounds, *J. Phys. Chem.* **102**, 7338 (1998).

73. D. A. Case, T. E. Cheathan III, T. Darden, H. Gohlke, R. Luo, K. M. Merz, A. Onufriev, C. Simmerling, B. Wang, and R. Woods, The amber biomolecular simulation programs, *J. Comput. Chem.* **26**, 1668 (2005).

74. J. A. D. Mackerell, B. R. Brooks, I. C. L. Brooks, L. Nilsson, B. Roux, Y. Won, and M. Karplus, in R. Scheleyer (Ed.), *The Encyclopedia of Computational Chemistry*, Wiley, New York, 1998, pp. 271–277.

75. W. L. Jorgensen, D. S. Maxwell, and J. Tirado-Rives, Development and testing of the OPLS all-atom force field on conformational energetics and properties of organic liquids, *J. Am. Chem. Soc.* **118**, 11225 (1996).

76. D. W. Brenner, Empirical potential for hydrocarbons for use in simulating the chemical vapor deposition of diamond films, *Phys. Rev. B* **42**, 9458 (1990).

77. A. C. van Duin, S. Dasgupta, F. Lorant, and W. A. Goddard III, *J. Phys. Chem. A* **105**, 9396 (2001).

78. M. S. Daw and M. Baskes, Embedded-atom method: derivation and applications to impurities, surfaces and other defects in metals, *Phys. Rev. B* **29**, 6433 (1984).

79. J. Tersoff, New empirical model for the structural properties of silicon, *Phys. Rev. Lett.* **56**, 632 (1986).

80. L. Verlet, Computer "experiments" on classical fluids: I. Themodynamical properties of Lennard Jones molecules, *Phys. Rev.* **159**, 98 (1967).

81. P. Ewald, Due Berechnung optischer und elektrostatischer Gitterpotentiale, *Ann. Phys.* **64**, 253 (1921).

82. L. Greengard, Fast algorithms for classical physics, *Science* **265**, 909 (1994).

83. S. Nose, A unified formulation of the constant temperature molecular dynamics methods, *J. Chem. Phys.* **81**, 511 (1984).

84. W. G. Hoover, Canonical dynamics: equilibration phase space distribution, *Phys. Rev. A* **31**, 1695 (1985).

85. H. J. C. Berendsen, J. P. M. Postma, W. F. van Gunsteren, A. DiNole, and J. R. Haak, Molecular dynamics with coupling to an external bath, *J. Chem. Phys.* **81**, 3684 (1984).

86. M. E. Clamp, P. G. Baker, C. J. Stirling, A. Brass, Hybrid Monte Carlo: an efficient algorithm for condensed matter simulation, *J. Comput. Chem.* **15**, 838 (1994).

87. K. Binder and D. W. Heerman, *Monte Carlo Simulation in Statistical Physics: An Introduction*, Springer-Verlag, Berlin, 2010.

88. J. I. McKehnie, R. N. Haward, D. Brown, and J. H. R. Clarke, *Macromolecules* **26**, 198 (1993).

89. A. V. Lyulin and M. A. J. Michels, Time scales and mechanism of relaxation in the energy landscape of polymer glass under deformation: direct atomistic modeling, *Phys. Rev. Lett.* **99**, 085504-1 (2007).

90. D. N. Theodorou and U. W. Suter, Atomistic modeling of mechanical properties of polymeric glasses, *Macromolecules* **19**, 139 (1986).

91. M. Hutnik, A. S. Argon, and U. W. Suter, Simulation of elastic and plastic response in the glassy polycarbonate of 4,4′-isopropylidenediphenol, *Macromolecules* **26**, 1097 (1993).

92. J. Che, T. Cagin, and W. A. Goddard, Thermal conductivity of carbon nanotubes, *Nanotechnology* **11**, 65 (2000).

93. P. K. Schelling, S. R. Phillpot, and P. Keblinski, Comparison of atomic-level simulation methods for computing thermal conductivity, *Phys. Rev. B* **65**, 144306 (2002).

94. V. Varshney, S. S. Patnaik, A. K. Roy, and B. L. Farmer, Heat transport in an epoxy network: a molecular dynamics study, *Polymer* **50**, 3378 (2009).

95. V. Varshney, S. S. Patnaik, A. K. Roy, and B. L. Farmer, Modeling of thermal conductance at transverse CNT-CNT interfaces, *J. Phys. Chem. C* (2010).

96. V. Varshney, S. S. Patnaik, A. K. Roy, G. E. Froudakis, and B. L. Farmer, Modeling of thermal transport in Pillard-graphene architectures, *ACS Nano* **4**, (2010).

97. M. Tsige and S. S. Patnaik, An all-atom simulation study of the ordering of liquid squalane near silicon dioxide surfaces, *Chemi. Phys. Lett.* **457**, 357 (2008).

98. M. Doi, Challenge in polymer physics, *Pure Appl. Chem.* **75**, 1395 (2003).

SECTION III

APPLICATIONS

8

NANOMATERIALS FOR ALTERNATIVE ENERGY

HONG HUANG AND BOR Z. JANG

College of Engineerig and Computer Science, Wright State University, Dayton, Ohio

Nanoscale Multifunctional Materials: Science and Applications, First Edition.
Edited by Sharmila M. Mukhopadhyay.
© 2012 John Wiley & Sons, Inc. Published 2012 by John Wiley & Sons, Inc.

1   BACKGROUND AND INTRODUCTION

The growing needs for energy and the increasing awareness of global environmental issues have spurred an intense search for alternatives to fossil fuels. Renewable and clean energy technologies, such as wind, solar, geothermal, and nuclear, are currently under extensive research and development. Electrochemical energy conversion and storage systems, including batteries, supercapacitors, and fuel cells, can store or convert chemical energies directly into electricity with high efficiency. These electrochemical energy systems are indispensable power supplies. Various types of batteries and supercapacitors are integral parts of large-scale energy storage systems for smart grid, wind, and solar energy.

At present, four electrochemical energy conversion and storage systems are under intense development: lithium-ion (Li-ion) batteries, supercapacitors, polymer electrolyte membrane fuel cell (PEMFCs), and solid oxide fuel cells (SOFCs) [1–6]. Li-ion rechargeable batteries are advantageous over conventional batteries such as alkaline and nickel hydride in view of high output voltage, high energy density, and good discharge–charge cycle life. Supercapacitors have high power density, extremely long cycle life (up to a million cycles), no or few safety issue, and rapid rechargeability. PEMFCs have better start–stop capabilities and fewer corrosion problems than other fuel cells and are well suited to portable and mobile applications. Pure hydrogen-fueled PEMFCs exhibit the highest power densities among all types of fuel cells at ambient conditions. SOFCs are promising stationery power generators because they are highly efficient, fuel-flexible (independent of purity of hydrogen fuel), low-noise, and have zero or little emission. At the typical high operating temperatures (600 to 1000°C), the electrical efficiency of an SOFC is about 50 to 60% and the total efficiency can reach as high as 90% at the heat cogeneration operation mode.

The performances of batteries, supercapacitors, and fuel cells depend intimately on the properties of electrode and electrolyte materials. Materials innovations and wise choice have led to technological breakthroughs and leaping advancements in the energy systems. Nanostructured materials, due to their unique and often significantly improved mechanical, electrical, thermal, and optical properties, have attracted great attention and interest in alternative energy applications [4–6]. In this chapter we focus our discussion on nanomaterials for Li-ion batteries, supercapacitors, PEMFCs, and SOFCs.

Our discussion will include nanoparticles, nano-thin films, nanowires, and nanotubes, with emphasis being placed on nano-graphene platelets (NGPs). NGPs herein collectively refer to one- or multiple-layered graphene platelets, which exhibit many astonishing properties: the highest intrinsic mechanical strength (130 GPa), the highest thermal conductivity (3000 to 5300 $Wm^{-1} K^{-1}$), high specific surface area (2675 m^2/g), and high electron mobility (10,000 $cm^2/V \cdot s$, 100 times faster than in Si). Graphene has become the most exciting and promising nanomaterial and has triggered a "gold rush" for exploiting its various possible applications [7–14]. Graphene-based materials have found extensive applications in alternative energy sources [15–24].

In this chapter, electrochemistry basics are introduced in Section 2 in consideration of the diverse technical backgrounds of the readers. In Sections 3 to 5 we present selected examples of nanocrystalline ionic conductors for SOFCs, nanostructured electrocatalysts for PEMFCs, polymer electrolytes with nanofillers for PEMFCs, and nanostructured electrodes for Li-ion batteries. Section 6 is devoted to a review of graphene-based materials for Li-ion batteries and supercapacitors.

2 ELECTROCHEMISTRY BASICS

2.1 Battery and Fuel Cell [25–27]

The building block of both batteries and fuel cells is a basic electrochemical unit cell made up of an electrolyte sandwiched between two electrodes. The electrolyte is an ionic conductor, serving as an ionic transport medium. The electrode is an electronic or mixed electronic and ionic conductor. The electrode where oxidation reaction occurs is termed the *anode* or negative electrode. Similarly, the electrode where reduction occurs is referred to as the *cathode* or positive electrode.

Such an electrochemical energy device can convert chemical energy stored in electroactive materials directly into electric energy via a pair of half electrochemical redox (reduction–oxidation) reactions, which involve both electrons and ions and take place at the two electrode–electrolyte interfaces. The electrochemical processes complete via separate electron and ion transport in two different paths: electrons are transported from one electrode to the other through an external electric circuit, and ions are transported through an internal electrolyte. The characteristics of electrochemical energy conversion, distinguished from conventional combustion, result in high-energy conversion efficiencies.

Despite these conceptual similarities, there exist clear distinctions between batteries and fuel cells:

- A battery is usually a closed system in which the cathode, anode, and electrolyte are enclosed. With appropriate material chemistry, reversible reactions can occur upon charging. Accordingly, batteries are categorized into nonrechargeable (primary) and rechargeable (secondary). The energy output of a battery is restricted by the predetermined amount of electroactive material in the electrode. A high storage capacity and output voltage from electrodes are essential to achieving the high energy density of a battery.
- A fuel cell, in contrast, is an open system. The active materials—the fuels and oxidants—are not confined in the system and can be supplied continuously to the electrode–electrolyte interface from a separate tank or the environment. A fuel cell system can deliver electricity as long as fuels and oxidants are supplied. The electrode in a fuel cell serves as a facilitator for the electrochemical reaction and is not consumed during the power-generating process. Therefore, high levels of catalytic activity with good stability to ensure a fast electrochemical reaction

over a long period are the key requirement for high-performance electrodes in fuel cells.

2.2   Theoretical Figures of Merit: Thermodynamics

To assess the performance of batteries and fuel cells, there are several basic figures of merit, such as open-circuit voltage, capacity, and energy, which are determined essentially by the nature and quality of electroactive materials. The theoretical values are governed by thermodynamics.

Theoretical Voltage   In an electrochemical cell, two half-reactions occur at the two electrodes and can be generalized as follows:

$$\text{Oxidation reaction at the anode: } m\text{M} \Leftrightarrow c\text{C} + n\text{e} \tag{1}$$

$$\text{Reduction reaction at the cathode: } b\text{B} + n\text{e} \Leftrightarrow d\text{D} \tag{2}$$

$$\text{The overall reaction: } m\text{M} + b\text{B} \Leftrightarrow c\text{C} + d\text{D} \tag{3}$$

where M, B, C, and D represent the formula of the reactants and products; m, b, c, and d are the corresponding numbers determined by the mass conservation; and n is the number of electrons e transferred in the reaction.

The total free-energy change ΔG of the reaction will be equal to the electrical work output (W_{electric}), provided that overall reaction (3) occurs spontaneously and there is no energy loss. Since the theoretical voltage of the electrochemical cell, E, is the electrical potential difference between the two electrodes, the electrical work is the product of E and the total charge nF (F is the Faraday constant) involved:

$$\Delta G = W_{\text{electric}} = -nFE \tag{4}$$

Hence, the theoretical voltage can then be calculated from the free-energy change, ΔG:

$$E = -\frac{\Delta G}{nF} \tag{5}$$

The theoretical voltage E° at standard conditions (1 atm, 25°C, all activities of relevant species are 1) in relation to the cell voltage E at nonstandard conditions is based on the Nernst equation:

$$E = E^\circ - \frac{RT}{nF} \ln \frac{a_{\text{C}}^c a_{\text{D}}^d}{a_{\text{M}}^m a_{\text{B}}^b} \tag{6}$$

where a is the activity of the relevant species, R the universal gas constant, and T the absolute temperature.

The theoretical voltage can also be calculated by subtracting the anode reduction potential (e_a) from the cathode reduction potential (e_c), (i.e., $E = e_c - e_a$). A number of electrode reduction potentials at the standard state are available in the electrochemical database.

Theoretical Capacity   Capacity is the total electricity that an electrochemical cell can deliver. The capacity of a battery is commonly expressed in terms of ampere-hours (Ah) instead of coulombs (C). As the maximum capacity is predetermined by the amount of active materials within the battery, a specific capacity (capacity per unit mass) as well as a volumetric capacity (capacity per unit volume) is generally used to assess the active electrode materials in batteries. For example, the specific capacity of the active material M in the half-reaction (1) can be calculated as

$$\text{specific capacity} = \frac{nF}{M_w} \tag{7}$$

where M_w is the molecular weight of material M, F is Faraday's constant (96,485 C = 26.8 Ah), and n is the number of electrons transferred per one formula M.

Theoretical Energy Density   Energy delivered by a cell is the integral of voltage and the quantity of electric charge. Theoretical gravimetric and volumetric energy densities (common engineering units are Wh/kg and Wh/L) are equal to the Gibbs free-energy change (ΔG) of the total electrochemical reaction per unit mass and volume, respectively.

2.3   Practical Output: Electrochemical Kinetics

Electrochemical reactions are generally complex and often involve many parallel and sequential steps. The electroactive species must be transported to the electrode–electrolyte interface via migration and diffusion to participate in charge-transfer reactions. Adsorption of electroactive materials may precede the charge-transfer step. The overall rate of the electrochemical process will be determined by the rate of the slowest step (rate-determining step) in the entire sequence of reactions.

The occurrence of an electrochemical reaction is always accompanied by various voltage loss when a load current i, which is a direct measure of the reaction rate, passes through an electrochemical cell. There are, in general, three primary voltage losses (electrochemically termed overvoltage, overpotential, or potential polarization) dictated by the electrochemical kinetics. These are ohmic loss (η_{ohm}), activation loss (η_{act}), and concentration loss (η_{conc}). All three losses are functions of the loading current i. The practical voltage output is less than the theoretical voltage E by the three losses:

$$V(i) = E - \eta_{ohm}(i) - \eta_{act}(i) - \eta_{conc}(i) \tag{8}$$

Ohmic Loss η_{ohm}   The ohmic loss η_{ohm} results from the internal resistance to both electronic and ionic conductions. The ohmic loss is proportional to the load current i and total resistance R,

$$\eta_{ohm}(i) = iR \qquad (9)$$

Here R is the sum of the electrolyte resistance; the electronic resistance of electrodes, current collectors, and electrical wires; and the contact resistance between electrolyte, electrode, and current collector. Usually, electrolyte resistance is predominant among all the others.

The electrolyte resistance R_i is determined fundamentally by the ionic conductivity σ_i of the electrolyte, and their relationship obeys Ohm's law:

$$R_i = \frac{1}{\sigma_i} \times \frac{L}{S} \qquad (10)$$

where L and S are the thickness and cross-sectional area of the electrolyte. In SOFCs, one of the major tasks is to increase the ionic conductivity of the solid oxide conductor at low temperatures. In PEMFCs, one research area is to increase the ionic conductivity of the polymeric electrolyte under a reduced water content environment.

Activation Loss η_{act}   When a current i flows through an electrode, the electrode potential will deviate away from the equilibrium electrode potential. The potential difference resulting from an activation-controlled charge-transfer process (e.g., Ox + $ne \Leftrightarrow$ Re d) is referred to as the activation overvoltage, η_{act}.

A general i–η_{act} relationship is represented by the Butler–Volmer equation:

$$i = i_0 \left[c_{ox} \exp\left(-\frac{\alpha n F \eta}{RT} \right) - c_{red} \exp\left(\frac{(1-\alpha)nF\eta}{RT} \right) \right] \qquad (11)$$

where i_0 is the exchange current, α the transfer coefficient, and η the overvoltage. It is noteworthy that c_{ox} and c_{red} are effective concentrations of oxidant (ox) and reductant (red) at the surface of the electrode, which are assumed to be the same as the bulk concentration. The exchange current i_0 equals the forward–backward reaction current at equilibrium.

When the forward reaction is dominant, the second term corresponding to the backward process is negligible. As a result, equation (11) can be simplified mathematically into the Tafel equation,

$$\eta_{act}(i) = x + b \log i \qquad (12)$$

Here x and b are constants in relation to the exchange current i_0 and transfer coefficient α. It can be deduced that increasing the exchange current i_0 at the operating temperature and current will decrease the activation loss. Therefore, reducing the activation loss can be realized by improving the catalytic activity at the electrode, which may increase the intrinsic reaction exchange current.

Concentration Loss η_{conc}   When the electrochemical reaction proceeds, the concentration [c_{ox} and c_{red} in equation (11)] at the electrode surface will differ from the equilibrium concentration. Reactants will be depleted and products will be accumulated near the electrode surface. The current in relation to the concentration gradient near the electrode surface can be expressed as

$$i = \frac{nFDA(c_B - c_S)}{\delta} \tag{13}$$

where A the electrode area, δ the thickness of the diffusion layer, D the diffusion coefficient of the electroactive species, and c_B and c_S the concentrations in the bulk and on the electrode surface, respectively. When the surface concentration c_S becomes zero, the current will reach the maximum value; that is the limiting current

$$i_L = \frac{nFDAc_B}{\delta} \tag{14}$$

The concentration difference due to diffusion results in the concentration overvoltage,

$$\eta_{conc} = \frac{RT}{nF} \ln \frac{c_B}{c_S} = \frac{RT}{nF} \ln \frac{i_L}{i_L - i} \tag{15}$$

Equations (14) and (15) indicate that increasing the limiting current and sustaining a high current loading at the low concentration loss can be realized via increasing the diffusion coefficient, electrode area, and bulk concentration. The diffusion process, sometimes accompanied by a phase transformation, is typically the rate-determining step in the battery systems. Improvement in the diffusion and/or phase transition rate is an appropriate direction for research to improve the battery rate performance.

2.4   Supercapacitor

Supercapacitors, also known as ultracapacitors, are electrochemical capacitors (ECs). A simple EC consists of an electrolyte and two electrodes, similar to a battery and a fuel cell. However, no significant electrochemical charge transfer occurs in an EC. Charges can be stored or delivered at the electric double layer (EDL), generated at the electrode–electrolyte interface. The capacitance of an EC, 10 to 100 times greater than that of conventional capacitors, derives from the large effective "plate area" of the porous electrodes and extremely small effective "plate separation" of the electric double layer, which is only about 1 nm thick. Energy storage of a capacitor is the integral of voltage difference and charge accumulation at the EDL. The charge is the product of voltage difference and capacitance.

The distinguished characteristic of an EC from a battery and a fuel cell is physical rather than chemical energy storage, giving ECs an extraordinarily long cycle life and high power density with no or few safety issues. ECs have a great potential for uses in hybrid electric vehicles (HEVs) to provide the bursts of power needed for rapid acceleration.

3 NANOCRYSTALLINE SOLID OXIDE CONDUCTORS FOR SOFCs

An SOFC employs a thin ceramic oxygen conductor as an electrolyte and porous mixed ion/electron-conducting ceramic composites as both electrodes. The temperatures for fabricating an SOFC are usually near or higher than 1000°C. SOFCs are usually operated above 800°C, due to the low oxygen ion conductivities of the electrolytes and high activation energies at the cathodes [28,29]. Present efforts have been dedicated to developing intermediate-temperature SOFCs (IT-SOFCs) in the range 500 to 700°C, which will mediate the sealing challenge and performance degradation commonly encountered in conventional high-temperature SOFCs. Even at such high temperatures, the potential for applying nanoengineering principles is enormous [28–35]. For example, utilizations of nanoparticles are beneficial to fuel cell fabrication procedures in terms of reducing the firing temperature and improving electrode microstructures; nanocrystalline ionic materials can deliver superior electrical properties suitable for IT-SOFCs. In this section we focus on discussing mechanisms of performance enhancement in nanocrystalline ionic conductors based on two exemplary cases.

3.1 Increased Ionic Conductivity in Nanocrystalline Ionic Conductors

When the grain sizes are reduced to the nanoscale, either increased or decreased ionic conductivities are reported in crystalline ionic materials. The confusing phenomena are governed by the characteristics of the space-charge region, naturally created at the grain boundaries or phase interface. On the one hand, the remarkably large areas of interface and/or grain boundaries in nanostructured materials will increase the density and conductivity of mobile defects *along* the space-charge region. For example, enhanced ionic conductivity was observed along the alternating layers of nanoscale CaF_2–BaF_2, which was attributed to the overlapping of space-charge layers [36]. On the other hand, ion conduction *across* the space-charge regions will be blocked, resulting in ionic conductivities in the vicinity of grain boundaries (σ_{gb}) several orders of magnitude lower than those in the bulk (σ_b) [36,38–41].

Statistically, polycrystalline films with thicknesses comparable to the grain sizes contain few grain boundaries parallel to the membrane surface (perpendicular to the ionic movement). Transport properties along the nanoscale grains may be enhanced and at the same time, blocking effects from crossing grain boundaries may be eliminated. Therefore, fast ionic transport properties are anticipated to achieve in nanoscale thin films with thicknesses that are comparable to the grain size.

Huang et al. [42] observed increased ionic conductivities by measuring the conduction directly across the ultrathin nanocrystalline gadolinia-doped ceria (GDC) films with thicknesses comparable to the grain size (20 to 50 nm) in the temperature range of 100 to 350°C. GDC (sometimes referred to as CGO in literatures) is a promising electrolyte candidate for IT-SOFCs because of its high oxide-ion conductivity relative to the standard yttria-stabilized zirconia (YSZ) electrolyte. It was found that grain conductivities in the ultrathin GDC films depended insignificantly on the dopant concentrations, and the effective values were higher in heavily doped samples at low temperatures. This characteristic was ascribed to the unusual dopant segregation and defect association present in nanocrystalline ionic conductors.

Dopant segregation is commonly observed in alloys and ceramics, and the width of the segregation zone is in the range 1 to 3 nm [37,43–46]. Dopant segregation may cause a compositional deviation from the nominal value in grain interior. In a grain of a few tens of nanometers, the compositional deviation may spread over a significant fraction of the grain. Due to the segregation of virtually all available Gd to the grain boundaries, leaving the grain interior undoped, 1.54 mol% nanocrystalline GDC exhibited unusual electronic conduction as observed by Chiang [37]. In contrast, the compositional deviation is negligible in micrometer-sized particles. Moreover, the oxide-ion conductivity is related directly to the dissociated oxide-ion vacancy concentration. At high dopant concentration levels and low temperatures, oxide-ion vacancies tend to associate with the dopants, resulting in decreased concentration in the free (dissociated) oxygen-ion vacancies, and hence decreased ionic conductivity. Huang [42] suggested the presence of three zones in a GDC nanograin: a grain boundary core (GB), a segregation–association zone (A-zone), and a free zone (F-zone: populated with free oxide-ion vacancies) (Figure 1). In a film whose thickness approached the grain size, the free zones were readily accessible to both electrodes and, consequently, served as conduits for fast ion transport from one electrode

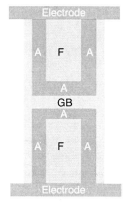
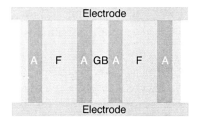

FIGURE 1   Ultrathin films (left: thickness equals two grain sizes; right: thickness equals one grain size) which contain three zones: a grain boundary core (GB), a segregation/association zone (A), and a free zone (F).

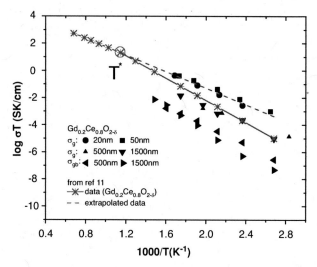

FIGURE 2 Ionic conductivity of GDC thin films of different thicknesses in comparison with the GDC of different dopant concentrations.

to the other. The experimental results and the segregation–association model corroborated very well, as shown in Figure 2. The conductivities in the 20- and 50-nm-thick $Gd_{0.2}Ce_{0.8}O_{1.9}$ films agreed with the line extrapolated from high-temperature data (fully dissociated) in reference data with lower Gd doping, $x = 0.10$ to 0.12 [47]. The enthalpies of migration and association in GDC were calculated to be 0.6 and 0.12 eV, respectively, similar to values reported previously [47–49].

It is noteworthy that the location of such fast ion conduits may regulate itself, depending on the variation in temperature and doping level. Similar phenomena are anticipated in other thin nanocrystalline films with thicknesses comparable to the grain size and with varying dopant concentrations.

3.2 Increased Cathodic Kinetics on a Nanocrystalline Ionic Conductor

Another crucial factor for improving IT-SOFC performance relies on reduction of the activation loss caused by the slow cathodic reaction kinetics. Oxides such as $La_{1-x}Sr_xMnO_3$ (LSM) and $La_{1-x}Sr_xCo_{1-y}Fe_yO_3$ (LSCF) are presently the representative cathode materials in SOFCs. The high activation energies (> 1.5 eV) for the oxygen reduction reaction (ORR) on these cathodes results in poor SOFC performances upon decreasing temperatures. Noble metallic catalysts such as Pt are superior to oxides at temperatures below 500°C [28,29,50–52]. However, the ORR at the metallic catalyst–electrolyte interface (e.g., Pt–YSZ) is constrained to the vicinity of the Pt–YSZ boundary due to the impermeability of oxygen, low surface oxygen diffusivity on Pt, and low ionic conductivity in YSZ.

Employing a thin functional interlayer between the electrolyte and the Pt cathode can extend the reaction zone away from the Pt electrode and hence increase the ORR reaction rate and reduce the activation loss. The criteria for a good interlayer with such functionality include one or more of the following properties: high ionic conductivity, high electronic conductivity, and high catalytic activity for oxygen dissociation.

Tanner and Chan [53,54] reduced activation loss by using a thin cathode with a fine microstructure and a highly ion-conducting electrolyte with small grain sizes. Tsai and Bernett [55] reported that cathode resistance was decreased 10 times via inserting yttria-doped ceria between YSZ and the cathode. Recently, Huang et al. [56] investigated the functions of the nanocrystalline interlayer via quantum simulation, electrochemical impedance analyses, and fuel cell performance assessment. Quantum simulation results suggested that a high oxygen vacancy density existed on the surface or near the grain boundary of YSZ, GDC, and undoped ceria. Experimentally, 50-nm-thick nanocrystalline YSZ, GDC, and pure ceria thin films were inserted between a Pt cathode and a microcrystalline YSZ electrolyte with the help of a sputtering technique. The electrode impedances were decreased several-fold by adding a nanocrystalline interlayer. This positive impact was found to be most significant using a GDC interlayer, a highly ionic-conductive nanocrystalline material. Characteristics of fuel cells with and without the GDC interlayer were investigated further on ultrathin SOFC architecture [56]. Figure 3(a) shows a scanning electron microscopic (SEM) cross section of the ultrathin SOFCs (UTSOFCs) with a 50-nm-thick GDC layer between a nanoporous Pt electrode and a 50-nm-thick YSZ electrolyte. Figure 3(b) shows the current–voltage (I–V) profiles of UTSOFCs composed of different cathodic interface configurations. The UTSOFCs with a higher ionic conductivity material adjacent to the Pt cathode exhibited a higher peak power density. The elevated mobility and density levels of oxide-ion vacancies in the vicinity of the Pt cathode–nanocrystalline electrolyte interface were ascribed to fast oxygen adsorption and incorporation. These results echoed De Souza's premise that high surface exchange rates correlate well with high oxygen diffusivity [57]. In sum, the total cathodic electrochemical kinetics was increased by using nanocrystalline, highly ion-conducting materials.

4   NANOSTRUCTURED ELECTROCATALYSTS AND ELECTROLYTES FOR PEMFCs

PEMFCs employ a thin porous polymeric membrane as the electrolyte and metallic catalysts supported on porous electron-conducting membrane as both the cathode and anode. Hydrogen, methanol, or formic acid is the choice of fuels, and oxygen in air serves as the oxidant. PEMFCs have attracted great interest in portable and transport applications and have been utilized in various military missions. Before PEMFCs can be widely commercialized, several technical obstacles need to be circumvented, such as low durability of the electrocatalysts, decreased conductivity of the electrolyte at low humidity and high temperatures, and high material cost. In this section,

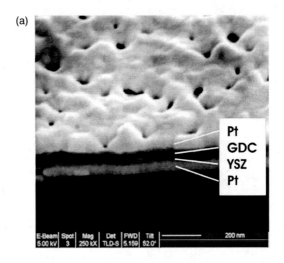

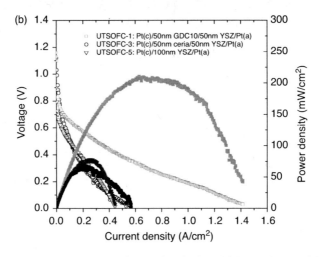

FIGURE 3 (a) SEM cross section of ultrathin SOFCs, which are 50-nm-thick GDC layers between nanoporous Pt electrodes and 50-nm-thick YSZ electrolytes; (b) voltage–current profile of UTSOFCs, composed of different cathodic interface configuration.

developments in nanostructured electrocatalysts and nanocomposite conductive electrolytes are addressed.

4.1   Nanostructured Electrocatalysts

In hydrogen-based PEMFCs, the ORR reaction is the rate-determining step and contributes the major activation loss even though the expensive platinum has been the most active electrocatalyst. To improve the catalytic activity as well as to reduce

Pt alloy   Pt skeleton   Pt skin

FIGURE 4 Three different nanostructures of Pt₃M alloys.

the material cost, the primary routes include tailoring the Pt particle size, alloying Pt with other metals, and searching for alternative nonnoble catalytic systems [1,58].

Pt-based alloys were often reported to improve the catalytic activity several fold relative to pure Pt catalyst [59–62]. Stamenkovic et al. [62] systematically studied the trends in ORR electrocatalysis on Pt alloyed with 3d transition metals using a combination of ex situ and in situ surface-sensitive probes and density functional theory (DFT) calculations. A fundamental "volcano-type" relationship was discovered on Pt_3M (M = Ni, Co, Fe, Ti, V) surfaces between the surface electronic structure (the d-band center) and the ORR activity. Moreover, three different nanostructures in Pt_3Ni alloy (i.e., Pt alloy, Pt skeleton, and Pt skin; see Figure 4) were achieved through controlling the alloy preparation procedures. Pt alloy, the as-sputtered sample, was a relatively homogeneous composition distribution from surface to bulk. A "corrugated" Pt-skeleton surface sample resulted from Ni dissolving after acid leaching. The Pt-skeleton sample has a bulklike alloy concentration profile beneath the subsurface atomic layer. Pt-skin structure was achieved after a 400°C annealing treatment. According to Stamenkovic et al. [62], a Pt-skin sample had a unique compositional arrangement in the near-surface region. The first layer of the alloy particle was composed entirely of Pt; the second atomic layer is Ni-rich (52% Ni compared to 25% Ni in the bulk); and the third layer was Pt-enriched (87%). The unique Pt-skin Pt_3Ni structure and composition distribution resulted in distinctive electronic properties which affected the Pt–OH chemicals significantly and enabled unusually high catalytic activity. The Pt_3Ni (111)-skin surface was 10-fold more active for the ORR than the Pt(111) surface and 90-fold more active than the current state-of-the-art Pt/C catalysts for PEMFCs. However, it is still technically challenging to mass-produce nanocatalysts that mimic Pt_3Ni(111) Pt-skin structures and have similar electronic and morphological properties.

4.2 Polymeric Electrolyte Membrane with Inorganic Nanofillers

Nafion, a Dupont-brand sulfonated tetrafluroethylene-based polymer, is the most common electrolyte membrane in PEMFCs. However, Nafion functions as a good protonic conductor only in the presence of sufficient water and contributes to a significant proportion of the total PEMFC cost [1]. Nafion-based fuel cells are restricted to operate below 80°C and at high relative humidity levels. For transport

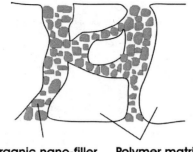

Inorganic nano-filler   Polymer matrix

FIGURE 5   Polymer membrane with inorganic nanofillers.

applications, low humidity levels and operating temperature in the range 90 to 160°C are desired. The benefits at such operating temperatures include fast electrode kinetics, high power density, improved catalyst tolerance to fuel impurities, and simple water management and cooling system structure [63,64].

Various polymeric nanocomposite membranes have been investigated to enhance water retention and protonic conductivity at low relative humidity and desired high temperatures [65–72]. An additional advantage of the composite membrane is the prevention of the crossover problem related to liquid fuel such as methanol, which becomes more severe at high temperatures.

Polymeric electrolyte composites feature inorganic particles (fillers) dispersed homogeneously in a polymeric matrix (Figure 5). The higher the interaction between the polymer and the filler particles, the greater the filler influence on the original property of the polymers. Due to their high specific surface, nanofillers modify the polymer characteristics to a greater extent than microfillers do.

Increased protonic conductivities at high temperatures and low relative humidity were reported in metal oxide nanoparticles (SiO_2, Al_2O_3, and TiO_2) filled in Nafion matrix. Surface functional groups on these oxide nanoparticles acted as water concentration centers to facilitate proton conduction. One typical synthesis approach of the polymeric composite was to impregnate SiO_2, TiO_2, and ZrO_2 nanoparticles in the ionomeric solution followed by drying and casting. The other approach was to grow in situ the fillers in the ionomer solution via hydrolysis of tetraethoxysilane (TEOS) or metal alkoxides such as $Ti(OEt)_4$ and $Zr(OPr)_4$. The conductivity of composite membranes loaded with 2.5 to 5 wt% silica was 2.7 to 5.8 times higher than that of Nafion 117 at 80°C and 100% relative humidity [65]. Direct methanol fuel cells based on this type of membrane were characterized as low methanol crossover and were able to operate up to 145°C with open-circuit voltages of 0.82 to 0.95 V and peak power densities of 150 to 240 mW/cm^2.

Alternative nanofillers are heteropolyacids (HPAs). HPAs have strong acidity and high-proton conductivity in their hydrated forms. Nafion membranes loaded with silicotungstic acid (STA), phosphotungstic acid (PTA), and phosphomolybdic acid (PMA) were systematically investigated on ionic conductivity, water uptake, tensile strength, and thermal behavior determinations [70–72]. At 80°C the current density at 0.600 V increased from 640 mA/cm^2 for Nafion 117 up to a maximum of 940 mA/cm^2 for PMA–Nafion 117 [70].

5   NANOSTRUCTURED ELECTRODE MATERIALS IN LI-ION BATTERIES

The Li-ion battery demonstrates a wise choice of materials, leading to a successful high-performance electrochemical energy storage system [2]. The first-generation Li-ion battery employed a graphitic carbon-based anode and a rock salt–structured lithium cobalt oxide (LiCoO$_2$) cathode. At the anode side, lithium ion can be reversibly inserted into graphite at a low electrode potential (less than 0.2 V vs. the Li$^+$/Li electrode potential) with a lithium storage capacity up to 350 mAh/g. The intercalation chemistry of lithium in the graphite prevented the formation of lithium dendrite (nonuniform lithium deposition) during charging. This significantly alleviated hazardous issues in rechargeable lithium batteries. At the cathode side, lithium can be reversibly inserted into lithium cobalt oxide around 4 V vs. Li$^+$/Li potential. The crystal structure of the cathode retained reasonable stability upon lithium removal and storage. The wise material selection and modern electrochemistry led to many advantages for Li-ion batteries: from high working voltage, high capacity, high energy density, to good discharge–charge cycle stability [2,73].

Although Li-ion batteries succeeded commercially in the portable electronics market, it is still pivotal to utilize Li-ion batteries in plug-in/hybrid electric vehicles (PHEVs). Improvements in electrode materials are in high demand to meet the target for transportation applications. Traditional approaches, relying on modifying bulk structures by tuning dopant compositions, are not able to bring about great advances, owing to thermodynamic and kinetic constraints. Exploiting nanostructural features in electrode materials in the emerging research area. In the following we provide a few examples of improved lithium storage and removal characteristics based on using nanostructured and/or nanocomposite electrode materials.

5.1   Nanostructured Cathode Materials

In the past 20 years, cathode materials for Li-ion batteries evolved from rock salt–structured LiMO$_2$ to spinel LiM$_2$O$_4$ to olivine phosphates LiMPO$_4$ (M = Fe, Co, Ni, Mn, etc.) toward the increasingly stringent demands for performance and safety [2,73–75].

Reducing the particle size to nanoscale has both negative and positive impacts on rock salt and spinel cathode. Microcrystalline LiCoO$_2$ has a capacity of around 140 mAh/g and good cyclability in the standard electrolyte and was used successfully in commercial Li-ion batteries. Nano-LiCoO$_2$, however, was excluded because it exhibited greater irreversible reaction with the electrolyte, worse stability on overcharge, and ultimately more safety problems than the same formula in micrometer forms [73]. LiMn$_2$O$_4$ is advantageous over LiCoO$_2$ in terms of low cost and low toxicity. In micro - Li$_x$Mn$_2$O$_4$, the range of x was confined to $0.5 < x < 1$, resulting in the lower capacity of 135 mAh/g. Beyond this capacity range the materials experienced a cubic-to-tetragonal phase transformation, together with significant anisotropic lattice change, and hence a dramatic capacity fading occurred [76,77]. Contrastingly, improved capacity and cyclability at room temperatures were observed in nanoscale $LiMn_2O_4$. Nanoscale $LiMn_2O_4$ could be cycled in the full capacity range ($0.5 < x < 2$) with over 99% capacity retention, because the nanodomains existing in the nanoparticles accommodated the slippage strain at the domain boundaries [78,79]. However, nano-$LiMn_2O_4$ showed a rapid capacity fade with discharge–charge cycling at 55°C, since the high surface area caused a significant increase in undesirable Mn dissolution [80].

Olivine-structured phosphate $LiMPO_4$ has a three-dimensional framework stabilized by the strong covalent bonds in PO_4^{3-} polyanions and therefore overcomes the structural weaknesses in the rock salt and spinel cathodes. The olivine-structured cathode has also exhibited better resistance against overcharge and thermal degradation than other oxide cathodes.

Olivine-structured phosphates $LiMPO_4$ experienced long-term development before it recently gained great momentum. In 1997, Padhi et al. [81,82] first reported a high theoretical capacity (170 mAh/g) via chemical removal of lithium from $LiFePO_4$. However, only 60% of the full capacity (around 100 mAh/g) was achieved in micro-$LiFePO_4$ and its capacity decreased remarkably at high rates. The very low intrinsic ionic and electronic conductivities in the olivine structure made it difficult to retrieve the full capacity from micro-$LiFePO_4$ electrochemically. Later, synthesis routes in the direction of reducing particle size, doping, or compositing with other elements had led to improvement in the electrochemical performances of $LiFePO_4$ [2,83–88]. A capacity of 156 mAh/g was achieved in nano-$LiFePO_4$ synthesized via the sol–gel approach [86]. To date, a capacity close to the theoretical value (170 mAh/g) and a high rate capability were reported consistently in $LiFePO_4$ coated with a few nanometer thick carbon layer [84,85,87,88]. The thin carbon layer, permeable for lithium ions, provides an efficient electronically conductive network.

Compared with other cathode materials, $LiFePO_4$/C-based nanocomposites exhibit a high reversible capacity, flat voltage, excellent cyclability, and a high rate capacity. In addition, $LiFePO_4$ is low-cost, non-toxic, and safe. These appealing properties make the $LiFePO_4$ family one of the most promising cathodes at present.

5.2 Nanostructured Carbon Anodes

There exist various forms of carbon: graphite, graphitizable soft carbon, nongraphitizable hard carbon, carbon nanotubes, and the monolayer graphene. Reversible Li-ion storage capacity and the electrode potential depend mainly on the carbon crystalline structure. Both theoretical and experimental studies have confirmed that the maximum amount of lithium that can be stored in the crystal structure of graphitic carbons is one lithium per six carbons (LiC_6), equivalent to a theoretical capacity of 372 mAh/g.

It was discovered that hard carbon fabricated by low-temperature thermal pyrolysis (500 to 1000°C) from various organics and polymers exhibited reversible capacities

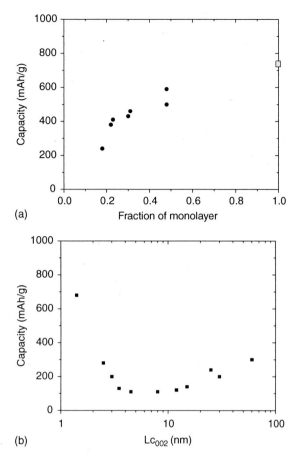

(a)

(b)

FIGURE 6 (a) Reversible capacity of hard carbon prepared between 900 and 1100°C as a function of their single-layer fraction. The square data at fraction value 1.0 are hypothesized for Li_2C_6 or 740 mAh/g. (Replotted from [98].) (b) Reversible capacity of various types of carbon electrodes as a function of crystallite thickness L_{c002}, determined by an x-ray (002) diffraction peak. (Replotted from [97].)

of 400 to 1000 mAh/g in the potential range of 0 to 2.5V [89–100]. According to Dahn et al. [97], hard carbons were made up predominantly of monolayered, bilayered, and trilayered graphene sheets arranged like a "house of cards." Lithium could adsorb on both sides of monolayer graphene sheets and onto internal surfaces of nanopores formed by the nano-graphene sheets, which would result in the formation of Li_2C_6 and hence a maximum capacity of 740 mAh/g. Figure 6 (a) presented capacities as a function of the fraction of monolayer graphene, estimated from small-angle x-ray scattering (SAXS) spectra, in hard carbons. The trend in capacity, changing with the single-layer fraction value, corroborated very well toward the hypothetical maximum value. Meanwhile, Endo et al. [98] studied the capacities of various hard carbons as a function of crystallite thickness L_{c002} [see Figure 6]. The lithium storage capacity increased to 700 mAh/g upon decreasing the carbon crystallite thickness to 1 nm. Endo suggested that a classical intercalation process occurred for L_{c002} greater than 10 nm, but a different doping and absorbing process occurred as the L_{c002} became smaller than 10 nm. The doping–absorbing process was enhanced upon decreasing the crystallite thickness to less than 1 nm, leading to the large capacity.

Lithium storage in carbon nanotubes was investigated extensively in the 1990s [101–105]. Capacities as high as $Li_{1.6}C_6$ were demonstrated in single-walled carbon nanotubes (SWNTs) and improved further, up to $Li_{2.7}C_6$, after ball-mill processing or chemical treatment. Shimoda et al. [103] systematically investigated lithium storage behaviors in closed and opened SWNTs of different lengths. It was observed that long closed SWNTs stored less than one lithium atom per six carbon atoms but short opened SWNTs could uptake more than two atoms lithium per six carbon atoms. First-principle computational studies [104,105] suggested that (1) the interaction between lithium and carbon nanotubes was mainly an ionic characteristic, due to the strong electron transfer; (2) the energy of lithium inside the tube was comparable to that outside the tube, implying that lithium could be stored in both nanotube exteriors and interiors leading to a high lithium storage density; and (3) lithium motion through perfect sidewalls was forbidden and lithium ions could enter tubes only through topological defects (at least nine-sided rings) or through the ends of open-ended nanotubes. Hence, the prolonged lithium diffusion length in long nanotubes determined the slow kinetics.

5.3 Alloy Anodes

Many metals or semiconductors, such as Si, Al, and Sn, can uptake lithium to form Li_xM alloys with very high stoichiometries [106–108]. Each Si can store up to 4.4 Li, rendering the highest theoretical specific capacity of 4200 mAh/g among all available materials. This value is more than 10 times higher than the theoretical limit of a graphitic carbon. However, these alloy anodes have difficulties in capacity retention upon charge–discharge cycling. Multiphase transitions and dramatic volume changes occur in the processes of alloying or dealloying with lithium. For examle, Si experiences 380% volume expansion to form $Li_{4.4}Si$. Repeated expansion and compression can generate enormous mechanical strains and eventually lead to pulverization and recrystallization of the host metal particles. Consequently, these active electrode particles lose electrical contacts with each other and with the current collector, resulting in dramatic capacity fading.

There have been many attempts to ameliorate the capacity–cycle life trade-off problems in the alloy anodes [106–120]. For the purpose of reducing the strain energy, reducing the size of the active material particle was investigated. However, particle reduction implied high surface area ready for side reaction with the liquid electrolyte. Nanocomposites composed of small electrode active particles supported with or protected by a less active or nonactive matrix were mostly adopted to preserve the integrity of the host structure matrix and to sustain continuity of electron-conducting paths [109–118]. The protective matrix provided a cushioning effect for

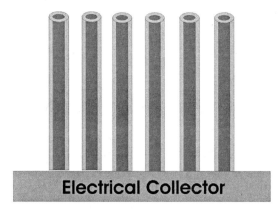

FIGURE 7   Hypothetical coated Si nanowire which may provide high reversible capacity and improve capacity retention.

particle expansion or shrinkage and prevented the electrolyte from reacting with the electrode active material. The matrix materials used to protect active particles (such as Si and Sn) included carbon, metal oxide, polymer, and ceramics. The Si–C nanocomposites, fabricated via pyrolysis of Si-containing resins, showed the capacity to be as high as 1000 mAh/g for more than 100 cycles [115]. Research has also been directed at alloying Si with Mg, Ni, Mn, and/or Cr, and so on. Si-based composite electrodes showed a considerable improvement in the cycling response. A nanostructured designed approach was demonstrated in silicon nanowires by Cui's group. Si nanowires reached a reversible capacity of 3000 mAh/g over 10 cycles [119,120]. The improved capacity retention was ascribed to the size confinement, which altered particle deformation and reduced fracturing. It was noted that significant degradation proceeded after 10 cycles, suggesting that Si nanowires could accommodate strain to a certain level but far from being adequate. Surface coating combining with atomic distribution of inactive metal may extend the cycle life of Si nanowires (see Figure 7).

5.4   Transition Metal Oxide Anodes

Lithium can be stored in transition metal oxide anodes via two different reaction mechanisms: insertion and displacement. The distinguished characteristic of the transition metal oxide from carbon and alloy anodes is the relatively high potential (1.0 to 2.0 V vs. lithium). The high anodic potential is beneficial in the following two aspects: (1) preventing the irreversible decomposition of electrolyte and polymer binder additive, which usually occurs at the voltage range of 0.6 to 0.8 V; and (2) avoiding the risk of lithium deposition at high discharge rates, which may lead to safety hazard.

Two common insertion oxide anodes are anatase TiO_2 and spinel $Li_4Ti_5O_{12}$. TiO_2 anatase has a higher lithium storage capacity of 220 mAh/g but a much lower rate

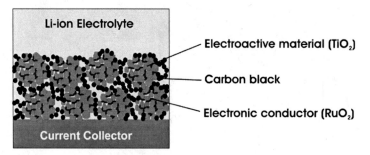

FIGURE 8   Model of a TiO_2–RuO_2 nanocomposite which has a three-dimensional nanonetwork superimposed on mesoporous particles [121].

capability than $Li_4Ti_5O_{12}$ (155 mAh/g), because TiO_2 is a poor electronic conductor and has a low lithium diffusion coefficient (ca. 2×10^{-15} cm^2/s). Efforts to facilitate lithium and electron transport in TiO_2 have been directed to nanostructured design [121–123]. An efficient mixed conducting three-dimensional nanonetwork superimposed on mesoporous particles was proposed by Guo and Maier (see Figure 8) [121]. The novel concept was demonstrated with TiO_2 : RuO_2 nanocomposite. Specifically, the mesoporous TiO_2 had an average pore size of a few nanometers. The pore channels were covered with nanocrystalline RuO_2, which allowed electronic conduction and quick lithium permeation. As the RuO_2 arrangement was highly porous but percolating, a large number of active triple-phase contacts were formed. The mixed conducting nanonetwork structure exhibited superior high rate capability.

Transition metal oxides such as CoO, CuO, Fe_3O_4, and MnO can be fully reduced by lithium to form metal and Li_2O

$$M_xO_y + 2y\mathrm{Li} \Leftrightarrow y\mathrm{Li}_2\mathrm{O} + x\mathrm{M} \qquad (16)$$

The displacement reaction results in a very high capacity [124–128]. For example, the theoretical capacity of Fe_3O_4 is 926 mAh/g. In microscale metal oxide the reverse reaction can hardly occur, dictated by thermodynamics. As the particle size was reduced to nanoscale, the products of the displacement reaction (i.e., metal nanocluster homogeneously dispersed in an amorphous Li_2O matrix) have much high surface energies, which facilitate the reverse reaction, according to Tarascon [124]. Similar phenomena were observed with sulfides, nitrides, and fluorides [129,130]. To provide further highly electron-conducting paths, nanoarchitectured anodes consisting of nanoparticle Fe_3O_4 coated on Cu nanorods were investigated by Tarascon's group [128]. Both improved cycle stability and rate capability were achieved.

5.5   Positive and Negative Impacts of Utilizing Nanostructured Electrodes

Moving toward nanostructured electrodes for Li-ion batteries brought about many advantageous aspects. These include (1) short path length for electron and Li-ion transports in insertion cathode or anode, leading to a high charge–discharge rate capability; (2) new storage mechanisms (e.g., displacement in oxide anode and adsorption in hard carbons) that did not occur in bulk materials; and (3) better accommodation of strain upon lithium insertion/removal, as in nano-silicon composites and nanowires, thereby improving cycle life.

The emergence of $LiFePO_4$ is a perfect example to illustrate the advantages of adopting nanomaterials processing. It opens up a new and wide route in the search for novel electrode materials. Nanoengineering will improve the dispensable properties extrinsically, such as high electronic conductivities and high lithium diffusion coefficients which are not possessed intrinsically in certain types of electrode materials [73].

Moving towards nanoscale may have some similar negative impacts. First, the inferior packing of particles resulting from the increased surface area will lower volumetric energy densities. Second, controlling the size and size distribution of the nanoparticles is still technically challenging. Third, fabrication of nanostructured composites means potentially more complex synthesis and expensive products. The greatest disadvantage of nanoparticulate electrode is the possibility of significant sidereactions with the electrolyte, leading to safety concerns and poor calendar life. Therefore, care must be taken in the development and application of nanostructured electrodes for Li-ion batteries.

6 GRAPHENE FOR ENERGY APPLICATIONS

6.1 Nano-Graphene Platelets: An Emerging Nanomaterial

Monolayer graphene is composed of carbon atoms forming a two-dimensional hexagonal lattice through strong in-plane sp^2 hybrid covalent bonds with a thickness of 0.335 nm. Multilayer graphenes are several graphene planes weakly bonded together through van der Waals forces in the thickness direction. Graphene may be viewed as a flattened sheet of a CNT, with a monolayer graphene corresponding to a SWCNT. Graphene may be oxidized to various extents, resulting in graphene oxide (GO) platelets. Nano-graphene platelets (NGPs), herein refer collectively to a platelet, sheet, or ribbon of monolayered or multilayered "pristine graphene" containing an insignificant amount of oxygen as well as GO of various oxygen contents.

For more than six decades, scientists have presumed that a monolayer graphene sheet could not exist in its free state, based on the assumption that its planar structure would be unstable thermodynamically. Surprisingly, several groups worldwide have recently succeeded in obtaining isolated graphene sheets [131–139]. In 2002, Jang et al. [131,132] reported a new class of nanomaterials now commonly referred to as NGPs. Monolayer graphene, discovered in 2004 by Novoselov et al., exhibited an electron mobility 100 times faster than that in silicon [9,10,133]. The major discovery has spurred rapidly growing global interests in studying graphene properties, developing production processes, and exploring novel applications.

TABLE 1   Comparison of Physical Properties of SWCNTs and NGPs

Property	Single-Walled CNTs	Nano-Graphene Platelets
Specific gravity	0.8 g/cm^3	2.2 g/cm^3
Elastic modulus	~1 TPa (axial direction)	~1 TPa (in-plane)
Intrinsic strength	100 GPa	Up to 130 GPa
Resistivity	5–50 $\mu\Omega \cdot$ cm	5–50 $\mu\Omega \cdot$ cm (in-plane)
Thermal conductivity	Up to 3000 W/m $\cdot$ K (axial)	5300 W/m $\cdot$ K (in-plane); 6–30 Wm $\cdot$ K (c-axis)
Magnetic susceptibility	22 $\times$ 10^6 emu/g (radial); 0.5 $\times$ 10^6 emu/g (axial)	22 $\times$ 10^6 emu/g ($\perp$ to plane); 0.5 $\times$ 10^6 emu/g (||to plane)
Thermal expansion	Negligible in the axial direction	-1×10^{-6} K^{-1} (in-plane); 29 $\times$ 10^{-6} K^{-1} (c-axis)
Thermal stability	>700°C (in air); 2800°C (in vacuum)	450–650°C (in air)
Specific surface area	Typically 10–200 m^2/g, up to 1300 m^2/g	Typically 100–1000 m^2/g, up to > 2600) m^2/g

NGPs are predicted to have a range of unique physical, chemical, and mechanical properties. For example, monolayer graphene was discovered to possess the highest intrinsic strength [133,134], thermal conductivity, and specific surface area [135] of all existing materials. Multiple-layer graphene platelets also exhibit unique and useful behaviors [140–142]. NGPs are superior to CNTs in several highly desirable properties: specific surface area, dispersability in polymers, thermal conductivity, and intrinsic strength. Selected properties of NGPs, in parallel with the values of CNTs, are listed in Table 1.

The dramatic advancements in graphene science and technology have opened up unlimited opportunities for the development of nano-enhanced products and other manufacturing innovations. In the microelectronics industry, graphene is considered the next-generation materials in replacement of silicon wafer for high-speed electronic devices. In chemical and materials manufacturing industries, NGP is a much lower-cost alternative to CNT for highly functional composites. Graphene-based materials have found extensive applications in alternative energy. Graphene-based materials have been investigated to serve as an electrode in Li-ion batteries [15–17]. The functionized graphene nanosheets exhibited high specific capacitances in electrochemical double-layer capacitors [18,19]. Graphene films, with a high conductivity of 550 S/cm and a transparency of more than 70% in the wavelength range 1000 to 3000 nm, were demonstrated successfully as the window electrodes for solid-state dye-sensitized solar cells [20]. NGPs also found their potential applications in fuel cells via serving as bipolar plates and active electrode catalyst supporters embedded with Pt nanoparticles [21–24]. Extensive investigations on graphene-based nanocomposites with improved functionalities are necessary to realize our clean energy dream.

6.2   A Brief Overview of Nano-Graphene Production Methods

There are four major different approaches to produce NGPs, reviewed recently by Jang and Zhamu [7] and by Park and Ruoff [143].

Approach 1: Chemical Formation and Reduction of Graphite Oxide Platelets   The first approach basically entails three distinct procedures: first expansion (oxidation or intercalation), further expansion (or "exfoliation"), and separation. Natural graphite is treated with an intercalant and an oxidant (e.g., concentrated sulfuric acid and nitric acid, respectively) to obtain a graphite intercalation compound (GIC). With an intercalation or oxidation treatment, the inter-graphene spacing is increased from 0.335 nm to >0.6 nm. This is the *first expansion* stage experienced by the graphite material. The GIC obtained is then subjected to exfoliation using either a thermal shock exposure or a solution-based exfoliation approach. In the thermal shock exposure approach, the GIC is exposed to a high temperature (typically, 800 to 1050°C) for 15 to 60 s for the formation of further expanded graphite worm. In both the solution and heat-induced exfoliation approaches, the resulting products are graphite oxide or graphene oxide platelets, not pristine graphene. Both steps require very tedious washing and purification steps [144–157]. Graphene dispersions with concentrations up to about 0.01 mg/mL may be produced by dispersion and exfoliation of graphite in highly selective solvents such as N-methylpyrrolidone [157]. The expanded graphite worm is then subjected to a flake *separation* treatment using mechanical shearing or ultrasonication in water.

Approach 2: Direct Formation of Pristine Nano-Graphene Platelets   Single- and multilayer graphene structures can be obtained from a polymer or pitch precursor via carbonization [131,132]. Novoselov and co-workers [9] prepared single-sheet graphene by removing graphene from a graphite sample one sheet at a time using a "Scotch-tape" method. A scanning probe microscope was used [158] to manipulate graphene layers at the step edges of graphite and etched HOPG, respectively, with the goal of fabricating ultrathin nanostructures. Mack et al. [159,160] developed a NGP production process via intercalating graphite with potassium melt followed by exfoliating the K-graphite intercalation compound in alcohol. The process must be carefully conducted in a vacuum or an extremely dry glove box environment, since pure alkali metals such as potassium and sodium are extremely sensitive to moisture and pose an explosion danger.

Approach 3: Epitaxial Growth and Chemical Vapor Deposition on Inorganic Surfaces   Small-scale production of ultrathin graphene sheets on a substrate can be obtained by thermal decomposition–based epitaxial growth [161], a laser desorption–ionization technique [162], and chemical vapor deposition (CVD) [163–165]. Epitaxial films of graphite with only one or a few atomic layers are of technological and scientific significance, due to their peculiar characteristics and great potential as a device substrate [166–168].

Approach 4: Synthesis of Graphene from Small Molecules Yang et al. [151] synthesized nano-graphene sheets with lengths of up to 12 nm using a method that began with Suzuki–Miyaura coupling of 1,4-diiodo-2,3,5,6-tetraphenylbenzene with 4-bromophenylboronic acid. The resulting hexaphenylbenzene derivative was further derivatized and ring-fused into small graphene sheets. This is a slow process that thus far has produced very small graphene sheets. Choucair et al. [169] reported the direct chemical synthesis of carbon nanosheets based on ethanol and sodium. The intermediate solid of the reaction was pyrolyzed followed by mild sonication, yielding a fused array of graphene sheets.

Major Issues Associated with the Current Graphene Production Processes It is of technological significance to develop an environmentally benign process for mass-producing pristine nano graphenes that are structurally smooth and have exceptional electrical and thermal conductivities. The processes described above are not amenable to the large-scale production. The GO nanoplatelets exhibit electrical conductivities several orders of magnitude lower than the conductivities of pristine NGP. Even after chemical reduction, the GO still has a much lower conductivity than pristine NGPs. The preparation of intercalated graphite usually involves undesirable chemicals such as sulfuric acid and potassium permanganate. The reduction procedure often involves hydrazine. Additionally, both heat- and solution-induced exfoliation approaches require a very tedious washing and purification step. The GO nanoplatelets, after a high degree of chemical reduction, are able to recover some of the properties of pristine graphite but are no longer dispersible in water and most organic solvents. Furthermore, their structures are severely damaged. The NGPs produced by approaches 2 and 3 are normally pristine graphene and highly conducting. Jang, Zhamu, and co-workers [170–193] have successfully developed several mass production processes that are capable of producing several kilograms per day of ultrathin NGPs (including single-layer graphene).

6.3 Nano-Graphene Platelets for Lithium-Ion Battery Applications

Lithium Storage Characteristics in NGPs Computations based on density function theory have been performed to elucidate the interaction between lithium and graphene and the nature of electron transfer in such a structure [194–200]. Depending on differences in detailed computational approaches, the binding distances varied from 1.60 to 2.10 nm, and binding energies were in the broad range 0.5 to 1.7 eV. Valencia et al. [196] reported a strong interaction between the lithium ion and the cloud π electrons in the graphene. Wang et al. [200] computed two lithium atoms on the each side of graphene manolayer structure. It was found that the most stable configuration was lithium at symmetric hydrogen-sites with graphene sheet as a mirror, resulting in the formation of LiC_3. The Fermi level of graphene for the two-side lithium configuration was higher than that for the one-side lithium adsorption resulting about 0.36 charge transfer from the two lithium atoms to every six carbon atoms. Li-ion diffusion on a model surface of $C_{96}H_{24}$ was investigated by means of a direct molecular orbital dynamics method. Tachikawa and Simigu [199] showed that the lithium ion became mobile on the surface above 250 K. The diffusion coefficient of Li^+ in the model graphene cluster was 9×10^{-11} m^2/s at 300 K, which was several orders of magnitude higher than on graphite (10^{-15} to 10^{-12} m^2/s).

The computational results and extensive experimental investigations on hard carbons and carbon nanotubes direct us to a reasonable hypothesis of high lithium storage capacity in NGPs. Further, NGPs circumvent cross-linkage in hard carbons and eliminate slow diffusion in the long tube interior. Therefore, graphene-based materials are promising anode alternatives for Li-ion batteries. A large reversible specific capacity up to 784 mAh/g was reported recently in graphene nanosheets incorporated with CNTs and C_{60} [15]. Cheekati et al. studied pristine NGP (less than 0.1 at% of oxygen) and oxide NGP (containing over 1 at% of oxygen) nanopowders [201]. The first discharge capacity was 890 mA/h and the charge capacity 560 mAh/g on pristine NGP. High discharge and charge capacities of 1086 and 780 mAh/g were achieved in oxide NGPs during the first cycle. The following 10 cycles showed a stable reversible capacity of 800 mAh/g with no significant loss.

To date, various approaches to fabricating graphene have been demonstrated, which were reviewed briefly in Section 6.2. For use as the anode in Li-ion batteries, graphene nanosheets were typically fabricated starting from graphite powders, followed by oxidation exfoliation and reduction (approach 1).

Si/Nano-Graphene Composite Anode   Si-based composite anodes, discussed in Section 5.3, usually have deficiencies: for example, less than satisfactory reversible capacity (per gram of the composite material), poor cycling stability, high irreversible capacity, ineffectiveness in reducing the internal stress or strain during Li-ion insertion and extraction steps, and other undesirable side effects. Ideally, the protective material should meet all the following requirements: (1) have high strength and stiffness so as to refrain the electrode active material particles from expanding to an excessive extent when lithiated; (2) have high fracture toughness to avoid disintegration during repeated cycling; (3) be inactive with the electrolyte but a good Li-ion conductor; and (4) providing an insignificant number of defect sites that trap lithium ions irreversibly.

Zhamu and Jang [202] developed hybrid graphene/Si nanostructured materials. In such a hybrid material, nano-Si was the primary lithium storage material that contributed to the high capacity. Although Si particles would expand and shrink during lithiation and delithiation, the strong but flexible graphene platelets surrounding the nano Si particles were capable of cushioning the stress and strain. Meanwhile, the graphene platelets ensured good electric contacts between Si particles as well as between Si particles and the current collector (e.g., Cu). With an ultrahigh length-to-thickness aspect ratio (up to 50,000) and low thickness (e.g., just one or a few atomic layers), a very small amount (in mass and volume) of NGPs was sufficient to provide an electrical network. In addition, through judicious design and preparation, the interaction between nano-Si particles and graphene sheets would help to prevent agglomeration and sintering of Si particles during the charge–discharge cycles.

Recently, an innovative anode material featuring spherical nanoparticle or nanowire electroactive material (e.g., Si with a diameter below 500 nm) dispersed in a protective carbon matrix reinforced with NGPs, was patented by Zhamu and Jang

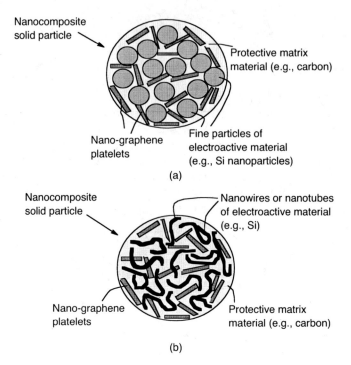

FIGURE 9   Si nanoparticles (a) or Si nanowires (b) dispersed in an NGP-reinforced carbon matrix.

[203], as shown in Figure 9. The nanocomposite itself was in a form of a fine spherical particle (<5 μm). Such a shape was conducive to the formation of an electrode with a high tap density, resulting in a greater amount of active material per unit volume under identical coating and laminating conditions during electrode fabrication. The NGP-reinforced carbon matrices protected Si nanoparticles from recrystallization and pulverization, rendering their capacity retention as high as 2000 mAh/g (per gram of the total composite weight) after 350 cycles.

Such a graphene-based nanocomposite platform technology has the following highly desirable advantages. NGPs have a high thermal conductivity ($\sim$5300 W/m · K), implying a high heat dissipation rate. This is an important feature since the charge–discharge operations of a battery produce a great amount of heat. Without a fast heat dissipation rate, the battery cannot be charged or discharged at a high rate. The high strength of NGPs ($\sim$130 GPa) significantly improves the overall strength and fracture toughness (resistance to cracking) of a protective matrix (carbon), which is otherwise weak and brittle. NGPs have an electrical conductivity value up to 20,000 S/cm which is several orders of magnitude higher than that of amorphous carbons (typically, 0.001 to 1 S/cm). Additionally, amorphous carbons intrinsically have an excessive number of defect sites that irreversibly trap or capture lithium atoms or ions, thereby significantly reducing the amount of lithium that can be shuttled back and forth between the anode and the cathode. By adding a certain number of NGPs, one can effectively reduce the proportion of amorphous carbon (hence reducing the level of irreversibility). Although conventional graphite particles can be added to reduce the amount of the carbon matrix, experimental data have demonstrated that these graphite particles do not improve the cracking resistance of carbon. Therefore, the resulting anode composite materials (i.e., Si and graphite particles dispersed in the amorphous carbon matrix) do not provide a long cycle life. In contrast, by replacing graphite particles with NGPs, which are of much higher strength, the useful cycle life was significantly increased and the reversible capacity was maintained at an unprecedented level. NGP-reinforced protective matrix materials effectively cushioned the large volume changes of electroactive materials such as Si. Also, the matrix itself becomes more resistant to crack initiation and propagation.

In sum, NGP-based nanocomposite particles impart the following highly desirable attributes: high reversible capacity, low irreversible capacity, small particle sizes for high-rate capacity, compatibility with commonly used electrolytes, and long charge–discharge cycle life.

Nano Graphene as a Supporting Substrate for an Anode Active Material   Concepts similar to those above were reported on combinations of graphene and an anode active material [204–215]. Some of the significant observations on graphene-supported anode active materials are summarized in the following examples. Paek et al. [204] studied the cyclic performance and lithium storage capacity of SnO_2–graphene nanoporous electrodes with a three-dimensionally delaminated flexible structure. The SnO_2–GNS (graphene nanosheets or NGPs) exhibited a reversible capacity of 810 mAh/g and enhanced cycling performance compared with that of the bare SnO_2 particles. Yao et al. [204,205] studied in situ chemical synthesis of SnO_2–graphene nanocomposite and achieved a reversible lithium storage capacity of 765 mAh/g in the first cycle and enhanced cyclability. Chou et al. [208] studied the Si–graphene composite, which maintained a capacity of 1168 mAh/g and an average coulombic efficiency of 93% up to 30 cycles.

Wang et al. [209] investigated self-assembled TiO_2–graphene hybrid nanostructures for enhanced Li-ion insertion. The results indicated that the specific capacity of the hybrid was more than doubled at high charge rates, compared with the pure TiO_2 phase. The improved capacity at high charge–discharge rates was attributed to the increased electrode conductivity afforded by a percolated graphene network embedded into the metal oxide electrodes. This result further demonstrated the superiority of using nano-graphene sheets as a supporting substrate, as they provide a effective network of electron-conducting paths. Wu et al. investigated three-dimensional flexible nanostructured cobalt oxide/graphene composites as an anode material [212]. It was concluded that ultrathin elastic graphene layers not only provided a carrier for the particles dispersed therein and high conductive backbone, but also effectively prevented the volume expansion or contraction and aggregation of nanoparticles during the lithium charge–discharge process. The resulting batteries exhibited significantly improved cycling performance.

Nano Graphene as a Supporting Substrate for a Cathode Active Material   Nano graphene for battery applications is not limited to serving as a supporting substrate for an anode active material. Actually, graphene has been used as a support for a cathode active material [213–216]. Ding et al. [216] reported a novel route for synthesizing $LiFePO_4$–graphene hybrids with high electrochemical performance. The graphene nanosheets were used as scaffolds with $LiFePO_4$ nanoparticles grown on the graphene surface. The hybrid material with only 1.5 wt% of graphene exhibited an initial discharge capacity of 160 mAh/g at 0.2 C and retained a capacity of 110 mAh/g even at a high rate of 10 C.

Some Thoughts on NGP-Based Electrodes   At this infancy stage, systematic experimental and fundamental mechanistic understanding of Lithium storage in NGPs has yet to be addressed. Many fundamental questions are unsolved, including (1) the maximum lithium storage capacities in mono-, bi-, and/or trilayered NGPs; (2) the rates of lithium absorption, diffusion, and desorption on NGPs; (3) the relationships between NGP dimensionalities (thickness and basal area) and lithium diffusion coefficients; (4) the formation mechanism and composition of the solid electrolyte interphase (SEI) layer on NGPs; (5) lithium storage characteristics and mechanism in nanostructured (e.g., NGP–supported or graphene-coated) NGP–composite anodes; (6) mass production of optimal nanostructured NGP–composite anodes at low cost; and so on. Answers to these questions are indispensable for gaining insights into high-capacity lithium storage mechanisms in the development of novel anode materials.

The current progress in graphene technologies facilitates our direct experimental investigations on lithium storage monolayer graphene, the building block of all types of carbons. Moreover, the bandgap of graphene can be tuned via adjusting the width of graphene ribbon, the number of layers, the electrical field, doping with foreign elements, and sublayer materials [217–220]. The well-controlled structure, dimensionality, and unique bandgap tunability grant graphene an ideal platform for accurate monitoring of lithium storage processes and investigation of lithium storage mechanisms, kinetics, and SEI formation as a function of layer number, basal area, foreign element, bandgap, and so on. The information may aid developments in anode material chemistry.

6.4   Nano Graphene for Supercapacitor Electrodes

Supercapacitor Electrode Materials   Supercapacitors have two energy storage mechanisms: electrical double-layer (EDL) capacitance and pseudocapacitance [3]. Current EDL capacitors contain carbon-based materials that have high surface areas as the electrode materials, with the capacitance coming from the charge accumulated at the electrode–electrolyte interface. Activated carbon, mesoporous carbon, and carbon nanotubes usually exhibit good stability, but the EDL capacitance values are limited by their microstructures [3,221]. For very high-surface-area carbons, typically only about 10 to 20% of the "theoretical" capacitance was observed. This disappointing performance is due to the presence of micropores that are inaccessible by the electrolyte, wetting deficiencies of electrolytes on the electrode surface, and/or the inability of a double layer to form successfully in pores. In contrast, the pseudocapacitors use conducting polymers or metal oxides as electrode materials, which undergo reversible faradaic redox reactions [222]. Conjugate-chain-conducting polymers such as polyanilines (PANIs), polypyrroles (PPYs), and polythiophenes [223] showed high pseudocapacitances but poor stabilities during the charge–discharge cycling. Further, the typical response times of pseudocapacitors are significantly longer than those of EDL capacitors and the capacitance decay rate is high upon charge–discharge cycling, which are undesirable in many supercapacitor applications [224–229].

Carbon nanotubes (CNTs) are being considered as feasible supercapacitor electrode materials, due to their unique properties and structure, which include high surface area, high conductivity, and chemical stability. Capacitance values from 20 to 180 F/g have been reported, depending on the CNT purity and the electrolyte [230–235] as well as on the specimen treatment, such as CO_2 physical activation [236], KOH chemical activation [237,238], or exposure to nitric acid [239–242], fluorine [243], or ammonia plasma [244]. The modification of CNTs with conducting polymers increased the capacitance of the composite resulting from the redox contribution of the conducting polymers [245–262]. In the CNT–conducting polymer composite, CNTs are electron acceptors while the conducting polymer serves as an electron donor. A charge-transfer complex is formed between CNTs in their ground state and aniline monomers [247]. The CNT–conducting polymer composite prepared by polymerization of pyrrole on the nanotubes with ammonium persulfate as an oxidant exhibited a capacitance value up to 180 F/g [248]. A SWCNT–polypyrrole composite exhibited a capacitance value up to 260 F/g [249]. The capacitance of an unoriented MWCNT–polypyrrole composite was found to be at least twice as high as that of either component alone [250,251]. The supercapacitor behavior of aligned arrays of CNTs coated with polypyrrole was also reported [252,253]. The capacitance of the composite was more than 200 F/g in potassium chloride and t-butylammonium bromide solutions [253]. CNT matrices provided a mesoporous scaffold on which a porous layer of polypyrrole was electrodeposited to achieve high charge dynamics for improved supercapacitor performance [254]. A MWCNT–poly(3-methylthylthiophene) composite–based supercapacitor in 1 M $LiClO_4$ acetonitrile solution exhibited a specific capacitance of about 80 F/g [255]. The capacitance of the SWCNT-PANI composite fabricated by in situ electrochemical polymerization was 310 F/g because the complex structure in this composite offered more active sites for faradaic reactions [256]. These recent studies clearly suggest that nanocomposites, composed of high-surface-area nanofillers and an intrinsically conductive polymer, provide an effective approach to producing superior supercapacitor electrodes.

Nano Graphene as a Supercapacitor Electrode Material An outstanding characteristic of graphene is its exceptionally high specific surface area; a single graphene sheet provides approximately 2675 m^2/g (that is accessible by liquid electrolyte). The intrinsic capacitance of graphene was recently found to be 21 $\mu F/cm^2$ [257], which sets the upper limit of double-layer capacitance for all carbon-based materials. This study asserted that graphene-based EDL supercapacitors would have a double-layer capacitance value of up to 550 F/g provided that the entire surface area was fully utilized.

The EDL capacitance of a mesoporous composite composed of NGPs (one to four graphene layers) bonded by a carbonized polyacrylonitrile (PAN) binder has already reached 150 F/g [258]. With a small amount of polypyrrole serving as a redox pair partner, the specific capacitance was increased to 200 F/g [259]. In early 2008, graphene-based supercapacitors were found by Vivekchand et al. [260] to exhibit an EDL-based specific capacitance of 117 F/g in aqueous H_2SO_4. By using an ionic liquid with an operating voltage of 3.5 V (instead of 1 V as in the case of aqulous H_2SO_4), the specific capacitance and energy density were 75 F/g and 31.9 Wh/kg, respectively. Stoller et al. [261] reported specific capacitances of 135 and 99 F/g in aqueous and organic electrolytes, respectively. A capacitance of 205 F/g, a power density of 10 kW/kg, and an energy density of 28.5 Wh/kg in an aqueous electrolyte solution were obtained by Wang et al. [262]. Further, the supercapacitor devices exhibit excellent long cycle life, exhibiting 90% specific capacitance retention after 1200 cycles. To further study the electrical contact between the graphene electrode and the current collector, Wang et al. [262] conducted electrochemical impedance spectroscopy (EIS) measurements of the supercapacitor. EIS analyses indicated that the graphene-based electrode has a short ion diffusion path, facilitating the efficient access of electrolyte ions to the graphene surface.

Zhang and Zhao [263] prepared chemically modified graphene and PANI nanofiber composites by in situ polymerization of aniline monomer in the presence of graphene oxide under acid conditions. The chemically modified graphene and PANI nanofibers were found to form a uniform nanocomposite with the PANI fibers absorbed on the graphene surface and/or filled between the graphene sheets. This uniform structure and the high electric conductivity of the nanocomposite, when used as supercapacitor electrodes, enabled high specific capacitance (480 F/g at a current density of 0.1 A/g) and good cycling stability during the charge–discharge process. When a higher current density of 1 A/g was used, the specific capacitance dropped to a typical range of 130 to 200 F/g in 2 M H_2SO_4 electrolyte. The cyclic voltammograms of the graphene–PANI nanocomposite show the current–voltage curves characteristic of a pseudocapacitor, suggesting that graphene and PANI form a redox pair.

Future Research Directions   Individual nano-graphene sheets have a great tendency to restack with each other, significantly reducing the specific surface area that is accessible to the electrolyte in a supercapacitor electrode. Most of the studies cited above ignored the significance of this graphene sheet overlap issue, which may be illustrated as follows. For a nano-graphene platelet with dimensions of l (length) $\times$ w (width) $\times$ t (thickness) and density ρ, the estimated surface area per unit mass is $S/m = (2/\rho)(1/l + 1/w + 1/t)$. For the monolayered graphene, with $\rho \cong 2.2$ g/cm^3, $l = 100$ nm, $w = 100$ nm, and $t = 0.34$ nm (single layer), the S/m value is 2675 m^2/g, which is much greater than that of most commercially available carbon black or activated carbon materials used in the state-of-the-art supercapacitors. If two single-layer graphene sheets stack to form a double-layer NGP, the specific surface area is reduced to 1345 m^2/g. For a three-layer NGP, $t = 1$ nm and $S/m = 906$ m^2/g. If more layers were stacked together, the specific surface area would be further significantly reduced. These calculations suggest that it is essential to find a way to prevent individual graphene sheets from fully restacking.

More important, if graphene sheets do partially restack, the resulting multilayer structure must have interlayer pores of adequate sizes. These pores must be sufficiently large to allow accessibility of the electrolyte and to enable the formation of double-layer charges. Earlier attempts to keep graphene sheets from fully restacking were not particularly effective and did not provide much insight into the double-layer formation mechanisms. The minimal pore sizes required in graphene-based electrodes for both aqueous and nonaqueous electrolytes remain unknown. No prior work was directed at addressing the issues of pore accessibility and surface compatibility of graphene by organic and ionic liquid electrolytes. This is a critically important issue, due to the fact that water-based electrolytes could not be operated at a voltage higher than 1.2 V. A higher working voltage implies a much higher specific energy density.

7 CONCLUDING REMARKS

This chapter demonstrates a few selected examples of nanomaterials used in electrochemical energy conversion and storage systems. Nano-thin films, nanocomposites, and nanostructured materials significantly change electrode and solid–polymer electrolyte properties, and consequently their performances in devices for energy storage and conversion. In some cases the effects may be the simple consequences of reduction in size. In others the effects may be more subtle, involving internally nanostructured materials or nanostructures with particular morphologies. Space-charge and dopant segregation effects result in substantial improvements in ionic conduction and can be utilized in ultrathin-film configurations. There is a profound effect from spatial confinement and contribution of surfaces, due to small particle size, on many of the properties of materials; this challenges us to develop new theory or at least to adapt and develop theories that have been established for bulk materials. We also foresee that this subject will bring together the disciplines of material chemistry and surface science, as both are necessary to understand nanomaterials. Nanomaterials offer various unique properties or combinations of properties as electrodes and electrolytes in a range of energy devices. Nanoscience holds the key to fundamental advances in energy conversion and storage.

REFERENCES

1. R. O'Hayre, S.-W. Cha, W. Colella, and F. B. Prinz, *Fuel Cell Fundamentals*, 2nd ed., Wiley, Hoboken, NJ, 2009.

2. G. A. Nazri and G. Pistoia, *Lithium Batteries: Science and Technology*, Kluwer Academic, Dordrecht, The Netherlands, 2004.

3. B. E. Conway, *Electrochemical Supercapacitors: Scientific Fundamentals and Technological Applications*, SpringerVerlag, New York, 1999.

4. A. S. Arico, P. Bruce, B. Acrosati, J.-M. Tarascon, and W. van Schalkwijk, Nanostructured materials for advanced energy conversion and storage devices, *Nat. Mat.*, 4, 366–377 (2005).

5. B. Peng, and J. Chen, Functional materials with high-efficiency energy storage and conversion for batteries and fuel cells, *Coord. Chem. Rev.*, 253, 2805–2813 (2009).

6. Y.-G. Guo, J.-S. Hu, and L.-J. Wan, Nanostructured materials for electrochemical energy conversion and storage devices, *Adv. Mater.*, 20, 2878–2887 (2008).

7. B. Z. Jang and A. Zhamu, Processing of nanographene plateltes (NGPs) and NGP nanocomposites: a review, *J. Mater. Sci.*, 43, 5092–5101 (2008).

8. S. Stankovich, D. A. Dikin, G. Dommett, K. Kohlhaas, E. J. Zimney, E. Stach, R. Piner, S. T. Nguyen, and R. S. Ruoff, Graphene-based composite materials, *Nature*, 442, 282–286 (2006).

9. K. S. Novoselov, A. K. Geim, S. V. Morozov, D. Jiang, S. V. Dubonos, I. V. Grigorieva, and A. A. Firsov, Electric field effect in atomically thin carbon films, *Science*, 306, 666–669 (2004).

10. A. K. Geim and K. S. Novoselov, The rise of graphene, *Nat. Mater.*, 6, 183–191 (2007).

11. C. M. Jannik, A. K. Geim, M. I. Katsnelson, K. S. Novoselov, T. J. Booth, and S. Roth, The structure of suspended graphene sheets, *Nature*, 446, 60–63 (2007).

12. A. A. Balandin, S. Ghosh, W. Bao, I. Calizo, D. Teweldebrhan, F. Miao, and C. N. Lau, Superior thermal conductivity of single-layer graphene, *Nano Lett.*, 8, 902–907 (2008).

13. X. Sun, Z. Liu, K. Welsher, J. T. Robinson, A. Goodwin, S. Zaric, and H. Dai, Nanographene oxide for cellular imaging and drug delivery, *Nano Res.*, 1, 203–212 (2008).

14. C. Stampfer, F. Schurtenberger, E. Molitor, J. Guttinger, T. Ihn, and K. Ensslin, Tunable graphene single electron transistor, *Nano Lett.*, 8, 2378–2383 (2008).

15. E. J. Yoo, J. Kim, E. Hosono, H. Zhou, T. Kudo, and I. Honma, Large reversible Li storage of graphene nanosheet families for use in lithium ion batteries, *Nano Lett.*, 8, 2277–2283 (2008).

16. G. Wang, X. Shen, J. Yao, and J. Park, Graphene nanosheets for enhanced lithium storage in lithium ion batteries, *Carbon*, 47, 2049–2053 (2009).

17. M. Liang and L. Zhi, Graphene-based electrode materials for rechargeable lithium batteries, *J. Mater. Chem.*, 19, 5871–5878 (2009).

18. M. D. Stoller, S. Park, Y. Zhu, J. An, and R. S. Ruoff, Graphene-based ultracapacitors, *Nano Lett.*, 8, 3498–3502 (2008).

19. S. R. C. Vivekchang, C. S. Rout, K. S. Subrahmanyam, A. Govindraraj, and C. N. Rao, Graphene-based electrochemical supercapacitor, *J. Chem. Sci.*, 120, 9–13 (2008).

20. X. Wang, L. Zhi, and K. Mullen, Transparent, conductive graphene electrodes for dye-sensitized solar cells, *Nano Lett.*, 8, 323–327 (2008).

21. Y. Si and E. T. Samulski, Exfoliated graphene separated by Pt nanoparticles, *Chem. Mater.*, 20, 6792–6797 (2008).

22. R. Kou, Y. Shao, D. Wang, M. H. Engelhard, J. H. Kwak, J. Wang, V. V. Viswanathan, C. Wang, Y. Lin, I. A. Aksay, and J. Liu, Enhanced activity and stability of Pt catalysts on functionalized graphene sheets for electrocatalytic oxygen reduction, *Electrochem. Commun.*, 11, 954–957 (2009).

23. B. Z. Jang, A. Zhamu, and J. Guo, Process for producing carbon-cladded composite bipolar plates for fuel cells, USPTO, Appl. 20080149900 (2008).

24. L. Song, J. Guo, A. Zhamu, and B. Z. Jang, Highly conductive nanoscale graphene plate nanocomposites and products, USPTO, Appl. 20070158618 (2007).

25. A. J. Bard and L. R. Faulkner, *Electrochemical Methods: Fundamentals and Applications*, Wiley, Hoboken, NJ, 2001.

26. D. Linden and T. B. Reddy, *Handbook of Batteries*, 3rd ed., McGraw-Hill, New York, 2001.

27. M. Winter, R. J. Brodd, What are batteries, fuel cells, and supercapacitors? *Chem. Rev.*, 104, 4245–4269 (2004).

28. B. C. H. Steele and A. Heinzel, Materials for fuel-cell technologies, *Nature (London)*, 414, 345–352 (2001).

29. B. P. Brandon, S. Skinner, and B. C. H. Steele, Recent advances in materials for fuel cells, *Annu. Rev. Mater. Res.*, 33, 183–213 (2003).

30. J. M. Ralph, A. C. Schoeler, and M. Krumpelt, Materials for lower temperature solid oxide fuel cells, *J. Mater. Sci.*, 36, 1161–1172 (2001).

31. T. Hibino, A. Hashimoto, T. Inoue, J. Tokuno, S. Yoshina, and M. Sano, A low-operating-temperature solid oxide fuel cell in hydrocarbon–air mixtures, *Science*, 288, 2031–2033 (2000).

32. S. Souza, S. J. Visco, and L. C. De Jonghe, Thin-film solid oxide fuel cell with high performance at low-temperature, *Solid State Ionics*, 98, 57 (1997).

33. X. Chen, N. J. Wu, L. Smith, and A. Ignatiev, Thin-film heterostructure solid oxide fuel cells, *Appl. Phys. Lett.*, 84, 2700 (2004).

34. H. Huang, M. Nakamura, P. Su, R. Fasching, Y. Saito, and F. Prinz, High-performance ultrathin solid oxide fuel cells for low-temperature operation, *J. Electrochem. Soc.*, 154, B20–B24 (2007).

35. J. H. Shim, C.-C. Chao, H. Huang, and F. B. Prinz, Atomic layer deposition of yttria-stabilized zirconia for solid oxide fuel cells, *Chem. Mater.*, 19, 3850–3854 (2007).

36. N. Sata, K. Eberman, K. Ebel, and J. Maier, Meoscopic fast ion conduction in nanometre-scale planar heterostructures, *Nature (London)*, 408, 946 (2000).

37. Y. M. Chiang, E. B. Lavik, I. Kosacki, H. L. Tuller, and J. Y. Ying, Defect and transport properties of nanocrystalline CeO_{2-x}, *Appl. Phys. Lett.*, 69, 185 (1996).

38. T. Suzuki, I. Ksacki, and U. H. Anderson, Defect and mixed conductivity in nanocrystalline doped cerium oxide, *J. Am. Ceram. Soc.*, 85, 1492–1498 (2002).

39. C. A. Leach, P. Tanev, and B. C. H. Steele, Effect of rapid cooling on the grain boundary conductivity of yttria partially stabilized zirconia, *J. Mater. Sci. Lett.*, 5, 893 (1986).

40. M. Aoki, Y.-M. Chiang, I. Kosacki, J. R. Lee, H. L. Tuller, and Y. P. Liu, Solute segregation and grain-boundary impedance in high-purity stabilized zirconia, *J. Am. Ceram, Soc.*, 79, 1169 (1996).

41. X. Guo and J. Maier, Grain boundary blocking effect in zirconia: a Schottky barrier analysis, *J. Electrochem. Soc.*, 148, E121 (2001).

42. H. Huang, T. M. Gür, Y. Saito, and F. Prinz, High ionic conductivity in ultrathin nanocrystalline gadolinia-doped ceria films, *Appl. Phys. Lett.*, 89, 143107 (2006).

43. J. Canabe and F. Canabe, in *Surface of Ceramic Interfaces II*, J. Nowotny (Ed.), Elsevier, New York, 1994.

44. Y. Lei, Y. Ito, and N. D. Browning, Segregation effects at grain boundaries in fluorite-structured ceramics, *J. Am. Ceram. Soc.*, 85, 2359 (2002).

45. Y.-M. Chiang and T. Takagi, Grain-boundary chemistry of barium titanate and strontium titanate: I., High-temperature equilibrium space charge, *J. Am. Ceram. Soc.*, 73, 3278 (1990).

46. D. K. Hohnke, Ionic conduction in doped oxides with the fluorite structure, *Solid State Ionics*, 5, 531 (1981).

47. J. B. Goodenough, Oxide-ion electrolyte, *Annu. Rev. Mater. Res.*, 33, 91 (2003).

48. D. Wang, D. S. Park, J. Griffith, and A. S. Nowick, Oxygen-ion conductivity and defect interactions in yttria-doped ceria, *Solid State Ionics*, 2, 95 (1981).

49. R. Gerhardt-Anderson and A. S. Mowick, Ionic conductivity of CeO_2 with trivalent dopants of different ionic radii, *Solid State Ionics*, 5, 547 (1981).

50. J. Fleig, Solid oxide fuel cell cathodes: polarization mechanisms and modeling of the electrochemical performance, *Annu. Rev. Mater. Res.*, 33, 361–382 (2003).

51. S. B. Adler, Factors governing oxygen reduction in solid oxide fuel cell cathodes, *Chem. Rev.*, 104, 4791–4844 (2004).

52. S. Wang, T. Kato, S. Nagata, T. Kaneko, N. Iwashita, T. Honda, and M. Dokiya, Electrodes and performance analysis of a ceria electrolyte SOFC, *Solid State Ionics*, 152–153, 477–484 (2002).

53. C. W. Tanner, K. Z. Fung, and A. V. Virkar, The effect of porous composite electrode structure on solid oxide fuel cell performance, *J. Electrochem. Soc.*, 144, 21–30 (1997).

54. S. H. Chan, X. J. Chen, and K. A. Khor, Cathode micromodel of solid oxide fuel cell, *J. Electrochem. Soc.*, 151, A164–A172 (2004).

55. T. Tsai and S. A. Barnett, Increased solid-oxide fuel cell power density using interfacial ceria layers, *Solid State Ionics*, 98, 191–196 (1997).

56. H. Huang, T. Holme, and F. B. Prinz, Increased cathodic kinetics on platinum in IT-SOFCs by inserting highly ionic-conducting nanocrystalline materials, *J. Fuel Cell Sci. Technol.*, 7, 041012 (2010).

57. R. A. De Souza, A universal empirical expression for the isotope surface exchange coefficients ($k*$) of acceptor-doped perovskite and fluorite oxides, *Phys. Chem. Chem. Phys.*, 8, 890–897 (2006).

58. N. M. Markovic, T. J. Schmidt, V. Stamenkovic, and P. N. Ross, Oxygen reduction reaction on Pt and Pt bimetallic surfaces: a selective review, *Fuel Cell*, 1, 105–116, (2001).

59. N. M. Markovic and P. N. Ross, Jr., Surface science studies of model fuel cell electrolcatalysts, *Surf. Sci. Rep.*, 45, 117–229 (2002).

60. M. Peuckert, T. Yoneda, R. A. Dalla Betta, and M. Boudart, Oxygen reduction on small supported platinum particles, *J. Electrochem. Soc.*, 133, 944–947 (1986).

61. V. R. Stamenkovic, B. S. Mun, M. Arenz, K. J. J. Mayrhofer, C. A. Lucas, G. Wand, P. N. Ross and N. M. Markovic, Trends in electrocatalysis on extended and nanoscale Pt-bimetallic alloy surfaces, *Nat. Mater.*, 6, 241–247 (2007).

62. V. R. Stamenkovic, B. Fowler, B. S. Mun, G. Wang, P. N. Ross, C. A. Lucas, N. M. Markovic, Improved oxygen reduction activity on Pt3Ni(111) via increased surface site availability, *Science*, 315, 493–497 (2007).

63. K.-D. Kreuer, Proton conductivity: materials and applications, *Chem. Mater.*, 8, 610–641 (1996).

64. G. Alberti and M. Casciola, Composite membranes for medium-temperature PEM fuel cells, *Annu. Rev. Mater. Res.*, 33, 129–154 (2003).

65. A. S. Aricò, V. Baglio, A. Di Blasi, P. Creti, P. L. Antonucci, and V. Antonucci, Influence of the acid–base characteristics of inorganic fillers on the high temperature performance of composite membranes in direct methanol fuel cells, *Solid State Ionics*, 161, 251–265 (2003).

66. S. Licoccia and E. Traversa, Increasing the operation temperature of polymer electrolyte membranes for fuel cells: from nanocomposites to hybrids, *J. Power Sources*, 159, 12–20 (2006).

67. O. Savadogo, Emerging membranes for electrochemical systems: II. High temperature composite membranes for polymer electrolyte fuel cell (PEFC) applications, *J. Power Sources*, 127, 135–161 (2004).

68. B. Smitha, S. Sridhar, and A. A. Khan, Solid polymer electrolyte membranes for fuel cell applications: a review, *J. Membrane Sci.*, 259, 10–26 (2005).

69. A. S. Aricò, V. Baglio, A. Di Blasi, E. Modica, P. L. Antonucci, and V. Antonucci, Surface properties of inorganic fillers for application in composite membranes—direct methanol fuel cells, *J. Power Sources*, 128, 113–118 (2004).

70. V. Ramani, H. R. Kunz, and J. M. Fenton, Investigation of Nafion$^{\circledR}$/HPA composite membranes for high temperature/low relative humidity PEMFC operation, *J. Membrane Sci.*, 232, 31–44 (2004).

71. Z.-G. Shao, H. Xu, M. Li, and I.-M. Hsing, Hybrid Nafion–inorganic oxides membrane doped with heteropolyacids for high temperature operation of proton exchange membrane fuel cell, *Solid State Ionics*, 177, 779–785 (2006).

72. H.-J. Kim, Y.-G. Shul, and H. Han, Sulfonic-functionalized heteropolyacid–silica nanoparticles for high temperature operation of a direct methanol fuel cell, *J. Power Sources*, 158, 137–142 (2006).

73. J.-M. Tarascon and M. Armand, Issues and challenges facing rechargeable lithium batteries, *Nature*, 414, 359–367 (2001).

74. M. S. Wittingham, Lithium batteries and cathode materials, *Chem. Rev.*, 104, 4271–4302 (2004).

75. T. Ohzuku and R. J. Brodd, An overview of positive-electrode materials for advanced lithium-ion batteries, *J. Power Sources*, 174, 449–456 (2007).

76. Y. Xia, Y. Zhou, and M. Yoshio, Capacity fading on cycling of 4 V $Li/LiMn_2O_4$ cells, *J. Electrochem. Soc.*, 144, 2593–2600 (1997).

77. S. J. Wen, T. J. Richarson, L. Ma, K. A. Striebel, P. N. Ross, Jr., and E. J. Cairins, FTIR spectroscopy of metal oxide insertion electrodes, *J. Eelectrochem. Soc.*, 143, L136–L138 (1996).

78. S. H. Kang, J. B. Goodenough, and L. K. Rabenberg, Effect of ball-milling on 3-V capacity of lithium−manganese oxospinel cathodes, *Chem. Mater.*, 13, 1758–1764 (2001).

79. S. H. Kang, J. B. Goodenough, and L. K. Rabenberg, Nanocrystalline lithium manganese oxide spinel cathode for rechargeable lithium batteries, *Electrochem. Solid State Lett.*, 4, A49–A51 (2001).

80. G. Amatucci, C. N. Schmutz, A. Blyr, C. Sigala, A. S. Gozdz, D. Larcher, and J.-M. Tarascon, Materials' effects on the elevated and room temperature performance of $C/LiMn_2O_4$ Li-ion batteries, *J. Power Sources*, 69, 11–25 (1997).

81. A. K. Padhi, K. S. Najundaswamy, and J. B. Goodenough, Phospho-olivines as positive-electrode materials for rechargeable lithium batteries, *J. Electrochem. Soc.*, 144, 1188–1194 (1997).

82. A. K. Padhi, K. S. Najundaswamy, C. Masquelier, S. Okada, and J. B. Goodenough, Effect of structure on the Fe^{3+}/Fe^{2+} redox couple in iron phosphates, *J. Electrochem. Soc.*, 144, 1609–1613 (1997).

83. S.-Y. Chung, J. T. Bloking, and Y.-M. Chiang, Electronically conductive phospho-olivines as lithium storage electrodes, *Nat. Mater.*, 1, 123 (2002).

84. H. Huang, S.-C. Yin, and L. F. Nazar, Approaching theoretical capacity of $LiFePO_4$ at room temperature at high rates, *Electrochem. Solid State Lett.*, 4, A170–A172 (2001).

85. A. Kumar, R. Thomas, N. K. Karan, J. J. Saavedra-Arias, M. K. Singh, S. B. Majumder, M. S. Tomar, and R. S. Katiyar, Structural and electrochemical characterization of pure $LiFePO_4$ and nanocomposite $C-LiFePO_4$ cathodes for lithium ion rechargeable batteries, *J. Nanotechnol.*, 176517, (2009), doi: 10.1155/2009/176517.

86. K.-F. Hsu, S.-Y. Tsay, and B.-J. Hwang, Synthesis and characterization of nano-sized $LiFePO_4$ cathode materials prepared by a citric acid–based sol–gel route, *J. Mater. Chem.*, 14, 2690–2695 (2004).

87. P. P. Prosini, D. Zane, and M. Pasquali, Improved electrochemical performance of a $LiFePO_4$-based composite cathode, *Electrochim. Acta*, 46, 3517–3523 (2001).

88. S. Yang, Y. Song, P. Y. Zavalij, and M. S. Whittingham, Reactivity, Stability and electrochemical behavior of lithium iron phosphates, *Electrochem. Commun.*, 4, 239–244 (2002).

89. S. Faldrois and B. Simon, Carbon materials for Li-ion batteries, *Carbon*, 37, 165–180 (1999).

90. Z. Wang, X. Huang, and L. Chen, Lithium insertion/extraction in pyrolyzed phenolic resin, *J. Power Sources*, **81–82**, 328–334 (1999).

91. J. Lee, K. An, J. Ju, B. Cho, W. Cho, D. Park, and K. S. Yun, Electrochemical properties of PAN-based carbon fibers as anodes for rechargeable lithium ion batteries, *Carbon*, 39, 1299–1305 (2001).

92. Y. Liu, J. S. Xue, T. Zheng, and J. R. Dahn, Mechanism of lithium insertion in hard carbons prepared by pyrolysis of epoxy resins, *Carbon*, 34, 193–200 (1996).

93. M. Noel and A. Suryanarayanan, Role of carbon host lattices in Li-ion intercalation/de-intercalation processes, *J. Power Sources*, 11, 193–209 (2002).

94. T. Zheng, Y. Liu, E. W. Fuller, S. Tseng, U. von Sacken, and J. R. Dahn, Lithium insertion in high capacity carbonaceous materials, *J. Electrochem. Soc.*, 142, 2581–2590 (1995).

95. T. Zheng, J. N. Reimers, and J. R. Dahn, The effect of turbostratic disorder in graphitic carbons on the intercalation of lithium, *Phys. Rev. B*, 51, 734–741 (1995).

96. K. Tokumitsu, H. Fujimoto, A. Mabuchi, and T. Kasuh, High capacity carbon anode for Li-ion battery: a theoretical explanation, *Carbon*, 37, 1599–1605 (1999).

97. J. R. Dahn, T. Zheng, Y. Liu, and J. Xue, Mechanism of lithium insertion in carbonaceous materials, *Science*, 270, 590–594 (1995).

98. M. Endo, Y. Nishimura, T. Takahashi, K. Takeuchi, and M. S. Dresselhaus, Lithium storage behavior for various kinds of carbon anodes in Li ion secondary battery, *J. Phys. Chem. Solids*, 57, 725–728 (1996).

99. K. Sato, M. Noguchi, A. Demachi, N. Oki, and M. Endo, A mechanism of lithium storage in disordered carbons, *Science*, 264, 556–558 (1994).

100. R. Yazami, Surface chemistry and lithium storage capability of the graphite–lithium electrode, *Electrochim. Acta*, 45, 87–97 (1999).

101. E. Frackowiak and F. Beguin, Electrochemical storage of energy in carbon nanotubes and nanostructured carbons, *Carbon*, 40, 1775–1787 (2002).

102. H. Zhang, G. Cao, Z. Wang, Y. Yang, Z. Shi, and Z. Gu, Carbon nanotube array anodes for high-rate Li-ion batteries, *Electrochim Acta*, 55, 2873–2877 (2010).

103. H. Shimoda, B. Gao, X. P. Tang, A. Kleinhammes, L. Fleming, Y. Wu, and O. Zhou, Lithium intercalation into opened single-wall carbon nanotubes: storage capacity and electronic properties, *Phy. Rev. B*, 88, 015502 (2002).

104. J. Zhao, A. Buldum, J. Han, and J. Lu, First-principles study of Li-intercalated carbon nanotube ropes, *Phys. Rev. Lett.*, 86, 1706–1709 (2002).

105. V. Meinier, J. Kephart, C. Roland, and J. Bernholc, Ab initio investigations of lithium diffusion in carbon nanotube systems, *Phys. Rev. B*, 88, 075506 (2002).

106. U. Kasavajjula, C. Wang, and A. J. Appleby, Nano- and bulk-silicon-based insertion anodes for lithium-ion secondary cells, *J. Power Sources*, 163, 1003–1039 (2007).

107. B. A. Boukamp, G. C. Lesh, and R. A. Huggins, All-solid lithium electrodes with mixed-conductor matrix, *J. Electrochem. Soc.*, 128, 725–729 (1981).

108. J. H. Ryu, J. W. Kim, Y. E. Sung, and S. M. Oh, Failure modes of silicon powder negative electrode in lithium secondary batteries, *Electrochem. Solid-State Lett.*, 7, A306–A309 (2004).

109. M. D. Fleischauer, J. M. Topple, and J. R. Dahn, Combinatorial investigations of Si-M (M = Cr, Ni, Fe, Mn) thin film negative electrode materials, *Electrochem. Solid-State Lett.*, 8, A137–A140 (2005).

110. M. Park, J. Lee, S. Rajendran, M. Song, H. Kim, and Y. Lee, Electrochemical properties of Si/Ni alloy–graphite composite as an anode material for Li-ion batteries, *Electrochim. Acta*, 50, 5561 (2005).

111. G. A. Roberts, E. J. Cairns, and J. A. Reimer, Magnesium silicide as a negative electrode material for lithium-ion batteries, *J. Power Sources*, 110, 424–429 (2002).

112. X. He, W. Pu, J. Ren, L. Wang, C. Jiang, and C. Wan, Synthesis of nanosized Si composite anode materials for Li-ion batteries, *Ionics*, 13, 51–54 (2007).

113. J. Niu and J. Y. Lee, Improvement of usable capacity and cyclability of Si-based anode materials for lithium batteries by sol–gel graphite matrix, *Electrochem. Solid-State Lett.*, 5, A107–A110 (2002).

114. J. Su, H. Li, R. Yang, Y. Shi, and X. Huang, Cage-like carbon nanotubes/Si composites as anode materials for lithium ion batteries, *Electrochem. Commun.*, 8, 51–54 (2006).

115. W. Xing, A. M. Wilson, K. Eguchi, G. Zank, and J. R. Dahn, Pyrolyzed polysiloxanes for use as anode materials in lithium-ion batteries, *J. Electrochem. Soc.*, 144, 2410–2416 (1997).

116. J. P. Guo, E. Milin, J. Z. Wang, J. Chen, and H. K. Liu, Silicon/disordered carbon nanocomposites for lithium-ion battery anodes, *J. Electrochem. Soc.*, 152, A2211–A2216 (2005).

117. G. X. Wang, L. Sun, D. H. Bradhurst, S. Zhong, S. X. Dou, and H. K. Liu, Nanocrystalline NiSi alloy as an anode material for lithium-ion batteries, *J. Alloys Compounds*, 306, 249–252 (2000).

118. Z. Wang, W. Tian, X. Liu, Y. Li and X. Li, Nanosized Si–Ni alloys anode prepared by hydrogen plasma–metal reaction for secondary lithium batteries, *Mater. Chem. Phys.*, 100, 92–97 (2006).

119. C. K. Chan, H. Peng, G. Liu, K. McIlwrath, X. F. Zhang, R. A. Huggins, and Y. Cui, High-performance lithium battery anodes using silicon nanowires, *Nat. Nanotechnol.*, 3, 31–35 (2008).

120. C. K. Chan, R. Ruffo, S. S. Hong, R. A. Huggins, and Y. Cui, Structural and electrochemical study of the reaction of lithium with silicon nanowires, *J. Power Sources*, 189, 34–39 (2009).

121. Y.-G. Guo, Y.-S. Hu, W. Sigle, and J. Maier, Superior electrode performance of nanostructured mesoporous TiO_2 (anatase) through efficient hierarchical mixed conducting networks, *Adv. Mater.*, 19, 2087–2091 (2007).

122. Z. Yang, D. Choi, S. Kerisit, K. M. Rosso, D. Wang, J. Zhang, G. Graff, and J. Liu, Nanostructures and lithium electrochemical reactivity of lithium titanites and titanium oxides: a review, *J. Power Sources*, 192, 588 (2009).

123. A. R. Armstrong G. Canales, J. R. Garcia, and P. G. Bruce, Lithium intercalation into TiO_2–B nanowires, *Adv. Mater.*, 17, 862–865 (2005).

124. P. Poizot, S. Laruelle, S. Grugeon, L. Dupont, and J.-M. Tarascon, Nano-sized transition metal oxides as negative electrode material for lithium-ion batteries, *Nature*, 407, 496–499 (2000).

125. F. Leroux, G. R. Coward, W. P. Power, and L. F. Nazar, Understanding the nature of low-potential Li uptake into high volumetric capacity molybedenum oxides, *Electrochem. Solid-State Lett.*, 1, 255–258 (1998).

126. D. Larcher, C. Masquelier, D. Bonnin, Y. Chabre, V. Masson, J.-B. Leriche, and J.-M. Tarascon, Effect of particle size on lithium intercalation into α-Fe_2O_3, *J. Electrochem. Soc.*, 150, A133–A139 (2003).

127. C. R. Sides, N. C. Li, C. J. Patrissi, B. Scrosati, and C. R. Martin, Nanoscale materials for lithium-ion batteries, *Mater. Res. Bull.*, 27, 604–607 (2002).

128. P. L. Taberna, S. Mitra, P. Poizot, P. Simon, and J.-M. Tarascon, High rate capabilities Fe_3O_4-based Cu nano-architectured electrodes for lithium-ion battery applications, *Nat. Mater.*, 5, 567 (2006).

129. F. Badway, F. Cosandey, N. Pereira, and G. G. Amatucci, Carbon metal fluoride nanocomposites: high capacity reversible metal fluoride conversion materials as rechargeable positive electrodes for Li batteries, *J. Electrochem. Soc.*, 150, A1318–A1327 (2003).

130. H. Li, G. Ritcher, and J. Maier, Reversibile formation and decomposition of LiF clusters using transition metal fluorides as precursors and their application in rechargeable Li batteries, *Adv. Mater.*, 15, 736–739 (2003).

131. B. Z. Jang and W. C. Huang, Nano-scaled graphene plates, U.S. patent 7,07,1258 (submitted Oct. 21, 2002; issued July 4, 2006).

132. B. Z. Jang, Process for nano-scaled graphene plates, U.S. Patent appl. 11/442,903 (June 20, 2006); a divisional of 10/274,473 (Oct. 21, 2002).

133. K. S. Novoselov, D. Jiang, F. Schedin, T. J. Booth, V. V. Khotkevich, S. V. Morozov, and A. K. Geim, Two dimensional atomic crystals, *Proc. Natl. Acad. Sci. USA*, 102, 10451–10453 (2005).

134. C. Lee, X. Wei, J. W. Kysar, and J. Hone, Measurement of the elastic properties and intrinsic strength of monolayer graphene, *Science*, 321, 385–388 (2008).

135. A. Balandin, S. Ghosh, W. Bao, I. Calizo, D. Teweldebrhan, F. Miao, and C. N. Lau, Superior thermal conductivity of single-layer graphene, *Nano Lett.*, 8, 902–907 (2008).

136. W. Schwalm, M. Schwalm, and B. Z. Jang, Local density of states for nanoscale graphene fragments, Paper C1.157, American Physics Society, Montreal, Canada, 2004.

137. S. C. Wong, E.M. Sutherland, and B. Z. Jang, Graphene nanoplatelet reinforced polymer coatings, *Proceedings of the 62nd SPE ANTEC*, Chicago, 2004.

138. S. C. Wong, E. Sutherland, and B. Z. Jang, Processing of graphene nanoplatelets, *Proceedings of the 2004 NSF Design and Manufacturing Grantees and Research Conference*, Dallas, TX, 2004.

139. M. J. McAllister, J. L. Li, D. H. Adamson, H. C. Schniepp, A. A. Abdala, J. Liu, M. Herrera-Alonso, D. L. Milius, R. Car, R. K. Prud'homme, and I. A. Aksay, Single sheet functionalized graphene by oxidation and thermal expansion of graphite, *Chem. Mater.*, 19, 4396–4404 (2007).

140. J.-L. Li, K. N. Kudin, M. J. McAllister, R. K. Prud'homme, I. A. Aksay, and R. Car, Oxygen-driven unzipping of graphitic materials, *Phys. Rev. Lett.*, 96, 176101–1-4 (2006).

141. H. C. Schniepp, J. L. Li, M. J. McAllister, H. Sai, M. Herrera-Alonso, D. H. Adamson, P. K. Prud'homme, R. Car, D. A. Saville, and I. A. Aksay, Functionalized single graphene sheets derived from splitting graphite oxide, J. Phys. Chem. *B*, 110, 8535–8547 (2006).

142. X. Li, X. Wang, L. Zhang, S. Lee, and H. Dai, Chemically derived, ultrasmooth graphene nanoribbon semiconductor, Science, 319, 1229–1232 (2008).

143. S. J. Park and R. S. Ruoff, Chemical methods for the production of graphenes, *Nat. Nanotechnol.*, 4, 217–224 (2009).

144. S. Horiuchi, T. Gotou, M. Fujiwara, T. Asaka, T. Yokosawa, and Y. Matsui, Single graphene sheet detected in a carbon nanofilm, *Appl. Phys. Lett.*, 84, 2403–2405 (2004).

145. M. Hirata, T. Gotou, and M. Ohba, Thin-film particles of graphite oxide: 2. Preliminary studies for internal micro fabrication of single particle and carbonaceous electronic circuits, *Carbon*, 43, 503–510 (2005).

146. M. Hirata, T. Gotou, S. Horiuchi, M. Fujiwara, and M. Ohba, Thin-film particles of graphite oxide: 1. High yield synthesis and flexibility of the particles, *Carbon*, 42, 2929–2937 (2004).

147. S. Stankovich, Stable aqueous dispersions of graphitic nanoplatelets via the reduction of exfoliated graphite oxide in the presence of poly(sodium 4-styrenesulfonate), *J. Mater. Chem.*, 16, 155–158 (2006).

148. S. Stankovich, R. D. Piner, S. T. Nguyen, and R. S. Ruoff, Synthesis and exfoliation of isocyanate-treated graphene oxide nanoplatelets, *Carbon*, 44, 3342–3347 (2006).

149. D. Li, M. C. Muller, S. Gilje, R. B. Kaner, and G. Wallace, Processable aqueous dispersions of graphene nanosheets, *Nat. Nanotechnol.*, 3, 101–105 (2008).

150. Y. Si and E. T. Samulski, Synthesis of water soluble graphene, *Nano Lett.*, 8, 1679–1682 (2008).

151. X. Yang, X. Dou, A. Rouhanipour, L. Zhi, H. J. Raider, and K. Mullen, Two-dimensional graphene nano-ribbons, *J. Am. Chem. Soc.*, 130, 4216–4217 (2008).

152. R. K. Prud'Homme, I. A. Aksay, D. Adamson, and A. Abdala, Thermally exfoliated graphite oxide. *U.S patent* appl. 11/249,404 (Oct. 14, 2005); pub. no. US 2007/0092432 (Apr. 26, 2007).

153. H. A. Becerril, J. Mao, Z. Liu, R. M. Stoltenberg, Z. Bao, and Y. Chen, Evaluation of solution-processed reduced graphene oxide films as transparent conductors, *ACS Nano*, 2, 463–470 (2008).

154. S. Gilje, S. Han, M. Wang, K. L. Wang, and R. B. Kaner, A chemical route to graphene for device applications, *Nano Lett.*, 7, 3394–3398 (2007).

155. K. S. Subrahmanyam, S. R. C. Vivekchand, A. Govindaraj and C. N. R. Rao, A study of graphenes prepared by different methods: characterization, properties and solubilization, *J. Mater. Chem.*, 18, 1517–1523 (2008).

156. V. C. Tung, M. J. Allen, Y. Yang, and R. B. Kaner, High throughput solution processing of large-scale graphene, *Nat. Nanotechnol.*, online, Nov. 9, 2008.

157. Y. Hernandez, V. Nicolosi, M. Lotya, F. Blighe, Z. Sun, S. De, I. T. McGovern, B. Holland, M. Byrne, Y. Gunko, J. Boland, P. Niraj, G. Duesberg, S. Krishnamurti, R. Goodhue, J. Hutchison, V. Scardaci, A. C. Ferrari, and J. N. Coleman, High-yield production of graphene by liquid-phase exfoliation of graphite, *Nat. Nanotechnol.*, 3, 563–568, (2008).

158. H. V. Roy, C. Kallinger, B. Marsen, and K. Sattler, Manipulation of graphitic sheets using a tunneling microscope, *J. Appl. Phys.*, 83, 4695–4699 (1998).

159. J. J. Mack, Chemical manufacture of nanostructured materials, U.S. patent. 6,872,330 (Mar. 29, 2005).

160. L. M. Viculis, J. J. Mack, and R. B. Kaner, A chemical route to carbon nanoscrolls, *Science*, 299, 1361 (2003).

161. C. Berger, Z. Song, T. Li, X. Li, A. Y. Ogbazghi, R. Feng, Z. Dai, A. N. Marchenkov, E. H. Conrad, P. N. First, and W. A. de Heer, Ultrathin epitaxial graphite: two-dimensional electron gas properties and a route toward graphene-based nanoelectronics, J. Phys. Chem. *B*, 108, 19912–19916 (2004).

162. J. D. Udy, Method of continuous, monoatomic thick structures, *U.S. patent* appl. 11/243,285 (Oct. 4, 2005); pub. no. 2006/0269740 (Nov. 30, 2006).

163. M. Zhu, J. Wang, R. A. Outlaw, K. Hou, D. M. Manos, and B. C. Holloway, Synthesis of carbon nanosheets and carbon nanotubes by radio frequency plasma enhanced chemical vapor deposition, *Diamond Relat. Mater.*, 16, 196–201 (2007).

164. B. L. French, J. J. Wang, M. Y. Zhu, and B. C. Holloway, Evolution of structure and morphology during plasma-enhanced chemical vapor deposition of carbon nanosheets, *Thin Solid Films*, 494, 105 (2006).

165. T. A. Land, T. Michely, R. J. Behm, J. C. Hemminger, and G. Comsa, STM investigation of single layer graphite structures produced on Pt(111) by hydrocarbon decomposition, *Surf. Sci.* 264, 261–270 (1992).

166. A. Nagashima, K. Nuka, H. Itoh, T. Ichinokawa, C. Oshima, and S. Otani, Electronic states of monolayer graphite formed on TiC(111) surface, *Surf. Sci.* 291, 93–98 (1993).

167. A. J. van Bommel, J. E. Crombeen, and A. van Tooren, LEED and Auger electron observations of the SiC(0001) surface, *Surf. Sci.*, 48, 463–472 (1995).

168. I. Forbeaux, J.-M. Themlin, and J. M. Debever, Heteroepitaxial graphite on 6H-SiC(0001): interface formation through conduction-band electronic structure, *Phys. Rev. B*, 58, 16396–16406 (1998).

169. M. Choucair, P. Thordarson, and J. A. Stride, Gram-scale production of graphene based on solvothermal synthesis and sonication, *Nat. Nanotechnol.*, 4, 30–33 (2009).

170. B. Z. Jang, S. C. Wong, and Y. Bai, Process for producing nano-scaled graphene plates, U.S. patent appl. 10/858,814 (June 3, 2004); pub. no. US 2005/0271574 (Dec. 8, 2005).

171. A. Zhamu, J. Shi, J. Guo, and Z. Jang, Method of producing exfoliated graphite, flexible graphite, and nano-scaled graphene plates, U.S. patent, pending, 11/800,728 (May 08, 2007).

172. B. Z. Jang, A. Zhamu, and J. Guo, Process for producing nano-scaled platelets and nanocomposites, U.S. patent pending, 11/509,424 (Aug. 25, 2006).

173. B. Z. Jang, A. Zhamu, and J. Guo, Mass production of nano-scaled platelets and products, U.S. patent pending, 11/526,489 (Sept. 26, 2006).

174. B. Z. Jang, A. Zhamu, and J. Guo, Method of producing nano-scaled graphene and inorganic platelets and their nanocomposites, U.S. patent pending, 11/709,274 (Feb. 22, 2007).

175. B. Z. Jang, A. Zhamu, and J. Guo, Nano-scaled graphene plate films and articles, U.S. patent pending, 11/784,606 (Apr. 9, 2007).

176. A. Zhamu, J. Shi, J. Guo, and B. Z. Jang, Low-temperature method of producing nano-scaled graphene platelets and their nanocomposites, U.S. patent pending, 11/787,442 (Apr. 17, 2007).

177. A. Zhamu, J. Shi, J. Guo, and B. Z. Jang, Method of producing exfoliated graphite, flexible graphite, and nano graphene plates, U.S. patent pending, 11/800,728 (May 8, 2007).

178. A. Zhamu, J. Jang, J. Shi, and B. Z. Jang, Method of producing ultra-thin nano-scaled graphene platelets, U.S. patent pending, 11/879,680 (July 19, 2007).

179. A. Zhamu, J. Jang, and B. Z. Jang, Electrochemical method of producing ultra-thin nano-scaled graphene platelets, U.S. patent pending, 11/881,388 (July 27, 2007).

180. B. Z. Jang, A. Zhamu, and L. Song, Method for producing highly conductive SMC, fuel cell flow field plate, and bipolar plate, U.S. patent pending, 11/293,541 (Dec. 5, 2005).

181. B. Z. Jang, A. Zhamu, and L. Song, Highly conductive composites for fuel cell flow field plates and bipolar plates, U.S. patent pending, 11/324,370 (Jan. 4, 2006).

182. L. Song, J. Guo, A. Zhamu, and B. Z. Jang, Highly conductive nano-scaled graphene plate nanocomposites and products, U.S. patent pending, 11/328,880 (Jan. 11, 2006).

183. A. Zhamu and B. Z. Jang, Hybrid anode compositions for lithium ion batteries, U.S. patent appl. 11/982,662 (Nov. 05, 2007).

184. A. Zhamu and B. Z. Jang, Nano graphene platelet-based composite anode compositions for lithium ion batteries, U.S. patent appl. 11/982,672 (Nov. 5, 2007).

185. J. Shi, A. Zhamu, and B. Z. Jang, Conductive nanocomposite-based electrodes for lithium batteries, U.S. patent appl. 12/156,644 (June 4, 2008).

186. A. Zhamu and B. Z. Jang, Graphene nanocomposites for electrochemical cell electrodes, U.S. patent appl. 12/220,651 (July 28, 2008).

187. L. Song, A. Zhamu, J. Guo, and B. Z. Jang, Nano-scaled graphene plate nanocomposites for supercapacitor electrodes, U.S. patent pending, 11/499,861 (Aug. 7, 2006).

188. A. Zhamu and B. Z. Jang, Process for producing nano graphene platelet nanocomposite electrodes for supercapacitors, U.S. patent pending, 11/906,786 (Oct. 4, 2007).

189. A. Zhamu and B. Z. Jang, Graphite–carbon composite electrodes for supercapacitors, U.S. patent pending 11/895,657 (Aug. 27, 2007).

190. A. Zhamu and B. Z. Jang, Method of producing graphite–carbon composite electrodes for supercapacitors, U.S. patent pending 11/895,588 (Aug. 27, 2007).

191. B. Z. Jang and A. Zhamu, Process for producing dispersible nano graphene platelets from non-oxidized graphitic materials, U.S. patent appl. 12/231,411 (Sept. 3, 2008).

192. B. Z. Jang and A. Zhamu, Process for producing dispersible nano graphene platelets from oxidized graphite, U.S. patent appl. 12/231,413 (Sept. 3, 2008).

193. B. Z. Jang and A. Zhamu, Dispersible nano graphene platelets, U.S. patent appl. 12/231,417 (Sept. 3, 2008).

194. C. K. Yang, A metallic graphene layer adsorbed with lithium, *Appl. Phys. Lett.*, 94, 162115 (2009).

195. M. Khantha, N. A. Cordero, L. M. Molina, J. A. Aloso, and L. A. Girifalco, Interaction of lithium with graphene: an ab initio study, *Phys. Rev. B*, 70, 1254202 (2004).

196. F. Valencia, A. H. Romero, F. Ancilotto, and P. L. Silverstrelli, Lithium adsorption on graphite from density functional theory calculation, *J. Phys. Chem. B*, 110, 14832–14841 (2006).

197. K. Rytkonen, J. Akola, and M. Manninen, Density functional study of alkali-metal atoms and monolayers on graphite(0001), *Phys. Rev. B*, 75, 075401 (2007).

198. K. T. Chan, J. B. Neaton, and M. L. Cohen, First principle study of metal adatom adsorption on graphene, *Phys. Rev B.*, 77, 235430 (2008).

199. H. Tachikawa and A. Simizu, Diffusion dynamics of the Li ion on a model surface of amorphous carbon: a direct molecular orbital dynamics study, *J. Phys. Chem.*, 109, 13255–13262 (2005).

200. G. Wang, B. Wang, X. Wang, J. Park, S. Dou, H. Ahn, and K. Kim, Sn/graphene nanocomposite with 3D architecture for enhanced reversible lithium storage in lithium ion batteries, *J. Mater. Sci.*, 19, 8378–8384 (2009).

201. S. Cheekati, Y. Xing, Y. Zhuang, and H. Huang, Lithium storage characteristics in nano graphene platelets based materials, *Ceram. Soc. Trans.*, 224, 117–127 (2010).

202. A. Zhamu and B. Z. Jang, Nano graphene platelet-based composite anode compositions for lithium ion batteries, U.S. patent appl. 11/982,672 (Nov. 5, 2007).

203. A. Zhamu and B. Z. Jang, Graphene nanocomposites for electrochemical cell electrodes, U.S. patent appl. 12/220,651 (July 28, 2008).

204. S.-M. Paek, E. Yoo, and I. Honma, Enhanced cyclic performance and lithium storage capacity of SnO$_2$/graphene nanoporous electrodes with three-dimensionally delaminated flexible structure, *Nano Lett.*, 9, 72–75 (2009).

205. J. Yao, X. Shen, B. Wang, H. Liu and G. Wang, In situ chemical synthesis of SnO$_2$–graphene nanocomposite as anode materials for lithium-ion batteries, *Electrochem. Commun.*, 11, 1849–1852 (2009).

206. G. Wang, B. Wang, X. Wang, J. Park, S. Dou, H. Ahn, and K. Kim, Sn/graphene nanocomposite with 3D architecture for enhanced reversible lithium storage batteries, *J. Mater. Chem.*, 19, 8378—8384 (2009).

207. G. Wang, X. Shen, J. Yao, and J. Park, Graphene nanosheets for enhanced lithium storage in lithium ion batteries, *Carbon*, 47, 2049–2053 (2009).

208. S.-L. Chou, J.-Z. Wang, M. Choucair, H.-K. Liu, J. A. Stride, and S.-X. Dou, Enhanced reversible lithium storage in a nanosize silicon/graphene composite, *Electrochem. Commun.*, 12, 303–306 (2010).

209. D. Wang, D. Choi, J. Li, Z. Yang, Z. Nie, R. Kou, D. Hu, C. Wang, L. V. Saraf, J. Zhang, I. A. Aksay, and J. Liu, Self-assembled TiO_2–graphene hybrid nanostructures for enhanced Li-ion insertion, *ACS Nano*, 3, 907–914 (2009).

210. M. Liang and L. Zhi, Graphene-based electrode materials for rechargeable lithium batteries, *J. Mater. Chem.*, 19, 5871–5878 (2009).

211. F. Ji, Y.-L. Li, J.-M. Feng, D. Su, Y.-Y. Wen, Y. Feng, and F. Hou, Electrochemical performance of graphene nanosheets and ceramic composites as anodes for lithium batteries, *J. Mater. Chem.*, 19, 9063–9067 (2009).

212. Z.-S. Wu, W. Ren, L. Wen, L. Gao, J. Zhao, Z. Chen, G. Zhou, F. Li, and H.-M. Cheng, Graphene anchored with Co_3O_4 nanoparticles as anode of lithium ion batteries with enhanced reversible capacity and cyclic performance, *ACS Nano*, 4, 3187–3194 (2010).

213. A. Zhamu and B. Z. Jang, Hybrid nano filament cathode compositions for lithium ion and lithium metal batteries, U.S. patent appl. 12/009,259 (Jan. 18, 2008).

214. A. Zhamu and B. Z. Jang, Method of producing hybrid nano filament electrodes for lithium metal or lithium ion batteries, U.S. patent appl. 12/077,520 (Mar. 20, 2008).

215. A. Zhamu and B. Z. Jang, Process for producing hybrid nano filament electrodes for lithium batteries, U.S. patent appl. 12/150,096 (Apr. 25, 2008).

216. Y. Ding, Y. Jiang, F. Xu, J. Yin, H. Ren, Q. Zhuo, Z. Long, P. Zhang, Preparation of nano-structured $LiFePO_4$/graphene composites by co-precipitation method, *Electrochem. Commun.*, 12, 10–13 (2010).

217. T. Ohta, A. Bostwick, T. Seyller, K. Horn, and E. Rotenberg, Controlling the electronic structure of bilayer graphene, *Science*, 313, 951–954 (2006).

218. S. Y. Zhou, G. H. Gweon, A. V. Fedorov, P. N. First, W. A. de Heer, D. H. Lee, F. Guinéa, A. H. Castro Neto, and A. Lanzara, Substrate-induced bandgap opening in epitaxial graphene, *Nat. Mater.*, 6, 770–775 (2007).

219. N. Gorjizadeh, A. A. Farajian, K. Esfarjani, and Y. Kawazoe, Spin and band-gap engineering in doped graphene nanoribbons, *Phys. Rev. B*, 78, 155427 (1–6) (2008).

220. M. Y. Han, B. Oezyilmaz, Y. Zhang, and P. Kim, Energy band-gap engineering of graphene nanoribbons, *Phys. Rev. Lett.*, 98, 206801–206804 (2007).

221. J. F. Byrne and H. Marsh, in *Porosity in Carbons: Characterization and Applications*, J. W. Patrick (Ed.), Halsted, Chichester, UK, 1995.

222. M. Mastragostino, C. Arbizzani, and F. Soavi, Polymer-based supercapacitors, *J. Power Sources*, 97–98, 812–815 (2001).

223. L. L. Zhang and X. S. Zhao, Carbon-based materials as supercapacitor electrodes, *Chem. Soc. Rev.*, 38, 2520–2531 (2009).

224. A. K. Shukla, S. Sampath, and K. Vijayamohanan, Electrochemical supercapacitors: energy storage beyond batteries, *Curr. Sci.*, 79, 1656–1661 (2000).

225. R. Kotz and M. Carlen, Principles and applications of electrochemical capacitors, *Electrochim. Acta*, 45, 2483–2498 (2000).

226. T. Broussea and D. Belangerb, A hybrid Fe_3O_4-MnO_2 capacitor in mild electrolyte, *Electrochem. Solid-State Lett.*, 6, A244–A248 (2003).

227. G. X. Wang, B. L. Zhang, Z. L. Yu, and M. Z. Qu, Manganese oxide/MWNTs composite electrodes for supercapacitors, *Solid State Ionics*, 176, 1169–1174 (2005).

228. S. Razoumov, A. Klementov, S. Litvinenko, and A. Beliakov, Asymmetric electrochemical capacitor and method of making, U.S. patent 6,222,723 (2001).

229. S. M. Lipka, J. R. Miller, T. D. Xiao, and D. E. Reisner, Asymmetric electrochemical supercapacitor and method of manufacture thereof, U.S. patent 7,199,997 (2007).

230. A. Peigney, Ch. Laurent, E. Flahaut, R. Bacsa, and A. Rousset, Specific surface area of carbon nanotubes and bundles of carbon nanotubes, *Carbon*, 39, 507 (2001).

231. J. E. Fischer, H. Dai, A. Thess, R. Lee, N. M. Hanjani, D. L. Dehaas, and R. E. Smalley, Metallic resistivity in crystalline ropes of single-wall carbon nanotubes, *Phys. Rev. B*, 55, R4921–R4924 (1997).

232. W. A. de Heer, W. S. Bacsa, A. Chatelain, T. Gerfin, R. Humphrey-Baker, L. Forro, and D. Ugarte, Aligned carbon nanotube films: production and optical and electronic properties, *Science*, 268, 845–847 (1995).

233. E. Frackowiak, K. Jurewicz, S. Delpeux, and F. Béguin, Nanotubular materials for supercapacitors, *J. Power Sources*, 97, 822 (2001).

234. J. N. Barisci, G. G. Wallace, and R. H. Baughman, Electrochemical properties of single-wall carbon nanotube electrodes, *J. Electrochem. Soc.*, 150, E409 (2003).

235. S. Shiraishi, H. Kurihara, K. Okabe, D. Hulicova, and A. Oya, Electric double layer capacitance of highly pure single-walled carbon nanotubes (HiPco™ Buckytubes™) in propylene carbonate electrolytes, *Electrochem. Commun.*, 4, 593 (2002).

236. C. S. Li, D. Z. Wang, T. X. Liang, G. T. Li, X. F. Wang, M. S. Cao, and Z. Liang, Oxidation behavior of CNTs and the electric double layer capacitor made of the CNT electrodes, *J. Sci. China E*, 46, 349 (2003).

237. Q. Jiang, M. Z. Qu, G. M. Zhou, B. L. Zhang, and Z. L. Yu, A study of activated carbon nanotubes as electrochemical super capacitors electrode materials, *Mater. Lett.*, 57, 988 (2002).

237a. E. Frackowiak, S. Delpeux, K. Jurewicz, K. Szostak, D. Cazorla-Amoros, and F. Béguin, Enhanced capacitance of carbon nanotubes through chemical activation, *Chem. Phys. Lett.*, 361, 35 (2002).

238. Y. H. Lee, K. H. An, S. C. Lim, W. S. Kim, H. J. Jeong, C. H. Doh, and S. I. Moon, Applications of carbon nanotubes to energy storage devices, *New Diamond Front. Carbon Technol.*, 12, 209–228 (2002).

239. E. Frackowiak, K. Metenier, V. Bertagna, and F. Béguin, Supercapacitor electrodes from multiwalled carbon nanotubes, *Appl. Phys. Lett.*, 77, 2421 (2000).

240. G. Niu, E. K. Sichel, R. Hoch, D. Moy, and H. Tennent, High power electrochemical capacitors based on carbon nanotube electrodes, *Appl. Phys. Lett.*, 70, 1480 (1997).

241. R.,Z. Ma, J. Liang, B., Q. Wei, B. Zhang, C. L. Xu, and D. H. Wu, Study of electrochemical capacitors utilizing carbon nanotube electrodes, *J. Power Source*, 84, 126 (1999).

242. J. Y. Lee, K. H. An, J. K. Heo, and Y. H. Lee, Fabrication of supercapacitor electrodes using fluorinated single-walled carbon nanotubes, *J. Phys. Chem. B*, 107, 8812–8815 (2003).

243. B. J. Yoon, S. H. Jeong, K. H. Lee, H. S. Kim, C. G. Park, and J. H. Han, Electrical properties of electrical double layer capacitors with integrated carbon nanotube electrodes, *Chem. Phys. Lett.*, 388, 170–174 (2004).

244. D. Belanger, X. Ren, J. Davey, F. Uribe, and S. Gottesfeld, Characterization and long-term performance of polyaniline-based electrochemical capacitors, *J. Electrochem. Soc.*, 147, 2923–2929 (2000).

245. E. Frackowiak, K. Jurewicz, S. Depleux, and F. Béguin, Nanotubular materials for supercapacitors, *J. Power Sources*, 97–98, 822–825 (2001).

246. J. H. Fan, M. X. Wan, D. B. Zhu, B. H. Chang, Z. W. Pan, and S. S. Xe, Synthesis, characterizations, and physical properties of carbon nanotubes coated by conducting polypyrrole, *J. Appl. Sci.*, 74, 2605–2610 (1999).

247. Y. Sun, S. R. Wilson, and D. I. Schuster, High dissolution and strong light emission of carbon nanotubes in aromatic amine solvents, *J. Am. Chem. Soc.*, 123, 5348–5349 (2001).

248. C. B. McCarthy and J. N. R. Czerw, Microscopy studies of nanotube-conjugated polymer interactions, *Synthetic Metals*, 121, 1225–1226 (2001).

249. K. H. An, K. K. Jeon, J. K. Heo, S. C. Lim, D. J. Bae, and Y. H. Lee, High-capacitance supercapacitor using a nanocomposite electrode of single-walled carbon nanotube and polypyrrole, *J. Electrochem. Soc.*, 149, A1058–A1062 (2002).

250. M. Hughes, G. Z. Chen, M. S. P. Shaffer, D. J. Fray, and A. H. Windle, Electrochemical capacitance of a nanoporous composite of carbon nanotubes and polypyrrole, *Chem. Mater.*, 14, 1610–1613 (2002).

251. C. Downs, J. Nugent, P. M. Ajayan, D. J. Duquette, and K. S. V. Santhanam, Efficient polymerization of aniline at carbon nanotube electrodes, *Adv. Mater.*, 11, 1028–1031 (1999).

252. M. Hughes, M. S. P. Shaffer, N. C. Renouf, C. Singh, G. Z. Chen, D. J. Fray, and A. H. Windle, Electrochemical capacitance of nanocomposite films formed by coating aligned arrays of carbon nanotubes with polypyrrole, *Adv. Mater.*, 14, 382–385 (2002).

253. G. A. Snook, G. Z. Chen, D. J. Fray, M. Hughes, and M. Shaffer, Studies of deposition of and charge storage in polypyrrole–chloride and polypyrrole–carbon nanotube composites with an electrochemical quartz crystal microbalance, *J. Electroanal. Chem.*, 568, 135–142 (2004).

254. E. Frackowiak, K. Jurewicz, K. Szostak, S. Delpeux, and F. Béguin, Nanotubular materials as electrodes for supercapacitors, *Fuel Process. Technol.*, **77–78**, 213–219 (2002).

255. Q. Xiao and X. Zhou, The study of multiwalled carbon nanotube deposited with conducting polymer for supercapacitor, *Electrochim. Acta*, 48, 575 (2003).

256. Y.-K. Zhou, B.-L. He, W.-J. Zhou, and H.-L. Li, Preparation and electrochemistry of SWNT/PANI composite films for electrochemical capacitors, *J. Electrochem. Soc.*, 151, A1052 (2004).

257. H. Wang, Q. Hao, X. Yang, L. Lu, and X. Wang, Graphene oxide doped polyaniline for supercapacitors. *Electrochem. Commun.*, 11, 1158–1161 (2009).

258. L. L. Song, A. Zhamu, J. S. Guo, and B. Z. Jang, Nano-scaled graphene plate nanocomposites for supercapacitor electrodes. U.S. patent appl. 11/328,880 (Jan. 11, 2006); now U.S. patient 7,623,340 (Nov. 24, 2009).

259. A. Zhamu and B. Z. Jang, Process for producing nano-scaled graphene platelet nanocomposite electrodes for supercapacitors. U.S. patent Appl. 11/906,786 (Oct. 4, 2007).

260. S. R. Vivekchand, S. R. Chandra, K. S. Subrahmanyam, A. Govindaraj, and C. N. Rao, Graphene-based electrochemical supercapacitor, *J. Chem. Sci.*. 120, 9–13 (2008).

261. M. D. Stoller, S. Park, Y. Zhu, J. H. An, and R. S. Ruoff, Graphene-based ultracapacitor, *Nano Lett.*, 8, 3498–3502 (2008).

262. Y. Wang, Z. Shi, Y. Huang, Y. Ma, C. Wang, M. Chen, and Y. Chen, Supercapacitor devices based on graphene materials, *J. Phys. Chem. C*, 113, 13103–13107 (2009).

263. L. L. Zhang and X. S. Zhao, Carbon-based materials as supercapacitor electrodes, *Chem. Soc. Rev.*, 38, 2520–2531 (2009).

9

ENHANCEMENT OF THROUGH-THICKNESS THERMAL CONDUCTIVITY IN ADHESIVELY BONDED JOINTS USING ALIGNED CARBON NANOTUBES

SANGWOOK SIHN AND SABYASACHI GANGULI
University of Dayton Research Institute, Dayton, Ohio

AJIT K. ROY
Air Force Research Laboratory, Wright-Patterson Air Force Base, Dayton, Ohio

LIANGTI QU
Beijing Institute of Technology, Beijing, China

LIMING DAI
Case Western Reserve University, Cleveland, Ohio

Nanoscale Multifunctional Materials: Science and Applications, First Edition.
Edited by Sharmila M. Mukhopadhyay.
© 2012 John Wiley & Sons, Inc. Published 2012 by John Wiley & Sons, Inc.

1   INTRODUCTION

Based on a current requirement of space industries, a through-thickness thermal conductivity (k_Z) of adhesive of about 7 to 10 W/m·K is expected to enable efficient multifunctionality and lean manufacturing of systems to numerous applications, ranging from electronic cooling, to efficient space structures and to dramatically improving the energy-conversion efficiency of the directed energy devices. The through-thickness thermal conductivity in adhesive joints currently in use severely lacks in meeting the requirement mentioned above. One of the barriers in achieving adequate through-thickness thermal conductivity in composite materials and also in composite joints is due to the extremely low thermal conductivity of matrix resins or adhesives (typically, $k_Z \sim 0.3$ W/m·K). The poor thermal conductivity of resin and/or adhesives fails to meet the needed k_Z at the system level, as structural components are assembled primarily through bonded joints. A possible concept of enhancing through-thickness thermal conductivity in the adhesive joints is to incorporate carbon nanotubes (CNTs) in the adhesive layer.

Various unique properties of the CNTs have generated interest among many researchers over the last decade [1]. These researchers have reported remarkable electrical [2], mechanical [3], and thermal properties [4] related to their unique structure and high aspect ratio. These unique properties make CNTs the material of choice for numerous applications, such as sensors [5], actuators [6], energy storage devices [7], and nanoelectronics [8]. The CNTs also have extremely high thermal conductivity in the axial direction [9]. According to molecular dynamics simulations, the value can reach as high as 6500 W/m·K at room temperature for single-walled CNTs (SWCNTs). Researchers have experimentally determined the thermal conductivity of multiwalled CNTs (MWCNTs) to be 2000 to 3000 W/m·K at room temperature [10]. This outstanding thermal property of nanotubes has made them target materials for thermal management applications for improving thermal properties of materials in suspensions to solid phases. Thermal conductivity enhancement has been observed in nanotube suspensions [11]. The results of this study were theoretically intriguing as the measured thermal conductivities were abnormally greater than theoretical predictions with conventional heat conduction models [12]. Biercuk et al. [13] measured thermal transport properties of industrial epoxy loaded with as-produced SWCNTs (>5 wt%) from 20 to 300 K. It was observed that samples with 1% unpurified SWCNT material showed a 125% increase in thermal conductivity at room temperature. Furthermore, the unpurified CNTs were dispersed in a silicone elastomer to investigate their effect on the thermal conductivity [14]. Microstructure studies by a scanning electron microscope (SEM) showed that the CNTs were well dispersed in the matrix by a "grinding method." The thermal conductivity of the composites was measured with the ASTM D5470 method. The measured values of the thermal conductivity were found to increase with the carbon amount. There was a 65% enhancement in the thermal conductivity with 3.8 wt% CNT loading. The enhancement by equal loading of carbon black was found to be a little lower than that by CNT loading. Meanwhile, the composites loaded with CNTs displayed an abrupt increase in the electrical conductivity, which suggests that the mechanism of electron transport is different from that of thermal (or phonon) transport.

Thermal interface materials (TIMs) used for dissipating heat efficiently from electronic components is also gaining increased attention. Thermal conducting pads made with various conductive fillers are widely used commercially. With their exceptional thermal properties, the CNTs are an ideal candidate for TIM applications. Huang et al. [15] developed TIM based on an aligned CNT embedded in an elastomer. Although they achieved a 120% enhancement of thermal conductivity using the nanocomposite film, the value of 1.21 W/m·K that was achieved was much less than the value predicted based on an assumption of total thermal transport through the CNTs and polymer interface (i.e., perfect thermal interface). Therefore, Huang's work reveals that the expected improvement in the thermal conductivity was not achieved because of phonon scattering at the CNTs and the polymer interface caused by the impedance mismatch. Except for the TIMs, the thermal conductivity in adhesive joints in the through-thickness direction is very low, which is limited by the low thermal conductivity of the adhesive resin and interfacial imperfections between the adhesive and the adherends. Prior efforts to improve thermal conductivity in polymer or adhesive through adding the CNTs, as reviewed above, indicate that special attention needs to be paid when terminating CNT ends to minimize impedance mismatch at the interface between the CNT tips and the host material. Therefore, in an effort to enhance the heat transfer efficiency in the TIM and the adhesive jointed structures, the present study used vertically aligned CNTs to improve the through-thickness thermal conductivity in the adhesive joints. In this respect, analytical modeling was used in the design of the joint configuration to identify key parameters influencing the thermal transport by incorporating aligned CNTs in the joint. The modeling for the design of the joint configuration also provided processing guidelines, as described below.

As stated above, the thermal conductivity of MWCNTs is on the order of 2000 W/m·K, which is larger compared to that of the adhesive ($\sim$0.3 W/m·K). Hence, the acoustic impedance, which is related proportionally to the thermal conductivity of MWCNTs is much higher than that of the adhesive. It is known that the phonon transmission coefficient (hence, the thermal conductivity) at the interface of dissimilar materials (MWCNTs and adhesive in this case) is critically affected by the acoustic impedance mismatch between the respective materials [16]. For this particular reason, the approach of mixing CNTs or nanofibers in adhesive, exhibiting a large impedance mismatch between the CNT ends and adhesive, only marginally improved its thermal conductivity to approximately 0.7 W/m·K [13–15], which fails to provide an adequate solution for this study. Furthermore, the interfacial imperfections between the adhesive material and adherends degrade the thermal transfer efficiency significantly. Thus, to improve the phonon transport through the adhesive joints, the mismatch of the acoustic impedance between the CNTs and the adhesive needs to be minimized.

In this study, an efficient adhesive joint system was developed to improve the through-thickness thermal conductivity. We used vertically aligned MWCNTs to improve the through-thickness thermal conductivity in adhesive joints. The thermal conductivity of the aligned MWCNTs is not known. However, it is expected that the thermal properties of the aligned MWCNTs are similar to those of individual

MWCNTs if the quality and chirality of the MWCNTs remain the same. Furthermore, we introduced a transition zone (TZ) at the interface between the MWCNTs and the surrounding material to minimize the impedance mismatch and thus to minimize the interfacial imperfections. First, a numerical analysis based on a finite element (FE) analysis was conducted to study the effect of the nanotube and the TZ on the thermal transfer performance. Both geometric and thermal property variation were considered in parametric studies. Second, for validation of the FE simulation, the adhesive joint system was fabricated with the vertically aligned MWCNTs as well as introduction of the TZ between the nanotube ends and the surrounding adhesive–adherend surfaces. The TZ was achieved experimentally between the nanotube ends and graphite adherend surfaces through a series of metallic coatings, after a suitable functionalization of nanotube ends and adherend surfaces.

2 DESIGN OF A JOINT CONFIGURATION

As described above, vertically aligned MWCNTs were incorporated in the adhesive joint configuration. To incorporate the MWCNTs in the joint, MWCNT films were grown on silicon substrates by chemical vapor deposition. As grown, the aligned nanotubes are not perfectly aligned vertically. Figure 1 shows the scanning electron micrograph of as-produced MWCNT film. Although the vertical alignment of the nanotubes is evident from the SEM image, there are some obvious regions of imperfections, specifically bent tubes and sparse forests. Furthermore, the thermal transport properties of the aligned CNTs are not certain. To design the adhesive device configuration, parametric study was conducted to determine how significantly the irregularity of the CNTs and the uncertainty of the thermal properties affect the overall performance of the adhesive joint device.

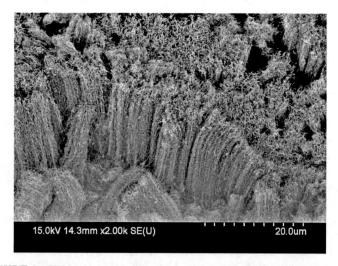

15.0kV 14.3mm x2.00k SE(U)   20.0um

FIGURE 1   SEM micrograph of a cross section of as-produced MWCNT film.

3   NUMERICAL THERMAL ANALYSIS

3.1   Steady-State and Transient Thermal Analysis

To understand the effect of material parameter on through-thickness thermal conductivity in an adhesive joint, a continuum-based FE thermal analysis of the adhesive joint was carried out. Figure 2 shows a schematic configuration of the present joint system. The vertically aligned MWCNTs (nanograss) are present in an adhesive layer that is placed between two adherent facesheets.

A representative unit-cell model was developed to pinpoint critical parameters that significantly affect the through-thickness conductivity in the adhesive joint. The unit-cell model considers a single NT that is embedded in the adhesive matrix at the ends. The connection between the NT and the matrix was achieved with a TZ. The FE analysis discretely modeled the nanotube, matrix material surrounding the nanotubes, and the TZ between the nanotube and matrix materials. Figure 3 shows FE meshes for both straight and curved nanotubes embedded in a matrix through the TZ. A typical diameter and length of the nanotube were selected to be 20 nm and 1 μm for the simulation, respectively. Steady-state and transient thermal analyses were conducted with the FE simulation using ANSYS, commercial FE software. Baseline properties of the thermal analysis were set as follows:

$$\text{Nanotube}: \quad k_{NT} = 500 \text{ W/m·K}, \quad \rho_{NT} = 1.5 \text{ g/cm}^3, \quad C_{p,NT} = 0.6 \text{ J/g·K}$$
$$\text{Matrix}: \quad k_m = 0.2 \text{ W/m·K}, \quad \rho_m = 1.6 \text{ g/cm}^3, \quad C_{p,m} = 1.3 \text{ J/g·K}$$
$$\text{TZ}: \quad k_{TZ} = 0.2 \text{ W/m·K}, \quad \rho_{TZ} = 1.6 \text{ g/cm}^3, \quad C_{p,TZ} = 1.3 \text{ J/g·K}$$

where k, ρ, and C_p are thermal conductivity, density, and heap capacity, respectively. The heat capacity of the nanotube was assumed to be similar to that of the graphite [17]. A boundary condition of fixed temperatures of 25 and 500°C was applied arbitrarily at the low- and high-temperature ends of the matrix, respectively.

Figure 4 shows the temperature profiles of the model after steady-state analysis. Two radii of the TZ were used for the straight nanotube, while a curved nanotube was modeled with a radius of curvature of 1.13 μm. In all cases, the temperature varies uniformly from the high temperature at 500°C to the low temperature at 25°C.

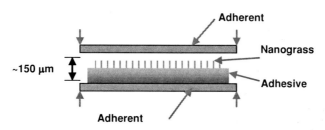

FIGURE 2   Configuration of nanograss (MWCNTs) infiltrated with adhesive in an adhesive joint.

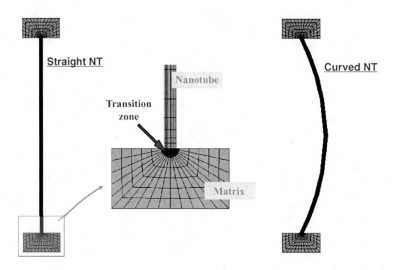

FIGURE 3 Finite-element model for straight and curved nanotubes embedded in matrix material through a transition zone.

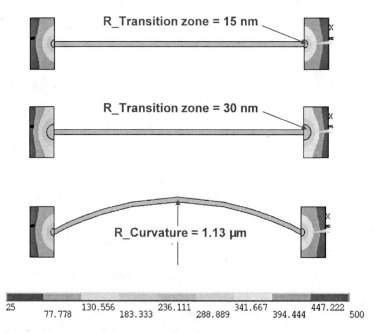

FIGURE 4 Uniform temperature profiles after steady-state analysis, regardless of the radius of TZ and radius of curvature of the nanotube.

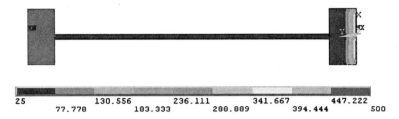

FIGURE 5   Temperature profile of a transient analysis after 3 ns.

Therefore, the thermal efficiency of either the TZ or the NT curvature could not be determined from steady-state analysis. Figure 5 shows a temperature distribution with the transient thermal analysis with the fixed temperature boundary conditions as stated above. In this model, the properties of the TZ were assumed to be the same as those of the surrounding matrix material. The temperature profile shows the instant temperature variation near the high-temperature end after a short duration of 3 ns, which indicated that the transient analysis could potentially be used to analyze the thermal efficiency of the TZ and NT conductivity and its curvature through a parametric study.

3.2 Parametric Study

Parametric studies were conducted using transient thermal analysis by varying parameters such as thermal conductivity of the nanotube and the TZ as well as the radius of curvature of the nanotube and the radius of the TZ. The baseline properties stated above were used if not stated explicitly in the parametric study. The parametric study was performed for an idealized single-nanotube scenario.

Effect of Thermal Conductivity of the Nanotube   First, the thermal conductivity of the nanotube was varied from 5 W/m·K to 500 W/m·K, while that of the TZ was fixed at 0.2 W/m·K. The temperatures at three points in the nanotube (A, B, and C in Figure 6) were calculated with an increase in the transient time history up to 0.3 μs. The lowest $k_{NT} = 5$ W/m·K results in the largest difference in the temperatures at these points, and the increase in k_{NT} gradually makes the difference smaller. With $k_{NT} = 500$ W/m·K, the temperature differences at the three points were negligible, which indicates extremely good heat transfer. Therefore, the conductivity of the nanotube, whether it is a single- or multiwalled nanotube, significantly affects the heat transfer efficiency.

Effect of the Radius of Curvature of the Nanotube   The transient thermal analysis was conducted for the curved nanotube with three different radii of curvature: $R_{NT} = 0.52$ μm, 1.13 μm, and ∞ (straight NT). With a nanotube radius of 10 nm, the ratios of the nanotube curvature to its radius were 52, 113, and ∞, respectively. Nanotubes having low (5 W/m·K) and high (500 W/m·K) thermal conductivities were used for the calculation, as shown in Figures 7 and 8, respectively. The temperatures at three

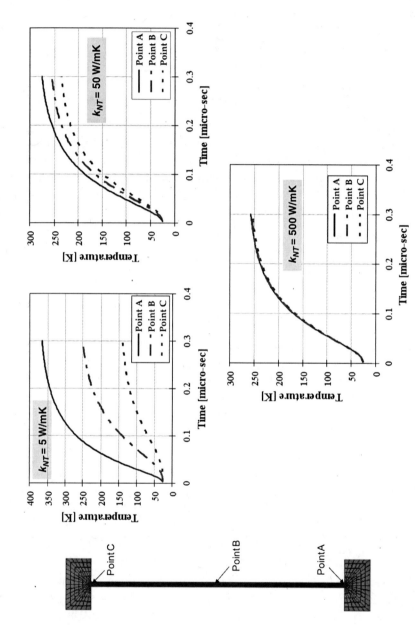

FIGURE 6 Temperature increase at points A, B, and C in the nanotube against transient time with various k_{NT} values.

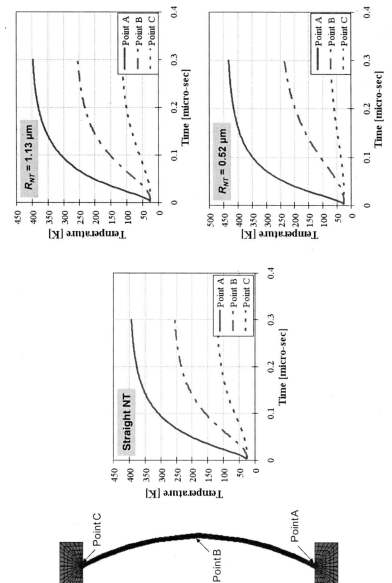

FIGURE 7 Temperature increase at points A, B, and C in straight and curved nanotubes against transient time with various R_{NT} values ($k_{NT} = 5$ W/m·K).

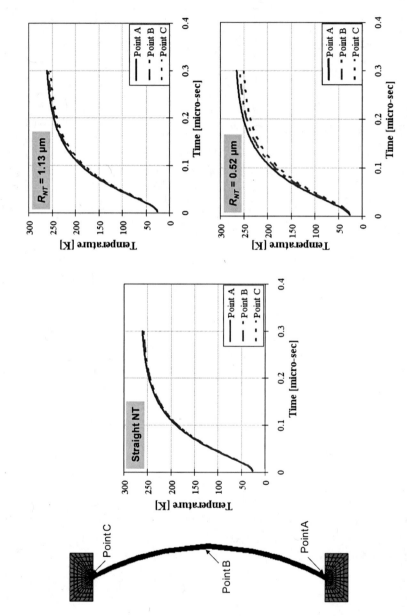

FIGURE 8 Temperature increase at points A, B and C in straight and curved nanotubes against transient time with various R_{NT} values ($k_{NT} = 500$ W/m·K).

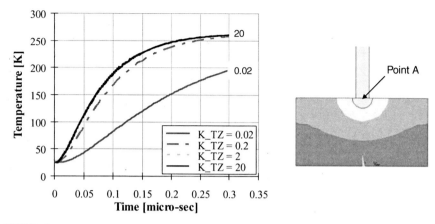

FIGURE 9   Temperature increase at a contact point between the nanotube and the TZ (point A) against transient time with various k_{TZ} values.

points (A, B, and C) were calculated with an increase in the transient time history up to 0.3 μs. Again, k_{TZ} was fixed at 0.2 W/m·K. Similar to the straight nanotube, curved nanotubes with low k_{NT} values result in a large difference in the temperatures at the three points, and those with high k_{NT} values result in a negligible difference. Furthermore, Figure 8 shows that the curviness of the nanotube has a negligible influence on the heat transfer with a high k_{NT} value.

Effect of the Thermal Conductivity of the Transition Zone   The parametric study was conducted by varying the thermal conductivity of the TZ from 0.02 to 20 W/m·K. The temperature at a contact point (point A in Figure 9) between the nanotube and the TZ was calculated with an increase in the transient time history up to 0.3 μs, as shown in Figure 9. The lowest k_{TZ} value, 0.02 W/m·K, results in the slowest heat transfer in the TZ, and the increase in k_{TZ} gradually increases the heat transfer with time. However, a k_{TZ} higher than 2 W/m·K results in a negligible temperature increase with time. Therefore, it is necessary to achieve a good thermal transition between the nanotube and the matrix material for better heat transfer, but the improvement has a diminishing return with improvement in k_{TZ}.

Effect of the Radius of the Transition Zone   The radius of the TZ was varied from 0 to 30 nm, and the temperature increase with time was calculated. The thermal conductivities of the nanotube and the TZ were set at 500 and 20 W/m·K, respectively. As Figure 10 shows, a larger radius of TZ with the higher thermal conductivity facilitates heat transfer (increases the temperature quickly). Therefore, it is desirable to introduce a large volume of high-conductive TZ for efficient thermal transfer.

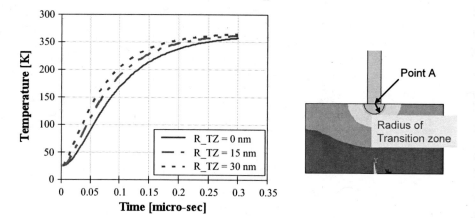

FIGURE 10   Temperature increase at a contact point between the nanotube and the TZ (point A) against transient time with various R_{TZ} values.

3.3   Analytical System Level Thermal Transport Study

An analytic unit-cell model was used to calculate the through-thickness thermal conductivity (k_{eff}) of the adhesive joint device with an adhesive layer reinforced with vertically aligned CNTs. The advantage of performing the analytical study was that it was system level, thus considering a more realistic system similar to the experimental setup. Primary parameters affecting the through-thickness conductivity in the adhesive layer reinforced with the vertically aligned CNTs are the density of CNT forest, thermal conductivities of CNT, adhesive, and adherent material. In addition, in fabricating the adhesive device, if the tips of the CNTs were not exposed through the adhesive layer or an excess of adhesive resin remain at the tips, the CNTs could not make a direct contact with the adherent facesheet and thus create the third material phase. This third phase is called a transition zone layer. The thickness and thermal conductivity of the TZ layer affect the through-thickness conductivity as well.

The measured thickness of the present adhesive joint device was 6 mm and that of the adhesive layer was 30 μm. Considering a radius of an individual nanotube of less than 100 nm and a device thickness of 6 mm, the aspect ratio of the total thickness of the system and the radius of the nanotube was greater than 100. Therefore, the present adhesive system can be analyzed with a one-dimensional model as shown in Figure 11. The effective through-thickness thermal conductivity can then be calculated by an isostress (Reuss) model as

$$k_{eff} = \frac{2t_{fs} + t_{TZ} + t_{ad}}{2t_{fs}/k_{fs} + t_{TZ}/k_{TZ} + t_{ad}/k_{ad}} \tag{1}$$

where k and t represent the thermal conductivity and the thickness of each phase of material, respectively. The subscripts fs, TZ, and ad represent the adherent facesheet,

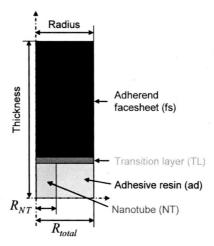

FIGURE 11 Axisymmetric unit-cell model for an adhesive joint with an adhesive layer reinforced with vertically aligned CNTs.

the TZ layer, and the adhesive layer, respectively. The effective thermal conductivity of the adhesive layer with the resin and the vertically aligned CNTs (k_{ad}) can be calculated further by an isostrain (Voight) model as

$$k_{ad} = \frac{k_{NT} A_{NT} + k_{ep} A_{ep}}{A_{NT} + A_{ep}} \tag{2}$$

where k and A represent the thermal conductivity and the surface area perpendicular to the through-thickness direction, respectively. The subscripts NT and ep represent the nanotube and the epoxy resin, respectively. The area can be calculated using the radius of the nanotube as $A_{NT} = \pi R_{NT}^2$ and $A_{ep} = \pi (R_{total}^2 - R_{NT}^2)$, where R_{NT} and R_{total} are the radii of the nanotube phase and the total phase, respectively. The thermal conductivities and the thicknesses of each phase of material are listed in Table 1.

Figure 12 shows the effective thermal conductivity with respect to the CNT density in the form of a ratio of the total radius to the CNT radius. Note that the k_{eff} is

TABLE 1 Thermal Conductivities and Thicknesses of Individual Phases of Material in Figure 11

Material	Conductivity, k (W/m·K)	Thickness, t (mm)
Facesheet	400	2984
Epoxy	0.3	30
Transition zone	0.3–50	1
Nanotube	0.3–2000	30

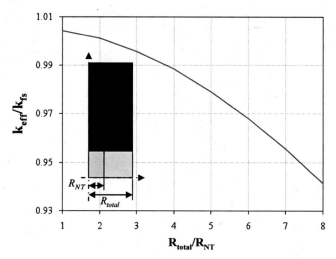

FIGURE 12 Effective through-thickness conductivity of adhesive versus a ratio of total radius to the radius of CNTs.

normalized by the k_{fs} hereafter. It was assumed that a TZ does not exist in this and later examples until stated explicitly. As the radius ratio, R_{total}/R_{NT}, increases, the CNT density decreases, resulting in reduction of k_{eff}. Note that it was observed from SEM images that the radius ratio ranges from approximately 4 to 8. In subsequent calculations, the radius ratio was set to 8.

Figure 13 shows the changes in k_{eff} versus the variation in k_{NT} ranging from the same value of the epoxy resin, 0.3 W/m·K, to a hypothetically achievable value of

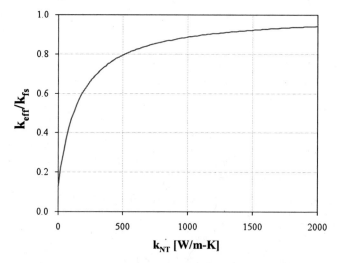

FIGURE 13 Effective through-thickness conductivity of adhesive versus the thermal conductivity of CNTs.

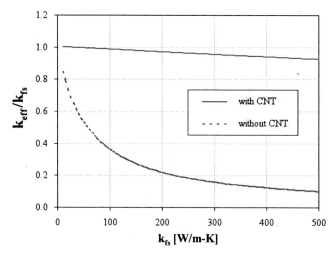

FIGURE 14 Effective through-thickness conductivity of adhesive against thermal conductivity of adherent facesheet with and without CNTs.

2000 W/m·K. All the other parameters remain the same. The calculation indicates that the increase in the thermal conductivity of the reinforcing material (CNT in this case) enhances k_{eff}. The rate of the increment is steeper with lower k_{NT} that with higher k_{NT}. k_{eff} becomes saturated to a value near k_{fs} for larger values of k_{NT}, so that further improvement in k_{NT} is not necessary for overall k_{eff} enhancement. The saturation limit is determined by k_{fs}.

Figure 14 shows the changes in k_{eff} versus the variation in k_{fs} without and with reinforcement of the CNTs. All other parameters remain the same. The calculation indicates that the increase in the thermal conductivity of the facesheet in the adhesive joint would not contribute efficiently to k_{eff} enhancement if the pristine adhesive was used without the CNTs, as the dotted line indicates. If vertically aligned CNTs were used to help the thermal transport in the through-thickness direction, the k_{eff} would be enhanced nearly proportionally to the increase in k_{fs}. It was found that enhancement of the k_{eff} is nearly linear with the increase in k_{fs} with the CNTs.

Figure 15 shows the changes in k_{eff} with respect to the variation in k_{NT} with and without a TZ. The thickness of the excessive adhesive layer (considered as the TZ) was assumed to be 1 μm. Solid and dotted lines represent the case with and without the TZ layer, respectively. Note that the dotted line in Figure 15 for the case without the TZ is the same as the line in Figure 13. As the figure shows, the excessive adhesive layer preventing direct contact between the CNTs and the adherent facesheets lowers k_{eff} significantly. No matter how the highly conductive facesheets are used, the maximum k_{eff} is limited to less than 70% of k_{fs} with the presence of adhesive TZ, compared to nearly 95% of k_{fs} with direct contact. Therefore, it is important to eliminate the excess adhesive resin near the CNT tips as much as possible.

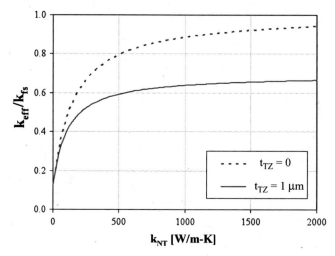

FIGURE 15   Effective through-thickness conductivity of adhesive versus thermal conductivity of CNT. Solid and dotted lines represent cases with and without a transition zone (TZ) between adherend and adhesive layer.

Although it is desirable that the CNT tips be exposed as much as possible for direct contact with the facesheet, it is still almost impossible to make all the tips contact the facesheet surfaces directly because of the difference in the nanotube height (length) and irregular surface morphology of the etched adhesive layer. The mismatch between the surface morphologies can be filled with the third TZ material, which has the highest thermal conductivity. In this case, the TZ is no longer the excessive adhesive layer but can be a highly conductive coating material phase. Figure 16 shows the changes in k_{eff} versus the variation in k_{TZ} with a TZ of 1 μm thickness. As the figure shows, the conductive coating layer improves k_{eff} significantly compared with the excessive adhesive layer. k_{eff} becomes saturated to the value of the perfect contact beyond approximately $k_{TZ} > 5$ W/m·K. Therefore, it is important to have the conductive coating layer in order to achieve the good thermal transport between the adhesive and adherent layers, but the thermal conductivity of the coating layer is not necessarily extremely high. This conductive layers were realized by metallic gold and indium coatings in the present study, which will be discussed later in this chapter.

In summary, the parametric study described above indicated that the irregularity due to the sparcity of the CNTs does not adversely affect the thermal conductivity of the adhesive joint device as far as the conductive phase of the CNTs is used, and good thermal contact between the CNT tips and the adherent facesheet are established through either direct contact and/or with a conductive TZ layer. Therefore, the most important point learned through the parametric study is that appropriate processing schemes need to be developed to ensure thermal contact of the CNT tips with the adherent facesheet through direct contact or a conductive TZ. The process steps developed to achieve that are discussed in the following section.

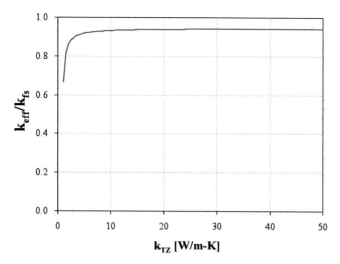

FIGURE 16 Effective through-thickness conductivity of adhesive versus thermal conductivity of TZ.

4 EXPERIMENTS

4.1 Processing

The finite element numerical study indicated that introduction of a small TZ at the MWCNT ends [i.e., at the interface of the MWCNTs and the surrounding matrix (adhesive and/or adherend materials)] of thermal conductivity higher than that of the matrix will enhance the phonon transport; hence, it is expected to improve through-thickness thermal conductivity in the adhesive joints. Among many possible methods, the TZ was introduced through suitable functionalization of nanotubes and a series of metallic coatings by gold and indium in this study.

First, MWCNT films were grown on quartz substrates by chemical vapor deposition. The aligned CNT films were prepared by pyrolyzing iron(II) phthalocyanine under Ar–H$_2$ at 900°C, as described in detail elsewhere [18]. The diameter of the tubes was 30 nm, and the length of the MWCNT film was 30 μm. Figure 17 shows the cross-sectional view of the as-produced MWCNT film. The vertical alignment of the nanotubes is evident from the SEM image. To infuse the epoxy (adhesive) in the CNT forest without disturbing its alignment, the wafer with the MWCNT side facing upward was dipped in a beaker containing a 10% Epon 862/W–acetone solution. The film was then kept in a vacuum oven at 60°C for 2 h for the solvent to escape. The epoxy was then cured at 177°C for 2 h. Figure 18 shows the cross-sectional view of the MWCNT film infused with the epoxy. The epoxy–MWCNT film was then peeled off the quartz substrate by etching with a 10% HF solution. The nanotube tips were exposed selectively by etching the film surface with 32-W RF oxygen plasma for 30 min. The SEM images of the film after the plasma etching are shown in Figure 19. It can be observed from the figure that the nanotube tips are clipped due to the

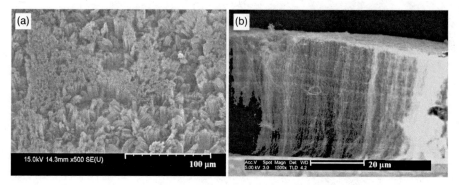

FIGURE 17 SEM micrograph of cross-sectional view of as-produced MWCNT film (nanograss).

FIGURE 18 SEM micrograph of a cross section of epoxy-infused MWCNT film before plasma etching.

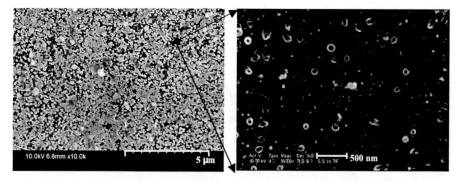

FIGURE 19 SEM micrograph of adhesive film with MWCNT tips exposed after plasma etching (view normal to the MWCNT film plane).

plasma etching. The side of the film that was previously anchored to the substrate was similarly etched in RF plasma under the conditions described above.

The parametric study indicated that MWCNT tips need to make good thermal contact with the adherent facesheet, either directly or through a conductive TZ. To demonstrate this through-thickness thermally conductive joint concept, we have chosen highly oriented pyrolytic graphite (HOPG) as the adherend, a conductive adherent facesheet. Due to the highly oriented graphene microstructure of HOPG, it is not simple to make a durable direct thermal contact between CNT tips and an HOPG surface. Thus, a thin gold layer was used to establish a conductive TZ to make thermal contact between CNT tips and a HOPG surface. A 900-Å layer of gold was then thermally evaporated on both sides of the film. In addition, the HOPG adherent facesheets were sputter-coated with gold and palladium for 3 min. Then a thin layer of indium metal was melt-coated on the graphite adherent surfaces and the epoxy–nanotube film to minimize the thermal resistance caused by surface roughness. Finally, the epoxy–nanotube film was sandwiched between the graphite facesheets and fused together by heating at 175°C to achieve the schematic configuration shown in Figure 2.

4.2   Measurement of Thermal Conductivity

The effective thermal conductivity of the adhesive joint system was measured by measuring thermal diffusivity using a laser flash (or heat pulse) technique. The measurement was made using a Netzsch laser flash apparatus under nitrogen purge. The laser flash technique allows measurement of the thermal diffusivity of solid materials over the temperature range −180 to 2000°C. The laser flash technique consists of applying a short-duration (less than 1 μs) heat pulse using the laser to one face of a parallel-sided sample and monitoring the temperature rise on the opposite face as a function of time. The temperature rise is monitored with an infrared detector. The thermal diffusivity (h) can then be calculated as

$$h = \frac{\varpi L^2}{\pi t_{1/2}}$$

(3)

where ϖ is a constant, L the thickness of the specimen, and $t_{1/2}$ the time for the rear surface temperature to reach one-half of its maximum value. The specific heat (C_p) of the samples can also be measured with the same laser flash apparatus by comparing the temperature rise of the sample to the temperature rise (ΔT) of a reference sample of known specific heat tested under the same conditions. Assuming that the laser pulse energy and its coupling to the sample remain unchanged between samples, the heat capacity can be obtained by

$$(C_p)_{\text{sample}} = \frac{(mC_p\Delta T)_{\text{ref}}}{(m\Delta T)_{\text{sample}}}$$

(4)

where m is mass and the subscripts ref and sample represent the reference sample and the sample of interest, respectively. We used a POCO graphite plate as the reference sample. The density (ρ) was calculated by measuring the weight and volume of the samples and taking a ratio of the measured weight to the measured volume.

Finally, with the measurement of the thermal diffusivity (h), the heat capacity (C_p) and the density (ρ), the thermal conductivity (k_z) of the samples can be calculated as

$$k_z = C_p \rho h \qquad (5)$$

The thermal conductivity of the graphite facesheet and the neat epoxy were also measured using the same method. Note that all the thermal conductivity measurements made in this study were performed at room temperature (24°C). Therefore, the values of the thermal conductivity (k_z) calculated using equation (5) is valid for the room temperature. The laser voltage used was 1826 V with a gain of 50 and an orifice area of 78.54 mm². A diffusivity model used for fitting the experimental data was a Cowan + pulse correction model, and the correlation coefficient for the data fit was 0.99864.

To determine quantitatively the effect of incorporating aligned MWCNTs on the thermal conductivity of the adhesive joint, four sets of samples were evaluated: HOPG facesheet, HOPG layers bonded by epoxy, HOPG layers bonded by indium, and HOPG layers bonded by indium with the modified MWCNT film in between. The measured values of the thermal conductivity of these samples are presented in Figures 20 and 21. As Figure 21 shows, the measured thermal conductivities of a graphite facesheet, a facesheet bonded by indium, a facesheet bonded by epoxy adhesive, and the actual device were 400, 9.2, 1.1, and 262 W/m·K, respectively. Compared to the present value, the study of mixing nanotubes in epoxy by Huang

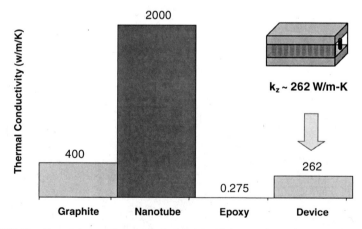

FIGURE 20 Comparison of measured through-thickness thermal conductivities of the graphite adherend (facesheet), an MWCNT (theoretical), epoxy adhesive, and the adhesive joint device with vertically aligned MWCNT.

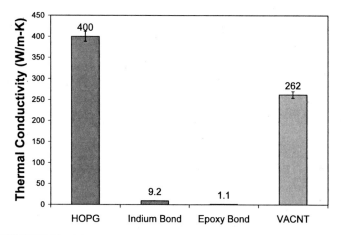

FIGURE 21   Measured thermal conductivities of bonded joint samples.

et al. [15] showed a considerably lower thermal conductivity: 1.21 W/m·K. Note that the joint device without using the vertically aligned CNTs yielded a thermal conductivity of 0.790 W/m·K. The low value of k_z in Huang's work and the present work without the CNTs was attributed to phonon scattering caused by significant acoustic impedance mismatch at the interface of the nanotube tips terminated in the epoxy (adhesive). It was possible to achieve the higher thermal conductivity value of 262 W/m·K in the present study compared to that of Huang because of reduced interface impedance mismatch through the use of the metallic interface as the TZ layer instead of a polymeric interface as revealed in the parametric study. The thermal conductivity of the device in this study is still lower than that of the pyrolytic graphite adherend (400 W/m·K), which may be due to the use of indium, whose thermal conductivity is about 70 W/m·K, much lower than that of pyrolytic graphite and MWCNTs.

5   SUMMARY AND DISCUSSION

A concept of incorporating aligned conductive phase (MWCNT) in an adhesive layer has been demonstrated to enhance the through-thickness thermal conductivity in adhesively bonded joints. The key parameters for improving the through-thickness thermal conductivity in an adhesive joint, identified though a numerical study, are (1) improvement in the thermal conductivity of the aligned nanotubes, and (2) the decent thermal conductivity and size of the TZ near the nanotube ends and the adherent surfaces. Therefore, good thermal contact of the conductive phase with adherent surfaces needs to be established in order to achieve the desirable thermal transport through the thickness. Further, acoustic impedance mismatch at the interface (TZ) needs to be minimized to minimize the phonon scattering to maximize thermal transport. This theoretical observation was realized experimentally using the aligned

MWCNTs (nanograss), surface functionalization by the gold on the nanotube ends, and by the gold-palladium on adherent surfaces and introduction of the TZ by indium.

Analytical modeling was used to identify the key parameters for inclusion of MWCNTs influencing the through-thickness thermal transport in joints. Parametric study of the analytical modeling demonstrated the need for a conductive TZ between the MWCNT tips and the adherent surface to efficiently utilize the superior thermal transport characteristics of the nanotubes as thermal interface materials. Highly oriented pyrolytic graphite (HOPG) facesheets were use in this study to demonstrate this conductive interface concept. Appropriate processing steps were developed to incorporate the aligned MWCNTs in the adhesive joint configuration with the conductive TZ. First, a self-standing film of the carbon nanotube film was formed by impregnating it with the epoxy resin. Selective etching of the epoxy matrix was then performed to expose the nanotube tips to the conductive TZ. The conductive TZ was established through metallization (gold sputter coating) of both the nanotube tips and graphite facesheets to ensure reduction in acoustic impedance mismatch between the nanotubes and the graphite faceplate.

The through-thickness thermal conductivity of the adhesive joint was measured by thermal diffusivity using the laser-flash method. The measured through-thickness thermal conductivity of the adhesive joint configuration of the aligned MWCNT infused with adhesive using the pyrolytic graphite facesheet was over 250 W/m·K, which is a significant improvement in the through-thickness thermal conductivity over that without aligned MWCNTs (<1 W/m·K). $k_z \sim 250$ W/m·K supersedes the through-thickness thermal conductivity requirement of the adhesive joints for space structures by an order of magnitude. However, it should be noted that since the k_z value of composite facesheets is much lower than that of pyrolitic graphite facesheets, the k_z value of the same joint configuration using composite facesheets is expected to be lower than 250 W/m·K.

Acknowledgment

This work was performed under U.S. Air Force Contract FA8650-05-D-5052.

REFERENCES

1. Iijima, S., Helical microtubules of graphitic carbon, *Nature (London)* 1991;354(6348): 56–58.

2. Frank, S., Poncharal, P., Wang Z. L., and De Heer, W. A., Carbon nanotube quantum resistors. *Science* 1998;280(5370):1744–1746.

3. Kim, P., and Lieber, C. M., Nanotube nanotweezers, *Science* 1999;286(5447):2148–2150.

4. Berber, S., Kwon, Y.-K., and Tomanek, D., Unusually high thermal conductivity of carbon nanotubes, *Phys. Rev. Lett.* 2000;84(20):4613–4616.

5. Ghosh, S., Sood, A. K., and Kumar, N., Carbon nanotube flow sensors, *Science* 2003;299(5609):1042–1044.

6. Baughman, R. H., Cui, C., Zakhidov, A. A., Iqbal, Z., Barisci, J. N., Spinks, G. M., Wallace, G. G., Mazzoldi, A., De Rossi, D., Rinzler, A. G., Jaschinski, O., Roth, S., and Kertesz, M., Carbon nanotube actuators, *Science* 1999;284(5418):1340–1344.

7. Frackowiak, E., and Béguin, F., Electrochemical storage of energy in carbon nanotubes and nanostructured carbons, *Carbon* 2002;40(10):1775–1787.

8. Clifford, J. P., John, D. L., and Castro, L. C., Pulfrey, D. L., Electrostatics of partially gated carbon nanotube FETS, *IEEE Trans. Nanotechnol.* 2004;3(2):281–286.

9. Che, J., Cagin, T., and Goddard, W. A., Thermal conductivity of carbon nanotubes, *Nanotechnology* 2000;11(2):65–69.

10. Kim, P., Shi, L., Majumdar, A., and McEuen, P. L., Thermal transport measurements of individual multiwalled nanotubes, *Phys. Rev. Lett.* 2001;87(21):215502.

11. Xie, H., Lee, H., Youn, W., and Choi, M., Nanofluids containing multiwalled carbon nanotubes and their enhanced thermal conductivities, *J. Appl. Phys.* 2003;94(8):4967–4971.

12. Xue, Q. Z., Model for effective thermal conductivity of nanofluids, *Phys. Lett. A* 2003;307(5–6):313–317.

13. Biercuk, M. J., Llaguno, M. C., Radosavljevic, M., Hyun, J. K., Johnson, A. T., and Fischer, J. E., Carbon nanotube composits for thermal management, *Appl. Phys. Lett.* 2002;80(15):2767–2769.

14. Liu, C. H., Huang, H., Wu, Y., and Fan, S. S., Thermal conductivity improvement of silicone elastomer with carbon nanotube loading, *Appl. Phys. Lett.* 2004;84(21):4248–4250.

15. Huang, H., Liu, C., Wu, Y., and Fan, S., Aligned carbon nanotube composite films for thermal management, *Adv. Mater.* 2005;17(13):1652–1656.

16. Cahill, D. G., Ford, W. K., Goodson, K. E., Mahan, G. D., Majumdar, A., Maris, H. J., Merlin, R., and Phillpot, S. R., Nanoscale thermal transport, *J. Appl. Phys.* 2003;93(2):793–818.

17. Masarapu, C., Henry, L. L., and Wei, B., Specific heat of aligned multiwalled carbon nanotubes, *Nanotechnology* 2005;16(9):1490–1494.

18. Wang, Q. H., Corrigan, T. D., Dai, J. Y., Chang, R. P. H., and Krauss, A. R., Field emission from nanotube bundle emitters at low fields, *Appl. Phys. Lett.* 1997;70(24):3308–3310.

10

USE OF METAL NANOPARTICLES IN ENVIRONMENTAL CLEANUP

SUSHIL R. KANEL

Department of Systems and Engineering Management, Air Force Institute of Technology, Wright-Patterson Air Force Base, Dayton, Ohio

CHUNMING SU

U.S. Environmental Protection Agency, Ada, Oklahoma

UPENDRA PATEL

Dr. Jivraj Mehta Institute of Technology, Gujarat, India

ABINASH AGRAWAL

Department of Earth and Environmental Sciences, Wright State University, Dayton, Ohio

Nanoscale Multifunctional Materials: Science and Applications, First Edition.
Edited by Sharmila M. Mukhopadhyay.
© 2012 John Wiley & Sons, Inc. Published 2012 by John Wiley & Sons, Inc.

1   INTRODUCTION: ZERO-VALENT IRON NANOPARTICLES (INPs) IN THE ENVIRONMENT

Iron exists in various forms, such as zero-valent iron, iron oxides, and oxyhydroxides. In this chapter, however, we focus on applications of zero-valent iron nanoparticles (INPs) for soil and water treatment. INPs are used to treat a variety of dissolved contaminants in the water, including organic and inorganic contaminants, heavy metals, metalloids, and actinides. Because of their high surface area/volume ratio (due to their very small size) and very high reactivity, INPs have been studied extensively for the treatment of contaminants in the soil, groundwater, and wastewater [1–4]. Recently, some excellent reviews have summarized the synthesis, properties, and environmental applications of INPs [3,5–9]. We emphasize here some of the important attributes of INPs, including their applications in soil and water treatment.

1.1   Synthesis of INPs

INPs have been synthesized in a variety of ways. A few recent methods of INP preparation are discussed briefly below.

Chemical Precipitation Method   Chemical precipitation is one of the most common approaches to INP synthesis [10]. Generally, a strong reducing agent such as $NaBH_4$ is used ferrous or ferric ion in solution to elemental iron; in this method INPs are synthesized by adding aqueous $NaBH_4$ solution slowly to a degassed aqueous solution of iron salt (ferric chloride or ferrous sulfate) at room temperature ($\sim$23°C) with simultaneous stirring [11]. Ferric or ferrous ion is reduced chemically according to the reaction [10]

$$Fe(H_2O)_6{}^{3+} + 3BH_4{}^- + 3H_2O = Fe^0 \downarrow + 3B(OH)_3 + 10.5H_2 \qquad (1)$$

After stirring the solution for 20 min, it is centrifuged at 6000 g for 2 min, and the supernatant is then replaced by acetone [2]. Acetone washing prevents the immediate rusting of INPs during synthesis and leads to a fine black powder product after freeze-drying. Ethanol washing was found to be more effective in terms of longer protection against corrosion and rusting [12]. Due to its potential for rapid oxidation, the synthesis of INPs is performed under constant N_2 atmosphere. INPs can also be synthesized in an anerobic chamber (glove box) at ambient temperature.

Heating Method   The nanosized Fe can also be synthesized that is supported on carbon particles. In this method, compressed carbon black is combined with aqueous solutions of iron(III) salts, either by adsorption or by impregnation [13]. In a typical synthesis procedure, 50 g of $Fe(NO_3)_3 \cdot 9H_2O$ and 5 g of carbon black (powder) are homogenized in 200 mL of deionized water for 30 min. The supernatant is then separated from the solid by vacuum filtration using a nylon membrane filter. The moist solid is then removed from the filter and placed in a vacuum chamber overnight to dry. After drying, the solid is transferred to an alumina boat and placed in a quartz tube inside a furnace; the solid is heated to 800°C at a ramp rate 4.5°C/min under Ar flow at 200 cm³/min and kept at 800°C for 3 hr. The tube containing the solid is purged with Ar for 1 hr before heating; further, the solid in the tube is kept under an Ar atmosphere for several hours while it is allowed to cool down to ambient temperature, before its removal from the furnace. The resulting carbon-supported nanosized iron ($C-Fe^0$) particles can then be handled in air to characterize the solid.

Sonolytic Method   Next we describe an approach for the synthesis of unagglomerated form of INPs. In this method, sonolysis is used in place of heating with constant stirring by the method described earlier. This method for the synthesis of INPs involves sonolytic breakdown of a solution of an organometallic, $Fe(CO)_5$, dissolved in anisol in the presence of poly(dimethylphenylene oxide) (PPO) that acts as a stabilizer for metal nanoparticles, and this leads to the formation of small nonagglomerated INPs (initial conditions: [Fe] = 0.3 mol/L; Fe/PPO = 170 wt%) [14]. High-resolution transmission electron microscopy (HRTEM) has shown that the median size of the INPs is centered around 3 nm; smaller INPs (diameter $\leq$ 2.5 nm) adopt the α-Fe (body-centered cubic) structure, whereas the larger ones (diameter $\geq$ 2.5 nm) adopt the γ-Fe (face-centered cubic) structure.

Electolytic–Sonolytic Hybrid Method   A new method for the synthesis of INPs was developed that combines electrochemical and ultrasonic techniques [15]. In this method, INPs are produced by reducing Fe^{3+} ions in a ferric chloride solution on a platinum cathode:

$$Fe^{3+} + 3e^- (+\text{stabilizer}) = Fe^0 (\text{or INPs}) \qquad (2)$$

In this synthesis, Fe^0 atoms developed on the platinum cathode instantaneously, and they grow to nanoscale range before the INPs were quickly harvested from the cathode into a surfactant solution. Ultrasonic vibrators (frequency: 20 kHz) were used during the synthesis to provide physical energy to help detach INPs from the cathode (see the experimental setup in Figure 1). The INPs produced by this method exhibited a diameter of 1 to 20 nm and a specific surface area of 25.4 m²/g.

Green Technique of INP Synthesis   INPs can also be synthesized with environmentally safe or green reagents, in which use of a strong reducing agent (such as

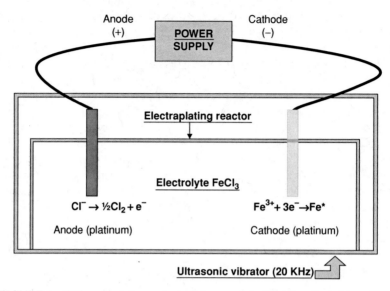

FIGURE 1   INP synthesis by an electrolytic–sonolytic hybrid method. (From [15].)

NaBH$_4$) is avoided. In this method, Fe(NO$_3$)$_3$ aqueous solution is reacted with tea extracts (polyphenols) to produce INPs [16] (Figure 2). This method does not form harmful by-products, such as boric acid, as in chemical precipitation methods, and it is hence considered a green technique. Large-scale synthesis of INPs using this method, however, may be a challenge.

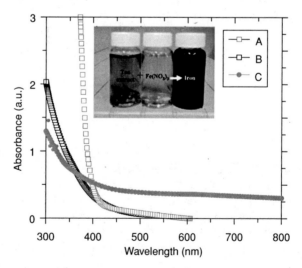

FIGURE 2   Ultraviolet spectra of (A) Fe(NO$_3$)$_3$, (B) tea extract (polyphenols), and (C) INPs that form as a reaction product of Fe(NO$_3$)$_3$ and tea extract. The inset shows the image of the reaction stages. (From [16].)

2   USES OF INPs IN ENVIRONMENTAL REMEDIATION

2.1   Treatment of Organic Contaminants

Overview of Organic Contaminant Remediation by INPs   Recent advances in nanotechnologies have shown a considerable potential for its use in environmental cleanup. Since the early 1990s, bench-scale studies and field-scale uses of INP-based technology have been investigated extensively for the mitigation of hazardous contaminants. Progress has been made in several areas, which include INP preparation by synthetic techniques, modification of the INP surface for improved stability and mobility, and improvement in field delivery techniques for subsurface destruction of environmental pollutants [17]. INPs are potent reductants, which makes them suitable for treating a wide range of hazardous aqueous-phase contaminants; INPs can assist in degrading many organic pollutants, such as chlorinated solvents, organochlorine pesticides, polychlorinated biphenyls (PCBs), organic dyes, and explosives (Table 1).

TABLE 1   Organic Contaminants Treatment by Metal Nanoparticles (Fe, Fe–Pd, Pd–Au, Fe–Ni, Fe–Cu, Fe–Ag, Fe–O$_2$, Ni–Fe–O$_2$, Fe–H$_2$O$_2$)

Chloromethanes	
Carbon tetrachloride, (CCl$_4$)	Fe [24,31]
Chloroform (CHCl$_3$)	Fe [24,31]
Dichloromethane (CH$_2$Cl$_2$)	Fe [24]
Chloromethane (CH$_3$Cl)	Fe [19]
Trihalomethanes	
Bromoform (CHBr$_3$)	Fe [19]
Dibromochloromethane (CHBr$_2$Cl)	Fe [19]
Dichlorobromomethane (CHBrCl$_2$)	Fe [19]
Chloroethanes	
Hexachloroethane (C$_2$Cl$_6$)	Fe [24,32]
Pentachloroethane (C$_2$HCl$_5$)	Fe [24,32]
1,1,2,2-Tetrachloroethane (C$_2$H$_2$Cl$_4$)	Fe–Pd [24]
1,1,1,2-Tetrachloroethane (C$_2$H$_2$Cl$_4$)	Fe–Pd and Fe [23,24]
1,1,2-Trichloroethane (C$_2$H$_3$Cl$_3$)	Fe [24]
1,1,1-Trichloroethane (C$_2$H$_3$Cl$_3$)	Fe–Pd–Fe [23,24]
1,2-Dichloroethane (C$_2$H$_4$Cl$_2$)	Fe [23,24]
1,1-Dichloroethane (C$_2$H$_4$Cl$_2$)	Fe [23,24]
Chloroethenes	
Tetrachloroethene (C$_2$Cl$_4$)	Fe, Fe–Pd [33]; Fe–Pd [34]
Trichloroethene (C$_2$HCl$_3$)	Pd–Fe, Pd–Zn, Pt–Fe, Ni–Fe, and Fe [33]; EZVI [35]; Pd–Au [36]; Fe–Pd [34]
cis-Dichloroethene (C$_2$H$_2$Cl$_2$)	[33]
trans-Dichloroethene (C$_2$H$_2$Cl$_2$)	Fe [19]
1,1-Dichloroethene (C$_2$H$_2$Cl$_2$)	Fe, Fe–Pd [33]
Vinyl chloride (C$_2$H$_3$Cl)	Fe, Fe–Pd [33]

(*Continued*)

TABLE 1   *(Continued)*

Chlorobenzenes	
Hexachlorobenzene (C_6Cl_6)	Fe–Ag [26]; Fe, Fe–Pd [33,37]
Pentachlorobenzene (C_6HCl_5)	Fe–Ag [26]
1,2,3,4-Tetrachlorobenzene ($C_6H_2Cl_4$)	Fe/Ag [26]
1,2,4,5-Tetrachlorobenzene ($C_6H_2Cl_4$)	Fe–Ag [26]
1,2,3-Trichlorobenzene ($C_6H_3Cl_3$)	Fe–Ag [26]
1,3,5-Trichlorobenzene ($C_6H_3Cl_3$)	Fe–Ag [26]
1,2,4-Trichlorobenzene ($C_6H_3Cl_3$)	Fe–Ag [26]; Fe–Pd [38]
1,3-Dichlorobenzene ($C_6H_2Cl_2$)	Fe–Ag [26]
1,2-Dichlorobenzene ($C_6H_2Cl_2$)	Fe–Ag [26]
1,4-Dichlorobenzene ($C_6H_2Cl_2$)	Fe–Ag [26]
Chlorobenzene (C_6H_5Cl)	Fe–Ag [26]
Herbicides	
Alachlor ($C_{14}H_{20}ClNO_2$)	Fe [28]; Fe [39]
Atrazine	Fe, Fe–Pd [29]
Molinate	Fe–O_2 [40]
Pesticides	
DDT ($C_{14}H_9Cl_5$)	Fe [19]
Lindane ($C_6H_6Cl_6$)	Fe [41]
Organic dyes	
Acid black 24 ($C_{36}H_{23}N_5Na_2O_6S_2$)	(Fe) [39,42]
Acid orange	Fe [19]
Acid red	Fe [19]
Chrysoidine ($C_{12}H_{13}ClN_4$)	Fe [19]
Indigo blue	Fe–Cu, C–Fe–Cu [43]
Orange G	Ni–Fe [44]
Orange II ($C_{16}H_{11}N_2NaO_4S$)	Fe [19]
Tropaeoline O ($C_{12}H_9N_2NaO_5S$)	Fe [19]
Other chlorinated hydrocarbons	
p-Chlorophenol (C_6H_5ClO)	Fe [45]; Fe–O_2 [46]; Fe–O_2 [47]
Polychlorinated biphenyls	Fe–Pd [18]; Fe–Pd [34]; Fe, 300°C [48]; Fe–Pd [49]; GAC–Fe–Pd [50]; Pd–Mg [51]
Polychlorinated dibenzo-*P*-dioxins	Fe, Fe–Pd [52]
Polychlorinated dibenzofurans	Fe, Fe–Pd [52]
Pentachlorophenol (C_6HCl_5O)	Fe–H_2O_2 [53]; Fe [54]
1,2,3-Trichloropropane ($C_3H_5Cl_3$)	Fe, Zn [25]
Explosives	
HMX ($C_4H_8N_8O_8$)	Fe [55]
RDX ($C_3H_6N_6O_6$)	Fe [55]; Fe [30]
N-Nitrosodimethylamine ($C_4H_{10}N_2O$)	Fe [19]
TNT ($C_7H_5N_3O_6$)	Fe [56]
Other organic contaminants	
Benzoic acid ($C_7H_6O_2$)	Fe–O_2 [57]; Fe–O_2 [58]
EDTA	Fe–O_2 [59]
Methanol	Fe–O_2 [60], Fe–Ni–O_2 [58]
2-Propanol	Fe–O_2 [60]
Pyrene ($C_{16}H_{10}$)	Fe [61]

Several performance-related issues have become evident with the use of commercially available zero-valent iron (ZVI) powder (range $\sim$30 to 200 mesh) for the treatment of groundwater pollutants. These include a decrease in ZVI reactivity during prolonged exposure to water and common aqueous solutes in the groundwater (a process referred to as *aging*); this occurs due to the metal corrosion and precipitation of metal hydroxides and carbonates on ZVI surface [18]. The discovery of INP applications in groundwater treatment can greatly minimize these problems; thus, the INPs typically used in emulsified and bimetallic forms are 100 to 200 nm in diameter [19]. In a recent study, the details of INP synthesis and its application have been reviewed [20]; the study also examined the case history of INP applications in aquifer cleanup at 22 North American and seven international field sites. The study [20] noted that the treatment time with INP was quite rapid and that over 99% of tetrachloroethene was removed in 1 hr and over 95% of *trans*-dichloroethene, *cis*-dichloroethene, 1,1,1-trichloroethane, trichloroethylene, and tetrachloromethane was removed in 120 hr. The INPs have also been quite effective in treating contaminated soils, sediments, and solid wastes in ex situ slurry reactors, and INPs were also successful in treating and stabilizing biosolids from domestic wastewater treatment plants [21], including the treatment of metal contaminants in the sludge. In summary, INPs have shown good promise thus far in treating both organic and inorganic contaminants and in neutralizing odorous sulfide compounds with a high reaction effectiveness and accessibility and in generating nonhazardous end products and compatibility with other treatment methods [22].

Reductive Transformation of Organic Contaminants   The potential of Fe–Pd bimetallic nanoparticles for the reductive dechlorination of seven chlorinated ethanes ($C_2H_{6-x}Cl_x$) was evaluated in batch experiments [23]. Hexachloroethane (HCA) (C_2Cl_6), pentachloroethane (PCA) (C_2HCl_5), 1,1,2,2-tetrachlorethane (1,1,2,2-TeCA, $C_2H_2Cl_4$), and 1,1,1,2-tetrachlorethane (1,1,1,2-TeCA, $C_2H_2Cl_4$) were rapidly dechlorinated (half-lives < 9 to 28 min) at a Fe–Pd nanoparticle loading of 5 g/L. The end products of degradation reactions were ethane and ethene. Only one chlorinated intermediate, a corresponding dichloro β-elimination product, formed briefly during the reactions. Further, the reductive dechlorination of 1,1,1-trichloroethane (1,1,1-TCA, $C_2H_3Cl_3$) to ethane was observed with Fe–Pd nanoparticles, yet at a relatively slower rate (half-life = 44.9 min). However, no significant degradation of dichloroethane ($C_2H_4Cl_2$) was observed within 24 hr. The Fe–Pd bimetallic nanoparticles generally exhibit much higher reactivity than that of conventional micro- and millimeter-scale ZVI powders. In comparison to ZVI, the dechlorination reactions with Fe–Pd are more complete, with a much higher yield of ethane as the final product, and a lower yield of chlorinated intermediates. Results from this work [23] suggest that the Fe–Pd bimetallic nanoparticles may represent a treatment alternative for in situ remediation of chlorinated ethanes.

Synthetic INPs (<100 nm in diameter) were investigated in bench-scale experiments to degrade eight chlorinated ethanes [hexachloroethane (HCA), pentachloroethane (PCA), 1,1,2,2-tetrachloroethane (1,1,2,2-TeCA), 1,1,1,2-tetrachloroethane (1,1,1,2-TeCA), 1,1,2-trichloroethane (1,1,2-TCA),

1,1,1-trichloroethane (1,1,1-TCA), 1,2-dichloroethane (1,2-DCA), and 1,1-dichloroethane (1,1-DCA)] in batch reactors [24]. Varying initial concentrations of PCA between 0.025 and 0.125 mM resulted in relatively similar pseudo-first-order rate constants, indicating that PCA removal conforms to pseudo-first-order kinetics. The reduction of 1,1,2,2-TeCA decreased with increasing pH; however, dehydrohalogenation of 1,1,2,2-TeCA became important at high pH. All chlorinated ethanes except 1,2-DCA were transformed to less chlorinated ethanes or ethenes. The surface area normalized rate constants from first-order kinetics ranged from $< 4 \times 10^{-6}$ to 0.80 L/m^2·h. In general, the reactivity increased with increasing chlorination. Among chlorinated hydrocarbon compounds with three and four organochlorines, the reactivity was higher for compounds with chlorines localized on a single carbon (e.g., 1,1,1-TCA $\geq$ 1,1,2-TCA). Reductive dichloro β-elimination was the major pathway for the chlorinated ethanes possessing α,β-pairs of chlorine atoms to form chlorinated ethenes, which in turn also reacted with INPs. Reductive α-elimination and hydrogenolysis were concurrent pathways for compounds possessing chlorine substitution on one carbon only, forming less chlorinated ethanes. Song and Carraway evaluated the dechlorination of three chlorinated methanes (CCl$_4$, CHCl$_3$, CH$_2$Cl$_2$) by INPs synthesized by borohydride reduction of Fe^{3+} under anaerobic conditions [24]. When reacted with chlorinated methanes in batch reactors, nanosized INPs transformed CCl$_4$ and CHCl$_3$ rapidly but showed negligible reactivity toward CH$_2$Cl$_2$, giving the reactivity order CCl$_4$ > CHCl$_3$ $\gg$ CH$_2$Cl$_2$. CH$_4$ was observed along with CH$_2$Cl$_2$ in the reduction of CCl$_4$ and CHCl$_3$, and they are presumed to be generated via concerted reductive elimination steps involving carbene and charged radical species. Pathways for CH$_4$ production from CCl$_4$ and CHCl$_3$ reductions are proposed. Evidence obtained from the study of several physicochemical factors, including pH, initial concentration, hydrogen concentration, and metal loading, indicates that reduction of chlorinated methanes occurs via a direct electron transfer reduction mechanism rather than an indirect mechanism involving reactive hydrogen species. In a comparative experiment with three types of commercial irons (Fisher, Connelly, and ARS) and CHCl$_3$, nanosized iron gave a surface area normalized rate constant (k_{SA}) value of 5.6 ($\pm$ 0.6) $\times$ 10^{-2} L/m^2·hr, which is one order of magnitude greater than Fisher iron and two orders of magnitude greater than ARS and Connelly irons. This comparison of k_{SA} values should be considered approximate, due to large differences in metal loadings.

1,2,3-Trichloropropane (TCP) is an emerging contaminant because of its occurrence in groundwater, potential carcinogenicity, and resistance to natural attenuation. The physical and chemical properties of TCP make it difficult to remediate, with all conventional options being relatively slow or inefficient [25]. Treatments that result in alkaline conditions (e.g., permeable reactive barriers containing ZVI) favor base-catalyzed hydrolysis of TCP, but high temperature (e.g., conditions of in situ thermal remediation) is necessary for this reaction to be significant. Common reductants (iron monosulfide, ferrous iron adsorbed to iron oxides, and most forms of construction-grade or nano Fe^0) may show insignificant degradation of TCP by reductive dechlorination. Quantifiable rates of TCP reduction were obtained with several types of activated INPs, but the surface area normalized rate contants (k_{SA}) for these reactions were lower than is generally considered useful for in situ remediation applications (10^{-4} L/m^2·hr). Much greater rate contants of TCE degradation were obtained with granular Zn^0 ($k_{SA} = 10^{-3}$ to 10^{-2} L/m^2·hr) without the production of potentially problematic dechlorination intermediates (e.g., 1,2- or 1,3-dichloropropane, 3-chloro-1-propene). The advantages of Zn^0 over Fe^0 are somewhat unique in terms of TCP treatment, which may offer promise for Zn^0 even though it has not yet found favor for remediation of other chlorinated solvents [25].

Subcolloidal (<0.1 μm) iron–silver bimetallic nanoparticles (Fe–Ag BMNPs) with 1% Ag were examined for the transformation of chlorinated benzenes in aqueous solution [26]. Hexachlorobenzene (HCB) (4 mg/L) was dechlorinated to tetra-, tri-, and dichlorobenzenes (TeCB, TCB, and DCB, respectively) within 24 hr at a metal loading of 25 g/L. Principal degradation products included 1,2,4,5-TeCB, 1,2,4-TCB, and 1,4-DCB. Continuous dechlorination was observed during a 57-day experiment. The rate of dechlorination was positively correlated to Ag loading of the Fe–Ag BMNPs. The bimetallic particles also effectively degraded penta- and tetrachlorobenzenes (PeCB and TeCB, respectively). Fe–Ag particles could become a cost-effective alternative to the previously reported Fe–Pd nanoparticles. The subcolloidal Fe–Ag bimetallic particles may be used in slurry reactors, in in situ applications, and in combination with biological treatment for the complete degradation of chlorinated benzenes and PCBs.

In a recent study [27], Fe–Ni BMNPs were prepared supported on carbon nanotubes and copolymerized with β-cyclodextrin; the resulting polymers were investigated for their potential in the degradation of organic pollutants in water. The Fe–Ni BMNPs were first embedded on functionalized carbon nanotubes before being copolymerized with β-cyclodextrin and hexamethylene diisocyanate, forming a water-insoluble polyurethane. The particle size and distribution of Fe–Ni BMNPs were determined by transmission electron microscopy (TEM), and the surface area of BMNPs was determined using the Brunauer–Emmett–Teller (BET) method. Energy-dispersive x-ray spectroscopy (EDXS) was used to confirm the formation of the Fe–Ni BMNPs. The degradation of trichloroethylene (TCE) as a model pollutant was studied, and more than a 98% reduction in TCE was achieved by polymers impregnated with the Fe–Ni BMNPs; the catalyst showed sustained TCE degradation in several cycles reproducibly. The degradation was monitored by gas chromatography–mass spectrometry (GC–MS), while the chloride produced during dechlorination was verified by ion chromatography (IC). Flame atomic absorption spectroscopy was employed to evaluate possible leaching of the BMNPs from the polymer, which was confirmed to be at a trace level.

The bench-scale degradation of the pesticides alachlor and atrazine in water with INPs (diameter < 90 nm, specific surface area = 25 m^2/g) was reported under anoxic conditions [28]. While alachlor (initial concentrations at 10, 20, 40 mg/L) degraded with INPs by 92 to 96% within 72 hr, degradation of atrazine with INPs was not observed. The rate constant of alachlor degradation with INPs observed (35.5×10^{-3} to 43.0×10^{-3} hr) increased with increasing alachlor concentration, and the degradation reaction was considered to follow pseudo-first-order kinetics. Above results are consistent with other reports on these pesticides with micro ZVI and iron filings. The authors [28] contend that the use of INPs may prove to be a simple method for on-site treatment of high concentrations of certain pesticides (in the range of 100 mg/L).

In contrast to the above report about the lack of atrazine degradation with INPs [28], another study has reported otherwise [29]. This study [29] examined the effects of Fe sources, solution pH, and the presence of Fe or Al sulfate salts on atrazine degradation with INPs in water and soil. Their results indicate that INPs can be employed to treat atrazine successfully in water and soil. A comparison of atrazine (30 mg/L) degradation treated with INPs ([Fe] = 2% w/v) and with commercial ZVI ([Fe] = 5% w/v) has demonstrated favorable reaction kinetics with INP treatment; the pseudo-first-order rate constant (k_{obs}) with INP treatment (1.39 day^{-1}) was about sevenfold greater than that with commercial ZVI treatment (0.18 day^{-1}). Reductive dechlorination is the major process in atrazine degradation by INPs, with 2-ethylamino-4-isopropylamino-1,3,5-triazine as the dechlorination product. Lowering the pH from 9 to 4 increased the destruction kinetic rates of atrazine by INPs. Moreover, INP–Pd enhanced the destruction kinetic rates of atrazine (3.36 day^{-1}). Atrazine destruction kinetics were greatly enhanced in both contaminated water and soil treatments by INPs when sulfate salts of Fe(II), Fe(III), or Al(III) were added with degradation rate contants, expressed in decreasing order: Al(III): 2.23 day^{-1} $\geq$ Fe(III): 2.04 day^{-1} $\geq$ Fe(II): 1.79 day^{-1}.

Explosive or munition compounds has been shown to degrade effectively by INPs. Hexahydro-1,3,5-trinitro-1,3,5-triazine (RDX) is a common contaminant of soil and water at military installations. RDX degradation has been reported with INPs in aqueous systems with or without a stabilizer additive [e.g., carboxymethyl cellulose (CMC), poly(acrylic acid) (PAA), etc.] [30]. The rate contants of RDX degradation in aqueous solution with INPs in decreasing order is reported to be CMC-INPs $\geq$ PAA-INPs $\geq$ INPs with k_1 values of 0.816 ± 0.067, 0.082 ± 0.002, and 0.019 ± 0.002 min^{-1}, respectively. The disappearance of RDX was accompanied by the formation of formaldehyde, nitrogen, nitrite, ammonium, nitrous oxide, and hydrazine by the intermediary formation of methylenedinitramine (MEDINA), MNX (hexahydro1-ntroso-,3,5-dintro-1,3,5-triazine), DNX (hexahydro-1,3-dinitroso-5-nitro-1,3,5-triazine), and TNX (hexahydro-1,3,5-trinitroso-1,3,5-triazine). When either of the reduced RDX products (MNX or TNX) was treated with INPs, the authors observed reaction products such as nitrite (only from MNX), NO (only from TNX), N_2O, NH_4^+, NH_2NH_2, and HCHO. In the case of TNX degradation a new reaction product, 1,3-dinitroso-5-hydro-1,3,5-triazacyclohexane, was tentatively identified. However, the equivalent denitrohydrogenated product of RDX and MNX degradation was not detected. Finally, during MNX degradation the authors observed a new intermediate, identified as N-nitrosomethylenenitramine ($ONNHCH_2NHNO_2$), which was perhaps equivalent to methylenedinitramine formed due to denitration of RDX. Experimental evidence gathered thus far suggested that

INPs may help in degradation of RDX and MNX via initial denitration and sequential reduction to the corresponding nitroso derivatives prior to completed decomposition, yet TNX degradation by INPs may occur exclusively via initial cleavage of the N—NO bond(s).

Fe–Pd bimetallic nanoparticles have been studied by several groups to characterize reductive degradation of chlorinated organics. Fe–Pd nanoparticles were prepared in three steps [49]: polymerization of acrylic acid (AA) in poly(vinylidene fluoride) (PVDF) microfiltration membrane pores, subsequent ion exchange of Fe^{2+}, followed by chemical reduction of ferrous ions (by borohydride) bound to the carboxylic acid groups. Fe–Pd bimetallic nanoparticles were formed by the partial reduction of Pd^{2+} with Fe^0 nanoparticles. The functionalized membrane and the nanoparticles were characterized by scanning electron microscopy (SEM) and transmission electron microscopy (TEM). The membrane-supported nanoparticles exhibited high reactivity in the dechlorination of 2,2-dichlorobiphenyl (DiCB) used as a model compound. The dechlorination mechanism and the role of water were probed by conducting the reaction in pure ethanol solution. Bulk Fe–Pd particles were also prepared to investigate the effect of particle size on catalytic activity. The effect of Pd content on catalytic activity was also studied to characterize the role of Pd in the Fe–Pd bimetallic nanoparticle system. The high catalytic activity of Pd was confirmed by the low activation energy compared to those other catalytic systems.

Dioxins, including polychlorinated dibenzo-p-dioxins (PCDDs) and polychlorinated dibenzofurans (PCDFs), are highly toxic and persistent compounds. Their detrimental health and environmental effects, low aqueous solubility, relatively high stability, and chlorinated nature contribute to their persistence and recalcitrance toward degradation. Hence, removal of dioxins is a global priority. Both micro-sized zero valent iron and INPs were examined for dechlorination of dioxins [52]; both forms of iron particles, without palladization were able to dechlorinate PCDD congeners with four chlorines in an aqueous system. Although dechlorination of dioxins is thermodynamically feasible, the focus of the above study was to examine the kinetics of dechlorination in the absence of any previous reports. It was observed that unamended INPs could dechlorinate PCDD congeners; however, the reaction proceeded slowly and complete dechlorination was not achieved within the duration of study. In contrast, palladized nanosized zero-valent iron (Pd–INPs) rapidly dechlorinated PCDDs, including the mono- to tetrachlorinated congeners. The rate of 1,2,3,4-tetrachlorodibenzo-p-dioxin (1,2,3,4-TeCDD) degradation using Pd–INPs was about three orders of magnitude greater than with INPs. The distribution of products obtained from dechlorination of 1,2,3,4-TeCDD suggests that palladization of INPs shifts the pathways of contaminant degradation toward a greater role of hydrogen atom transfer than electron transfer. The decision between choosing INPs or Pd–INPs for treatment of PCDD–Fs (polychlorinated dibenzodioxins or polychlorinated dibenzofurans) is still ambiguous because there are no reports for the toxic equivalent quantity (TEQ) changes during dechlorination. The study observed that dechlorination of higher chlorinated congeners such as octachlorodibenzo-p-dioxin (OCDD) increased the overall TEQ threefold with the formation of lower chlorinated compounds that had a higher TEQ [52]. Predictions made on the basis of the modeling studies show that it might take more than 100 years for complete and efficient removal of PCDDs from the environment by natural attenuation processes. Therefore, the choice of using INPs or Pd–INPs for dechlorination should not only be determined by the rate of dechlorination but also by the toxicity of the breakdown products.

There are many concerns and challenges in current remediation strategies for sediments contaminated with polychlorinated biphenyls (PCBs). Recent efforts have been geared toward the development of granular activated carbon (GAC) impregnated with reactive iron–palladium (Fe–Pd) bimetallic nanoparticles [reactive activated carbon (RAC)] [50]. The following reactions occurred either in parallel or consecutively: (1) 2Cl-biphenyl is promptly and completely sequestrated to RAC phase, (2) the adsorbed 2-Cl-biphenyl is dechlorinated almost simultaneously by Fe–Pd particles to form a reaction product biphenyl (BP), and (3) the BP formed is instantly and strongly adsorbed to RAC. The 2-Cl-biphenyl adsorption and dechlorination rate constants were estimated through simple first-order reaction kinetic models with an assumption for unextractable portion of carbon (the amount of carbon sorbed irreversibly) in RAC. The extent of 2-Cl-biphenyl accumulation and BP formation in RAC phase was explained by the kinetic model, and adsorption was found to be the rate-limiting step for overall reaction. On the basis of their observations, a new strategy and concept of "reactive" cap/barrier composed of RAC was proposed as a new environmental risk management option for PCB-contaminated sites.

Conflicting accounts are reported on the reactivity of substituted chlorines and the ensuing dechlorination pathway of PCBs through catalytic hydrodechlorination. To understand these relationships, intermediates and dechlorination pathways of 17 carefully selected congeners were investigated with reactive Pd–Mg systems to bring about their rapid and complete dechlorination [51]. The preferential site of electrophilic attack and its mechanistic aspects were interpreted in terms of steric, inductive, and resonance stabilization. The trends for electrophilic substitution were consistently p- $\geq$ m- $\geq$ o-positions, indicating that more toxic "coplanar" PCB congeners were easily reduced. The dechlorination rates and pathways were influenced both by the inductive effect of Cl, which probably governs the stability of the intermediate arenium ion and by steric effects affecting primarily the adsorption step (especially for the o-congeners). Electrophilic attack occurred preferentially on the less substituted phenyl ring in the absence of steric effects. A distinct correlation between the rate of hydrodechlorination and the degree of chlorination was not observed; rather, it depended on positions of organochlorine with respect to the biphenyl bond and the dominance between counteracting factors of deactivation by subsequent chlorinations and improvement in the probability of dechlorination through an increased number of chlorine atoms.

The destruction efficiency during the preliminary treatment (mixing of soil and INPs in water) can be increased by increasing the water temperature [48]. The maximum thermal destruction (pyrolysis or combustion of soil after preliminary treatment) of soil-bound PCBs occurs at 300°C in air. A minimum total PCB destruction efficiency of 95% can be achieved by this process. The effect of changing treatment parameters such as type of mixing, time of mixing and mixing conditions, and the application of other catalysts, such as iron oxide and V_2O_5–TiO_2, was also investigated. It was found that at 300°C in air, iron oxide and V_2O_5–TiO_2 are also good catalysts for remediating PCB-contaminated soils.

Oxidative Transformation of Organic Contaminants Addition of INPs to oxygen-containing water results in oxidation of organic compounds such as herbicide molinate [40], EDTA [59], and *p*-chlorophenol [46]. To assess the potential application of INPs for oxidative transformation of organic contaminants, the conversion of benzoic acid (BA) to *p*-hydroxybenzoic acid (*p*-HBA) was used as a probe reaction [57]. When INPs were added to BA-containing water, an initial pulse of *p*-HBA was detected during the first 30 min, followed by the slow generation of additional *p*-HBA over periods of at least 24 h. The yield of *p*-HBA increased with increasing BA concentration, presumably due to the increasing ability of BA to compete with alternative oxidant sinks, such as ferrous ion. At pH 3, during the initial phase of the reaction maximum yields of *p*-HBA of up to 25% were observed. The initial rate of INP-mediated oxidation of BA exhibited a marked reduction at pH values above 3. Despite the decrease in oxidant production rate, *p*-HBA was observed during the initial reaction phase at pH values up to 8. Competition experiments with probe compounds expected to exhibit different affinities for the INP surface (phenol, aniline, *o*-hydroxybenzoic acid, and synthetic humic acids) indicated relative rates of reaction that were similar to those observed in competitive experiments in which hydroxyl radicals were generated in solution. Examination of the oxidizing capacity of a range of Fe^0 particles reveals a capacity in all cases to induce oxidative transformation of benzoic acid, but the high surface areas that can be achieved with nanosized particles renders such particles particularly effective oxidants.

Although organic compounds can be oxidized by zero-valent iron or dissolved Fe(II) in the presence of oxygen, this process is not very effective for degrading contaminants because the yields of oxidants are usually low (i.e., typically less than 5% of the iron added is converted into oxidants capable of transforming organic compounds). The addition of polyoxometalate (POM) greatly increases the yield of oxidants in both systems [60]. The mechanism of POM enhancement depends on the solution pH. Under acidic conditions, POM mediates the electron transfer from INPs or Fe(II) to oxygen, increasing the production of hydrogen peroxide, which is subsequently converted to hydroxyl radical through the Fenton reaction. At neutral pH values, iron forms a complex with POM, preventing iron precipitation on the INP surface and in bulk solution. At pH 7, the yield of oxidant approaches the theoretical maximum in the INP–O_2 and the Fe(II)–O_2 systems when POM is present, suggesting that coordination of iron by POM alters the mechanism of the Fenton reaction by converting the active oxidant from ferryl ion to hydroxyl radical. Comparable enhancements in oxidant yields are also observed when INP or Fe(II) is exposed to oxygen in the presence of silica-immobilized POM.

Bimetallic nickel–iron nanoparticles (*n*Ni–Fe; i.e., Ni–Fe alloy and Ni-coated Fe nanoparticles) exhibit enhanced yields of oxidants compared to INPs. *n*Ni–Fe (Ni–Fe alloy nanoparticles with [Ni]/[Fe] = 0.28 and Ni-coated Fe nanoparticles with [Ni]/[Fe] = 0.035) produced approximately 40 and 85% higher yields of formaldehyde from the oxidation of methanol relative to INPs at pH 4 and 7, respectively [58].

Ni-coated Fe nanoparticles showed a higher efficiency for oxidant production relative to Ni–Fe alloy nanoparticles based on Ni content. Addition of Ni did not increase the oxidation of 2-propanol or benzoic acid, indicating that Ni addition did not enhance hydroxyl radical formation. The enhancement in oxidant yield was observed over the pH range 4 to 9. The enhanced production of oxidant by nNi–Fe appears to be attributable to two factors. First, the nNi–Fe surface is less reactive toward hydrogen peroxide (H_2O_2) than the INP surface, which favors the reaction of H_2O_2 with dissolved Fe(II) (the Fenton reaction). Second, the nNi–Fe surface promotes oxidant production from the oxidation of ferrous ion by oxygen at neutral pH values.

The oxidative degradation of organic compounds utilizing INPs is promoted further by the presence of natural organic matter (NOM) [as humic acid (HA) or fulvic acid (FA)] working as an electron shuttle [47]. The main target substrate used in the study was 4-chlorophenol. Both HA and FA can mediate electron transfer from the INP surface to O_2 while enhancing the production of Fe^{2+} and H_2O_2 that subsequently initiates the OH-radical-mediated oxidation of organic compounds through Fenton reaction. The electron transfer–mediating role of NOM was supported by the observation that higher concentrations of H_2O_2 and ferrous ion were generated in the presence of NOM. The NOM-induced enhancement in oxidation was observed with NOM concentrations ranging from 0.1 to 10 ppm. Since the reactive sites responsible for the electron transfer are likely to be the quinone moieties of NOM, benzoquinone that was tested as a proxy of NOM also enhanced the oxidative degradation of 4-chlorophenol with INP suspension. The NOM-mediated oxidation reaction with INPs was completely inhibited in the presence of methanol, an OH radical scavenger, and in the absence of dissolved oxygen.

INPs with hydrogen peroxide were used to treat pentachlorophenol-contaminated soil with the expectation of offering a new and effective solution for future treatment work [53]. When different amounts of INPs (0, 0.2, 0.5, and 1 wt%) are added to 1% H_2O_2, 1 wt% is found to be best for treating pentachlorophenol-contaminated soils. Among three types of sampling soils, the Pianchen soil is proven to be more effective because INPs release Fe^{2+} during oxidation. The addition of H_2O_2 promotes the generation of Fenton with OH radical and an oxidation–reduction reaction. Due to the dual mechanism, the phenol compounds that are otherwise difficult to be removed were effectively treated. This study also indicates that when 5% of calcium carbonate is added for 40 h, the decay rate on pentachlorophenol of Chengchun soil increases from 37% to 78%, and of Pianchen soil increases from 43% to 76%. The result of treatment of pentachlorophenol-contaminated soil with added calcium carbonate is a new treatment approach and serves as the reference for future on-site treatment.

Field Use of INPs for Organic Contaminant Remediation   Several studies have demonstrated that on an equal mass basis, INPs show higher reductive dechlorination rates for chlorinated hydrocarbons than those of granular iron [18,23,62–65], giving hope that faster site cleanup may be achievable at source zones by using INP-based remediation technologies. Previous laboratory batch and column tests [66,67] and a field test [68] show that the 'emulsified zero-valent iron' (EZVI) technology, patented by the National Aeronautics and Space Administration (NASA), represents such a

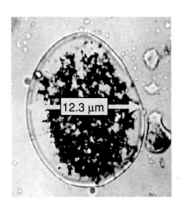 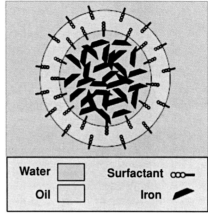

Water		Surfactant ∞⊸
Oil		Iron ◤

FIGURE 3 Optical microscope image (left) and schematic description (right) of an emulsified zero-valent iron or EZVI emulsion droplet containing nanoscale iron particles in water surrounded by an oil–liquid membrane. (From Jacqueline Quinn, NASA.)

promising approach to treat dense non-aqueous phase liquid (DNAPL) at the source zones. This technology creates surfactant-stabilized, biodegradable oil-water with water droplets containing the INPs (Figure 3). The vegetable oil emulsion can mix with the DNAPL to enhance contact between the INPs in EZVI and the DNAPL. The INPs can provide rapid abiotic degradation of the chlorinated hydrocarbons whereas the oil can serve as a long-term electron donor source to enhance microbial degradation. Additional advantages of the EZVI technology are that EZVI can be injected to areas where conventional PRB construction may not be feasible (e.g., due to depth restrictions or geological media) and that EZVI can be injected directly into DNAPL source zones. Regardless of the detailed steps of dechlorination reactions, the overall chemical reduction of tetrachloroethene (PCE) can be written as follows:

$$C_2Cl_4 + 6Fe^0 + 10H^+ \rightarrow C_2H_6 + 6Fe^{2+} + 4HCl \tag{3}$$

Based on a 100% iron use efficiency (defined as the percentage of the Fe^0 particles that is utilized to dechlorinate PCE), for every kilogram of PCE, 1.68 kg of Fe^0 is needed. The actual efficiency is lower. Liu et al. [65] reported an iron use efficiency of 52% for Toda iron (RNIP) to dechlorinate TCE in an anaerobic laboratory test. The lower efficiency is due to a competing Fe^0 corrosion reaction [65]:

$$\text{anaerobic corrosion: } Fe^0 + 2H_2O \rightarrow Fe^{2+} + H_2 + 2OH^- \tag{4}$$

In field applications, even lower Fe^0 use efficiency may occur because of other side reactions, such as

$$\text{aerobic corrosion: } Fe^0 + 2H_2O + O_2 \rightarrow 2Fe^{2+} + 4OH^- \tag{5}$$

may also occur and consume Fe^0. Furthermore, field injections may not result in 100% contact between the injected Fe^0 and the contaminants. Hence, extra Fe^0 is needed for successful treatment.

A pilot-scale study was reported using a palladium-catalyzed and polymer-coated nanoscale ZVI particle suspension at the Naval Air Station in Jacksonville, Florida [69]. A total of 300 lb of INP suspension was injected via a gravity feed and re-circulated through a source area containing chlorinated volatile organic compounds (CVOCs). The recirculation created favorable mixing and distribution of the INP suspension and enhanced the mass transfer of sorbed and non-aqueous phase con-stituents into the aqueous phase, where the contaminants could be reduced. Between 65 and 99% aqueous-phase CVOC concentration reduction occurred, due to abiotic degradation, within five weeks of the injection, as per the report. The rapid abiotic degradation processes then yielded to slower biological degradation as subsequent decreases in β-elimination reaction products were observed (Beta elimination is a reaction in which an alkyl group bonded to a metal center is converted into the corresponding metal-bonded hydride and an alkene.) However, favorable redox con-ditions were maintained as a result of the INP treatment. Post-treatment analyses revealed cumulative reduction of soil contaminant concentrations between 8 and 92%. Aqueous-phase VOC concentrations in the wells and the down-gradient of the source zone were reduced up to 99%, and became near or below applicable regulatory criteria. These reductions, coupled with the generation of innocuous by-products, in-dicate that INPs effectively degraded contamination and reduced the mass flux from the source, a critical metric identified for source treatment.

2.2   Knowledge Gaps

Many detailed mechanisms of organic degradation by INPs and bimetallic materials are not completely understood; hence, carefully designed lab experiments should be conducted employing a variety of techniques, such as spectroscopy methods and isotope analyses. Future studies should also address knowledge gaps in the fate and transport of INPs, economic analysis (cost vs. benefits), and public acceptance of the nanotechnology for site remediation.

2.3   Inorganic Contaminants

In this section, the use of INPs for the treatment of such inorganic contaminants, such as arsenic (As), lead (Pb), selenium (Se) and chromium (Cr), are reviewed that are significant threats to the environment and to human health. These contaminants are introduced into the environment both through natural processes (e.g., biogeochemical reactions, natural erosion, volcanic emissions) and human activities (e.g., mining, industrial disposal, coal burning, auto exhaust) [3]. Table 2 summarizes the treatment of inorganic contaminant by stabilized and unstabilized INPs.

INPs can adsorb or reduce heavy metals such as Cr(VI) [72,81], lead, nickel, or mercury [3], radionuclides such as U(VI) [82], metalloids [2,70], perchlorate [83],

TABLE 2   Summary of Inorganic Contaminant Treatment by INPs

Contaminants	Materials[a]	Findings	Refs.
As(III) and As(V)	INPs	INP $k_{obs} \gg$ ZVI k_{obs}	[2,70]
Cr(VI)	INPs and C–INPs	Cr(VI) is reduced and sorbed on INPs	[71–75]
Pb(II)	C–INPs	Pb(II) is sorbed onto INPs	[73]
Selenium	INPs and Fe–Ni		[76]
Cu(II)	Fe^0		[77]
Co(II)	Fe^0		[78]
Ba(II)	Fe^0		[79]
Ni(II)	Fe^0		[80]

[a]C–INPs = carbon-stabilized INPs.

humic acid, and nitrate [12,84–86]. In this section, the focus on the applications of INPs for the treatment of inorganic contaminants.

Cr(VI) Removal   Among the heavy metals, Cr(VI) is one of the contaminants most commonly treated using INPs. Several studies have examined the reduction of Cr(VI) by INPs. Recent investigations of synthetic INPs have shown that this material is effective in the treatment of Cr(VI) contaminations [71,74,75,87–90]. Ponder et al. [73] investigated the kinetics of Cr(VI) reduction by both unsupported nano Fe^0 and resin-supported nano Fe^0 (Ferragel) and concluded that the Cr(VI) removal rate for INPs was up to 30 times greater than conventional Fe^0 filings on a mass basis. The mechanisms of Cr(VI) reduction by both INPs and Fe^0 filings are similar, in which both involved complex surface chemistry during in situ corrosion of the Fe^0 surface, reduction of Cr(VI) to Cr(III) through coupled oxidation of Fe^0 and Fe^{2+}, and coprecipitation of Cr(III) with Fe(III) in a poorly defined Cr(III)–Fe(III) mixed hydroxide phase [72]. Another complicating factor uncovered during electrochemical investigations of Cr(VI) reduction by Fe^0 wire involves surface passivation of the corroding Fe^0 surface, where absolute, first-order Cr(VI) removal rates decrease with increasing Cr(VI) concentration [91]. This phenomenon has also been reported for Cr(VI) reduction by the Fe^{2+}-bearing mineral magnetite (Fe_3O_4) [92], which is a corrosion product on Fe^0 filing surfaces [93].

Although several studies have investigated Cr(VI) reduction by various Fe^0 materials, the identity of the solid-phase reaction product remains inconclusive. Several investigations have proposed formation of a mixed Fe(III)–Cr(III) hydroxide solid [93] according to the following reaction:

$$xCr^{3+} + (1-x)Fe^{3+} + 3OH^- = Cr_xFe_{1-x}(OH)_3 \qquad (6)$$

The Cr(VI) removal from simulative contaminated groundwater using ZVI (Fe^0) filings, Fe^0 powder, and INPs was investigated in batch experiments [94]. ZVI transformed Cr(VI) to Cr(III), which is less toxic and immobile. The Cr(VI) removal percentage was 87% at a metal/solution ratio of 6 g/L for commercial iron powder (200 mesh) in 120 min, and 100% Cr(VI) was removed when the metal/solution

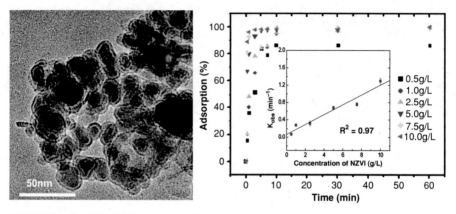

FIGURE 4   (Left) TEM micrograph of INPs; (right) removal of As(III) by INPs. (From [2].)

ratio was 10 g/L. The results demonstrate that the Cr(VI) removal percentage was apparently affected by pH, the amount of Fe powder, and the reaction temperature [94].

Arsenic Removal   Zero valent iron or ZVI has become one of the most common adsorbents for the rapid removal of metalloids such as arsenic [As(III) and As(V)] in the subsurface environment [63,95–97]. The kinetics of As removal by granular ZVI is moderate, with half-lives on the order of hours to days. However, due to the large size, lower surface area, and lack of mobility of ZVI, its use was limited to shallow groundwater treatment. To overcome these problems and to take the best advantage of good redox properties and the sorption capacity of iron, INPs has been applied for its treatment in the subsurface. Due to high surface area and reactivity, it has already shown great potential for the treatment of groundwater contaminants. Recently, removal of As(III) and As(V) by INPs has been reported, and the application of INPs for As remediation from the groundwater has been demonstrated in developing countries [2,98]. Figure 4 shows a TEM image of INPs and INPs' fast reaction for As(III) treatment [1].

Based on synthesis techniques, INP's surface properties and its porosity can vary, which affects INP's adsorption properties. For example, Kanel et al. prepared INP using the $NaBH_4$ reduction method under an inert atmosphere and obtained INPs with diameters of 5 to 100 nm and a surface area of 24.4 m^2/g [1]. It is observed clearly from the TEM picture and x-ray diffraction (XRD) that INPs remain as a core–shell structure, with the outer layer being iron oxide and the inner core, a zero-valent state. This makes INPs a versatile adsorbent to interact with various contaminants. There have been several successful reports of applications of the INPs for water treatment using different model contaminants, including As(III), As(V), Cr(III), TCE, PCE, NO_3^-, humic acids, and U(VI). INPs have been found to degrade chlorinated hydrocarbons such as trichloroethylene (TCE), tetrachloroethene (PCE), and carbon tetrachloride [4,98].

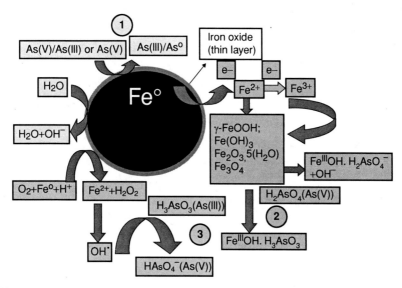

FIGURE 5   Mechanisms for INPs reacting with arsenic (1, 2, and 3 show three different mechanisms of arsenic reaction with INPs).

INPs can also produce hydroxide radicals in the presence of oxygen to oxidize a variety of organic contaminants, such as carbothioate herbicide/molinate and benzoic acid [99]. For example, the reactivity of these INPs was found to be three orders of magnitude greater than micrometer-sized iron for As(III). This is due primarily to versatile reactivity of INPs induced by continuous generation of the corrosion surface by ZVI from the core. Due to such characteristics, the interaction of INPs with a model contaminant can occur by multiple mechanisms (e.g., adsorption, oxidation, reduction). Typically, TCE and PCE can be reduced by INPs, whereas Cr(VI) can be both reduced and adsorbed by INPs. However, in the case of As(III), an oxidation and adsorption mechanism will be dominant (Figure 5).

Kanel et al. investigated the long-term interaction between INPs and As(III) in a batch study using solid samples collected from pristine INPs and INPs treated by 100 mg/L As(III) for 7, 30, and 60 days [2]. Figure 6 shows different surface textures and different pore sizes with respect to the adsorption time and precipitation of As onto INPs. Aggregation of particles increases with reaction time due to Fe(III) oxide/hydroxide precipitation. SEM pictures clearly show the growth of a fine urchin-like crystallie form (in 7 days), which may lead to an apparently amorphous phase (in 60 days). The very thin crystallites (about 100 nm long by 20 nm wide) are expected to be energetically unstable and should in turn disappear and get replaced by more stable phases, according to the Gay-Lussac/Oswald ripening rule [2].

In a recent work, INPs were supported on activated carbon (INP–AC) by impregnating carbon with ferrous sulfate followed by chemical reduction with NaBH4 [100]. Approximately 8.2 wt% of iron was loaded onto carbon. The adsorption capacity of the synthesized sorbent for arsenite and arsenate at pH 6.5 calculated from Langmuir

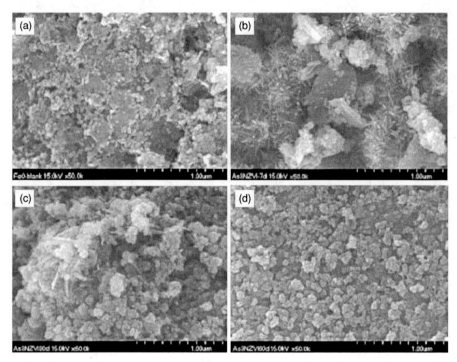

FIGURE 6   (a) SEM image of pristine INPs and As(III) sorbed on INPs for (b) 7, (c) 30, and (d) 60 days, respectively. Reaction conditions: 100 mg/L As(III) adsorbed on 50 g/L INPs in 0.01 M NaCl at pH 7, 25°C. (From [2].)

adsorption isotherms in batch experiments was 18.2 and 12.0 mg/g, respectively. Phosphate and silicate markedly decreased the removal of both arsenite and arsenate, whereas the effect of other anions and humic acid was insignificant. Common metal cations (Ca^{2+}, Mg^{2+}) enhanced arsenate adsorption, but ferrous iron (Fe^{2+}) was found to suppress arsenite adsorption. INP–AC can be regenerated effectively by reaction with 0.1 M NaOH.

Removal of Selenium   Selenium-contaminated water is also harmful to human health. According to U.S. Environmental Protection Agency (EPA), long-term exposure by selenium (Se) above the maximum contaminant levels has the potential to cause hair and fingernail loss, as well as damage to kidneys, livers, and the nervous and circulatory systems. The EPA has set a maximum contaminant level of 0.05 mg/L for Se as the drinking water standard (from the US EPA web site). Previous studies reported that use of ZVI is an attractive method for removing selenium from aqueous solutions, by reducing selenium to elemental form. Laboratory-synthesized INPs and Fe–Ni BMNPs were investigated for the removal of Se [76]. During their five-hour experiment, nearly 100% of selenate was removed by Fe and Fe–Ni nano particles (although removal by Fe–Ni nano particles was greater).

At a particle concentration of 0.1 g/L, Se removal by INPs was 155 mg/g, whereas removal by Fe–Ni BMNPs was 225 mg/g. Experimental data at pH 7.7 showed that for certain Se(VI) concentrations the removal percentage of both INPs and Fe–Ni BMNPs increased with nano particle concentration. Specific removal of Se increased with concentrations below 1 g/L of Se. For Fe–Ni BMNPs, Ni content also affected Se removal. Under conditions of 0.5 g/L Fe–Ni BMNPs and a 50.04 mg/L initial concentration of Se(VI), maximum selenate reduction occurred when the Ni content in the bimetallic powder was between 30 and 50%. As for pH, results indicated that high pH condition caused to decrease in Se removal. At pH 11, almost no Se(IV) removal was observed, whereas at pH 3.5 and pH 7.7 the total Se(IV) removal was 77.4 and 90%, respectively. The kinetics of removal was pseudo first-order at low Se(IV) concentration, shifting to zero order at higher Se(IV) concentrations. Many suboxic sediments and soils contain an Fe(II,III) oxide called *green rust*. Spectroscopic evidence showed that selenium is reduced from an oxidation state of $+VI$ to 0 in the presence of green rust at rates comparable with those found in sediments [101]. Selenium speciation was different in solid and aqueous phases. These redox reactions represent an abiotic pathway for selenium cycling in natural environments, which has previously been considered to be mediated principally by microorganisms. Similar green rust–mediated abiotic redox reactions are likely to be involved in the mobility of several other trace elements and contaminants in the environment.

Removal of Lead   Lead is a significant environmental hazard in drinking water. The EPA has set a maximum limit of 0.015 mg/L for Pb. While some studies have examined the reduction of aqueous Pb(II) by ZVI, Ponder et al. proved that supported nanoscale ZVI (Ferragel) is more effective [73]. The supported INPs rapidly separated and immobilized Pb(II) from aqueous solutions, reducing Pb(II) to Pb(0) while oxidizing the Fe to goethite. Based on tests of 0.5 g INPs in contact with 100 mL of 50 mM solutions of Pb(II) in 8 days, INPs removed and immobilized 0.0018 mmoles of Pb(II)/g of INPs. The removal and immobilization of aqueous Pb(II) showed pseudo-first-order reaction kinetics. The pseudo first-order rate constants for Pb(II) removal was more than five times greater for INPs than for the commercial microscale ZVI particles. There was a rapid disappearance of Pb(II) during the initial stage which appeared to be complete after 10 min. The initially high removal rate and the following slower rate indicated that the removal mechanism may have been physical rather than chemical.

The effects of heavy metals (Pb) on the dechlorination of carbon tetrachloride by INPs were investigated in terms of reaction kinetics and product distribution in batch systems [102]. Removal of Pb and its interactions with INPs were also examined. It was found that Pb(II) increased the CT reduction rate slightly but also increased the production of more toxic intermediates, such as dichloromethane.

Removal of Copper   Effects of Cu on the dechlorination of carbon tetrachloride by iron nanoparticles were investigated in terms of reaction kinetics and product distribution using batch systems [102]. Removal of heavy metals and the interaction between heavy metals and iron nanoparticles at the iron surface were also examined.

It was found that Cu(II) enhanced the carbon tetrachloride dechlorination by iron nanoparticles and led to the production of more benign products (i.e., methane). XRD analysis showed that Cu(II) was reduced to metallic copper and cuprite (Cu_2O) at the iron surface, whereas no reduced lead species was observed from the Pb(II)-treated iron nanoparticles. Limited data suggested that an oxidized lead species formed. The enhanced dechlorination by Cu(II) can be attributed to the deposition of metallic copper and cuprite at the iron surface [77]. As Cu(II) precipitates from solution, various other mechanisms, including coprecipitation, sorption, and ion exchange, may also enhance the removal of metals from solution.

Removal of Cobalt   INPs were synthesized by the borohydride reduction method, characterized and then examined for the removal of aqueous Co^{2+} ions over a wide range of concentrations (1 to 1000 mg/L) [103]. The experiments investigated the effects of the volume/mass ratio, concentration, contact time, repetitive loading, pH, and aging on the extent of retardation of Co^{2+} ions. INPs demonstrated very rapid uptake and large capacity for the removal of Co^{2+} ions. The extent of Co^{2+} ion uptake increased with increasing pH. X-ray photoelectron spectroscopy (XPS) indicates that the fixation of Co^{2+} ions takes place through the interaction of these ions with the oxohydroxyl groups on INP surface in addition to spontaneous precipitate at high Co loadings.

Removal of Barium   INPs are being tested increasingly as adsorbents for various types of inorganic pollutants. In this study, INPs synthesized under atmospheric conditions were employed for the removal of Ba^{2+} ions in the concentration range 10^{-3} to 10^{-6} M. Pseudo-second-order kinetics and the Dubinin–Radushkevich isotherm model provided the best correlation with the data obtained. The thermodynamic parameters observed showed that the process is exothermic and hence enthalpy-driven [79].

Removal of Ni   It is demonstrated that INPs function as a sorbent and a reductant for the sequestration of Ni(II) in water [104]. A relatively high capacity for nickel removal is observed (0.13 g Ni/g Fe), which is over 100% higher than the best inorganic sorbents available. High-resolution x-ray photoelectron spectroscopy confirmed that INPs have a core–shell structure and exhibit characteristics of both hydrous iron oxides (i.e., as a sorbent) and metallic iron (i.e., as a reductant). Ni(II) quickly forms a surface complex and is then reduced to metallic nickel on the nanoparticle surface. Surface reactions of nickel removal by iron nanoparticles (Figure 7) may be described by the following equations:

$$\equiv FeOH + Ni^{2+} = \equiv FeO\text{–}Ni^+ + H^+ \qquad (7)$$

$$\equiv FeONi^+ \cdot H_2O = \equiv FeONi\text{–}OH + H^+ \qquad (8)$$

$$\equiv FeONi^+ \cdot Fe^0 + H^+ = \equiv FeOH\text{–}Ni + Fe^{2+} \qquad (9)$$

Equations (7) and (8) illustrate the surface complex formation, and reaction (a) depicts the surface reduction of nickel [104].

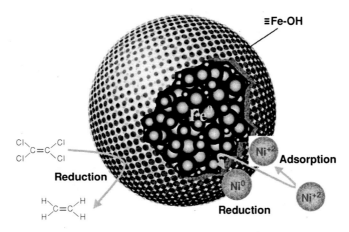

FIGURE 7   Sorption of Ni by INPs. (From [105].)

3   FATE AND TRANSPORT OF INPs IN POROUS MEDIA

Groundwater is a major source of drinking water and can be vulnerable to both natural and anthropogenic contaminations. Remediation techniques have been introduced, including soil vapor extraction (for volatile organic compounds), pump-and-treat methods, heat treatment, bioremediation, electroosmosis, use of a permeable reactive barrier (PRB), injection of reactive materials (e.g., oxidizers), and the in situ placement of chemical reactive barriers (CRBs) [105]. Among them, the use of CRBs is considered as one of the most cost-effective and innovative techniques for in situ groundwater remediation [106]. This is because CRBs can be injected as mobile slurry to reach the contaminated zone.

INPs can be used as a CRB material, and therefore have a potential for developing novel in situ groundwater remediation methods. For example, INPs are about one to three orders of magnitude more efficient than micro- to millimeter-sized iron particles for As treatment [2] and have been shown having higher reactivity for other contaminants (e.g., TCE, PCE, nitrates) than for micrometer-sized ZVI particles [3]. Successful application of these INPs for water treatment using different model contaminants was reported, including As(III), As(V), Cr(III), TCE, PCE, nitrate, humic acids [107], and U(VI) [7,40]. Despite its excellent reactivity, the natural tendency for INPs to aggregate limits their transport as well as their reactivity in porous media [108–110]. For example, iron cannot reach the contaminated plume in the deep groundwater aquifer (generally, ≥ 30 m below the surface). Even though INPs have been shown to be suitable for remediating contaminated aquifers, they usually aggregate rapidly, and this results in a very limited migration distance that inhibits their usefulness [111]. To overcome this critical limitation, various surface modification and particle stabilization strategies have been developed using various stabilizers, such as surfactant (Tween-20) [110], polymer (polyacrylic acid) [109,112], carboxymethyl cellulose (CMC) [108], starch [113], noble metals [62], and oil [114]. These surface-modified INPs (S-INPs) can be mobile and can theoretically reach the contaminated area in the deep groundwater aquifer even though their practical application still needs to be evaluated.

With the increased use of INPs for waste treatment at superfund sites in the last five years, the environmental fate and consequences of NPs during disposal has been examined. Remediation field studies have shown that INPs remained active for 4 to 8 weeks in groundwater flows to distances up to 20 m [19]. The mobility of NPs in a porous medium depends on the potential for their collision with the medium. Greater potential for collision increases the chance for their removal from flow and retention in the porous medium [7]. Movement of the NPs depends on their Brownian diffusion, interception potential, and gravitational sedimentation; Brownian diffusion has been found to be the dominant factor.

Only a few studies have reported the transport behavior of S-INPs. Schrick et al. studied polyacrylic acid–stabilized INP in a glass burette and transport of S-INPs (stabilized on surfactant and polymer) in a one-dimensional column packed with sand was reported [112]. In another study, polyacrylic acid was used to modify INPs [115] and demonstrated improved elution as a result of particle aggregation. The use of poly(vinyl alcohol)-co-vinyl acetate-co-itaconic acid (PV3A), a nontoxic biodegradable surfactant, to synthesize INPs substantially increases its stability in suspension to six months for PV3A-coated INPs [63]. Saleh et al. compared three different S-INPs and demonstrated far greater elution characteristics of modified NPs, with the triblock copolymer coating offering the best performance among the three. Thus, S-INPs offered the best treatment potential, whereas the leaching potential of the NPs in the environment remains unpredictable [116].

Arsenic removal using S-INPs (in S-INP-pretreated 10-cm sand-packed columns containing about 2 g of S-INPs at a flow rate 1.8 mL/min) showed that 100% of As(III) was removed from influent solutions containing 0.2, 0.5, and 1.0 mg/L As(III) for 9, 7, and 4 days, providing 23.3, 20.7, and 10.4 L of arsenic-free water, respectively. In addition, it was found that As(III) in 0.5 mg/L feed solution at a flow rate of 1.8 mL/min was removed by 100%, for more than 2.5 months, with an S-INPs pretreated 50-cm sand-packed column containing 12 g of S-INPs, providing 194.4 L of arsenic-free water [110]. Figure 8 shows a typical experimental setup and the breakthrough curve; the S-INPs are much more mobile than unmodified INPs.

It is important to note that one-dimensional transport studies cannot represent more complex transport scenarios occurring in multidimensional groundwater aquifers. Recently, Kanel et al. investigated the transport properties of polyarylic acid–stabilized INPs (S-INPs) through porous media in a two-dimensional physical model that was constructed under saturated, steady-state flow conditions [110]. Transport data for INPs, S-INPs, and a nonreactive tracer were collected under the similar flow conditions. The results show that unstabilized INPs cannot travel a long distance in the groundwater aquifer, while the S-INPs can travel like a tracer without any significant retardation. However, the S-INP plume migrates (vertically) downward as it transits horizontally in the physical model, indicating that the density of the S-INPs

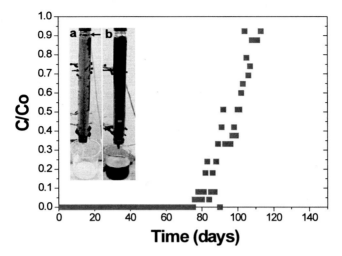

FIGURE 8 Breakthrough curves of As(III) (500 μg/L) passed through S-INP (12 g) anchored sand (425 to 600 μm) packed columns at a flow rate of 1.8 mL/min in an upward direction. The inset shows a comparison of effluents collected (a) without and (b) with S-INPs. The arrow in (a) shows INPs stuck at the top. (From [110].)

would affect the transport of S-INPs [117]. Nevertheless, S-INPs have a great potential to stabilize a variety of contaminants under in situ conditions. More basic and fundamental studies are needed for the use of this promising technology for in situ groundwater remediation.

4 BIMETALLIC NANOPARTICLES

4.1 Introduction

Bimetallic nanoparticles (BMNPs) are synthesized by coating INPs with palladium, gold, platinum, silver, nickel, cobalt, or copper. Experiments have shown that the chemical degradation resulting from the use of such BMNPs are 10 to 100 times faster than that of the microparticles [19]. In fact, Pd–Au BMNPs have a reaction rate 100 times greater than that of Pd alone; furthermore, the Au or gold does not react with the organic compound and serves merely as a catalyst and hence could be recycled [118]. Bimetallic NPs are more expensive to develop than the INPs; however, they serve as a catalyst instead of as a reactant and thus can be reused [36]. The Pd–Au BMNPs have been studied in the laboratory to catalyze the dechlorination of tetrachloroethene in water in the presence of hydrogen. Pilot-scale reactor testing of the process is under consideration at the DuPont facility [118].

Zero-valent metals (ZVMs) and catalytic reduction systems involving Pd^0, Ni^0, Pt^0, and so on, as catalysts are widely used for remediation of many types of environmental pollutants. The mechanisms of pollutant reduction by ZVMs and

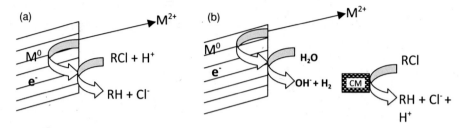

FIGURE 9 Proposed pathways for reductive dechlorination: (a) direct electron transfer; (b) catalyzed dechlorination. RCl and RH represent chlorinated and dechlorinated organic compounds, respectively. (From [119].)

reductive hydrogenation can be explained as shown below. Zero-oxidation-state metal (Fe^0, Mg^0) form a redox couple with the aqueous ions:

$$Fe^0 \rightarrow Fe^{2+} + 2e^- \qquad E^\circ = -0.44 \text{ V vs. S.H.E. (standard hydrogen electrode)} \tag{10}$$

$$Mg^0 \rightarrow Mg^{2+} + 2e^- \qquad E^\circ = -2.2 \text{ V vs. S.H.E.} \tag{11}$$

Reduction occurs by direct electron transfer from the metal surface to the adsorbed chlorinated organic compound (RX):

$$M^0 + RX + H^+ \rightarrow M^{2+} + RH + X^- \tag{12}$$

The reaction mechanism is shown schematically in Figure 9(a).

4.2 Reduction by Catalytic System–Mediated Catalytic Hydrogenation

The underlying principle of catalytic hydrogenation was discovered by Paul Sabatier in 1912. This type of reductive reaction involves a source of hydrogen in the reactor such as a more basic metal in the presence of acid [reaction (13)], formic acid, molecular hydrogen, and so on. The hydrogen subsequently forms a metal–hydride complex (CM–Hx) by intercalation on a catalyst metal (CM) such as Pd^0 or Ni^0, Pt^0 [reaction (14)]. The metal–hydride complexes reductively dehalogenate the halo-organic compound, or cleave the diazo (N=N) bond in a dye, or reduce an inorganic heavy metal ion to its lower valence state. Catalyst metal is concomitantly regenerated [reaction (15)]. This reaction mechanism is shown schematically in Figure 9(b).

$$M^0 + 2H^+ \rightarrow H_2 + M^{2+} \tag{13}$$

$$2CM + H_2 \rightarrow 2CM-H \tag{14}$$

$$CM-H + RX \rightarrow RH + X^- + H^+ + CM \tag{15}$$

4.3   Reduction of Organic Compounds by Pd-Catalyzed Systems

Palladium is one of the platinum group metals (group 8 of periodic table) capable of interacting with molecular hydrogen by intercalating it in its lattice structure. The intercalation of hydrogen in palladium lattice is believed to form palladium hydride; a potent reducing agent. The extent of intercalation of hydrogen depends on many factors, such as temperature, size of palladium particles, and effective concentration of hydrogen [120]. Palladium is a strong reducing catalyst capable of dechlorinating a chlorinated organic compound all the way to its hydrocarbon skeleton. Catalytic systems involving nano palladium can be divided in the following subgroups:

1. Bimetallic systems containing a palladized base metal
2. Nano palladium supported on inert supports
3. Nano palladium modified by incorporation of an other metal (palladium–metal alloy)
4. Catalytic systems containing nano-palladium particles under supercritical reaction conditions

4.4   Reaction Mechanism of Bimetallic Systems

Bimetallic systems consist of two metals: a catalyst such as palladium and a base metal such as Mg^0, Zn^0, or Fe^0 that can undergo corrosion in the presence of aqueous hydrogen ions. The anodic corrosion of base metal releases electrons, resulting in the formation of molecular hydrogen and base metal ions. Palladization of base metal can be achieved by adding a solution of a metal salt such as potassium hexachloropalladate (in Pd^{4+} oxidation state). Palladium ions can readily be reduced to their metallic (zero-valent) form by electrons released by anodic corrosion of base metal and deposited on the surface of base metal. Molecular hydrogen is intercalated on palladium particles, producing palladium–hydride complex. Aqueous anodic corrosion of base metal also increases the pH of the reaction medium. A high enough concentration of acid must be added to maintain the pH at such a level that metal hydroxide formation can be avoided. Metal hydroxide formation (and if molecular oxygen is not removed from the reaction system, metal oxide formation) can interfere with catalytic activity and prevent or reduce further corrosion of base metal by covering the bimetallic surface. Figure 10 (a) and (b) show, respectively, scanning electron microscopic images of an Mg^0 surface deposited with silver and the presence of oxides and hydroxides of magnesium at the end of the reaction where molecular oxygen is not excluded.

It may be noted from equations (13) and (14) that the rate of corrosion and hence formation of oxides and hydroxide for magnesium will be much higher than that of iron. Thus, bimetallic systems involving Fe^0 as a base metal may not require the addition of acid to maintain the pH of the reaction medium. In the absence of acid addition, the pH of a reaction solution containing 10 g/L Pd–Fe particles increased from 7 to 7.6 after 6 h of reaction [121]. On the other hand, the pH of a reaction solution containing 5 g/L Ag–Mg or Mg^0 particles increased from 4 to 10 after 10 min of reaction [122].

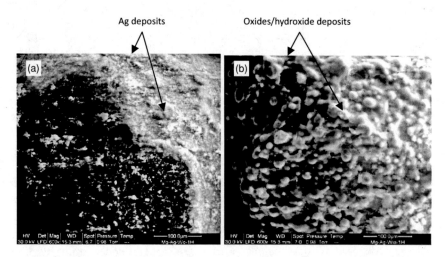

FIGURE 10   Deposition of (a) catalyst metal (silver) and (b) magnesium oxides and hydroxide on magnesium surface. (From [122].)

In a heterogeneous reaction system such as a bimetallic system, reactions are surface-mediated. Some of the several reaction steps in such a reaction could be (1) diffusion of the reactant to the surface, (2) a chemical reaction on the surface, and (3) diffusion of the product to the bulk solution [123]. Graham and Jovanovic used a Pd–Fe bimetallic system for dechlorination of p-chlorophenol and reported that the reduction of a target compound occurs at the interface of Fe and Pd, as shown in Figure 11 [124]. Later it was demonstrated by Patel and Suresh that the reduction of target compound need not necessarily occur at the metal–catalyst interface [125].

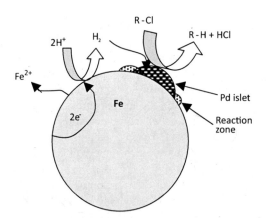

FIGURE 11   Mechanism of dechlorination on a Pd—Fe bimetallic system. R—Cl and R—H represent chlorinated and dechlorinated organic compounds, respectively. (From [124] and [121].)

In this study pentachlorophenol is degraded by a Pd–Mg bimetallic system and found that acid addition was essential to start the reaction and that Pd deposits on a magnesium surface were sloughed off due to the addition of acid [125]. The Pd deposits were found to be spatially separated, and reductive dechlorination of PCP occurred on such deposits, away from the Mg^0 surfaces.

4.5   Bimetallic Systems Containing Nano Palladium

Nanoscale bimetallic systems possess a much higher surface area (10 to 60 m^2/g) and hence much higher reactivity than those of microscale systems ($\sim$1 m^2/g) [121,126,127]. It was also shown by Lieu et al. that the surface area of nano Pd–Fe particles increased with an increase in Pd loading on Fe [128]. Due to their small size, nanoparticles remain suspended and move mainly by brownian motion when kept in a gently mixed reaction medium [128]. Thus, they can be delivered to deep contamination zones [121,129].

The extent and rate of dechlorination of chlorinated organic compounds (COC) by Fe–Pd BMNP systems have been reported to be affected by (1) loading of Pd on base metal, (2) concentration of Fe–Pd BMNPs with respect to concentration of the target compound, and (3) the initial pH of the reaction medium [121,126,127].

Wang et al. studied dechlorination of carbon tetrachloride (CT), chloroform (CF), and dichloromethane (DCM) using with Fe–Pd BMNPs, with Pd loading varying from 0.04 to 3 wt%. It was found that the optimum Pd loading was 0.2 wt% [121]. Lien and Zhang reported that higher reaction rates for TCE dechlorination were observed for palladium loading of 1 to 5% of Fe mass. When palladium loading was increased to 50% of Fe mass, reaction could not occur [126]. The decrease in removal efficiency due to increased loading of catalyst metal has been attributed to suppression of hydrogen evolution by excessive coverage of base metal [26,76,124]. Wang et al. also reported that efficiency of removal of chloromethanes increased with increase in concentration of Fe–Pd BMNPs (0.0 g/L to 20 g/L); however, the increase in removal efficiency was negligible for Fe–Pd BMNP concentrations greater than 10 g/L [121]. Increased concentration of BMNPs generates excess hydrogen, which may hinder further corrosion of base metal and suppress number of reaction sites [119].

Lien and Zhang also studied dechlorination of chlorinated ethanes containing two to six chlorine atoms using Fe–Pd BMNPs [123]. Authors reported that the observed reaction rate constant decreased with decrease in the number of organo chlorines on ethane. Attempts were made to develop a correlation between rate constants and thermodynamic properties, such as one electron reduction potential and C—Cl bond strength of chlorinated ethanes. The decrease in rate constants correlated well with the C—Cl bond strength. In a similar study by Patel and Suresh, decrease in second-order reaction rate constant values for dechlorination of penta-, tetra-, and trichlorophenols by Ag–Mg bimetallic particles correlated well with dipole moments of these chlorophenols [122].

Nagpal et al. compared INPs, Fe–Pd BMNPs, and granular ZVI for dechlorination of lindane. It was found that at Fe–Pd BMNPs at a concentration of 0.5 g/L achieved almost complete removal of lindane after 5 min of reaction [127]. On the other hand, granular ZVI could not achieve any removal and INPs achieved partial ($\sim$60%) removal of lindane under identical reaction conditions. These results were attributed to superior catalytic activity of Fe–Pd BMNPs due to the presence of Pd. Lien and Zhang reported that the activation energy for dechlorination of PCE was lower for Fe–Pd BMNPs (31.1 kJ/mol) than that of INPs (44.9 kJ/mol) for dechlorination of tetrachloroethene [126]. Moreover, Nagpal et al. found that the time-dependent profile of lindane removal was biphasic: initial rapid removal up to 1 min due to increased surface area of BMNPs, followed by slower removal, up to 5 min, due to rapid passivation of the reactive surface by the deposition of oxides [127].

BMNPs can be immobilized on resins or zeolites advantageously to achieve (1) a reduction in the loss of particles during reaction, (2) prevention of particle agglomeration, and (3) a stationary phase for treatment of a flowing medium [129,130]. Lin et al. employed resin-supported Sn–Pd BMNPs to dechlorinate tetrachloroethene [130]. Authors indicated that cation-exchange resins can exchange metal ions produced by corrosion of base metal with Na^+ or H^+ thereby preventing precipitation of metal hydroxide and oxides. Use of such immobilized BMNPs for groundwater remediation may be affected by the presence of constituents such as nitrate, sulfide, and dissolved oxygen. Lin et al. found that the presence of 10 mg/L of HS^- poisoned palladium and reduced the reaction rate constant by almost half. The poisoned Sn–Pd BMNPs could be regenerated by sodium borohydride but had reduced catalytic efficiency [130]. Zhu et al. demonstrated dechlorination of 1,2,4-trichlorobenzene (TCB) using Fe–Pd BMNPs supported on chitosan and silica. Complete dechlorination of TCB, all the way to benzene, was reported. The Pd–Fe–chitosan system was superior to the Pd–Fe–silica system [129].

As discussed earlier, the transport of the target compound to the surface of BMNPs is an important step in surface reactions. Thus, it may be possible that sorption of the target compound during the reaction may contribute significantly to the overall removal of that compound. It has been shown that sorption of phenolates on metal oxides occurs through inner-sphere coordination; and it may not be possible to remove such sorbed compounds simply by washing with water [131]. Morales et al. reported more than 95% removal of pentachlorophenol (PCP, initial concentration 2.5 mM) using a Pd–Mg system [132]. Repetition of this experimental protocol by Patel and Suresh [125] revealed that almost 10% of initial PCP was sorbed to residual surfaces, which remained unaccounted for by Morales et al. in the absence of an effective procedure to quantify sorption. Similar results were also obtained by Kim and Carraway in a study to dechlorinate PCP by INP, Fe–Pd, Fe–Pt, Fe–Cu, and Fe–Ni BMNP systems [133]. Recent studies revealed that PCP dechlorination rates obtained using ZVI were higher than that obtained using bimetallic systems, where a significant fraction of the apparent removal of PCP by bimetallic systems was attributed to sorption. From these results, Kim and Carraway concluded that sampling of the aqueous phase alone to determine the extent of removal of a target compound may lead to greatly overestimated reaction rates [133]. It must be kept in mind that the presence of a catalyst such as Pd in a bimetallic system lowers the activation energy and therefore improves the rate and extent of removal of a target compound.

Contradictory results obtained by Kim and Carraway may be attributed to the fact that they used previously prepared, stored, air-dried BMNPs [133]. Morales et al. indicated that such particles may be much less effective than freshly prepared nano particles [132]. Nonetheless, sorption of target compounds on residual surfaces at the end of the reaction time may not be neglected, and a procedure must be developed to account for the sorption.

4.6 Nano Palladium Supported Using Inert Supports

In the systems described above, catalysts such as Pd, Pt, and Ni are supported on an inert support. Reduction of a target compound is carried out in the presence of a hydrogen donor such as molecular hydrogen, formic acid, sodium borohydride, ethanol, isopropanol, or formate. Some of the advantages of supported catalytic systems over nonsupported systems are (1) easy separation and recycling of catalyst from and to the reaction medium, (2) reduction in loss of precious catalyst along with products, and (3) easy regeneration. Various supports reported in the published literature include carbon, metal oxides, chitosan, chitin, resin, silica, and bacterial cellulose [134–136].

Most common method of producing Pd NPs is using sodium borohydride to reduce a salt of palladium to metallic palladium. To prevent agglomeration of NPs, stabilizers may subsequently be added in a dispersing medium, which can be evaporated to obtain dried NPs [137,138]. Stabilizers prevent agglomeration by inducing a steric and/or electrostatic barrier between the particles. In the case of supported catalyst preparation, a support such as alumina or carbon can be impregnated with palladium ions and subsequently reduced using sodium borohydride in aqueous condition or dried and reduced under hydrogen stream at elevated temperature [136]. Some natural supports such as bacterial cellulose possess reducing sugars in their matrix. Tetravalent Pd can readily be reduced to its metallic form by these reducing sugars [139,140]. Pd NPs can also be electrodeposited on a support [141,142]. Recently, biologically reduced Pd NPs have been developed for a variety of catalytic reduction applications. In the presence of a hydrogen donor, bacterial strains can reduce Pd^{2+} and subsequently precipitate Pd NPs on their cell walls and in periplasmatic space [143–148]. After initial precipitation of Pd^0 by cells, further reduction of Pd^{2+} is sustained by absorption of hydrogen by existing Pd^0 and subsequent catalytic reduction of Pd^{2+} [148]. Lloyd et al. reported biological reduction of Pd^{2+} to Pd^0 by *Desulfovibrio desulfuricans* using pyruvate, formate, or H_2 as an electron donor. The authors also demonstrated that reduction of Pd^{2+} was insensitive to the presence of O_2, indicating potential for alternate ways of synthesizing biogenic Pd NPs [149].

Effect of Pd Catalyst Supports on Dechlorination   It has been reported that the selectivity of target compound, catalytic activity, and stability of immobilized catalyst are strongly influenced by the type of support [135,150]. A good catalyst support is expected to resist the corrosive environment developed due to the formation of hydrogen halides produced as a result of dehalogenation. In addition to this, a good support must also possess good selectivity toward the target compound and undergo minimal ion-exchange reactions with the halides [135].

Hennebel et al. compared catalytic activities of freely suspended and encapsulated [in poly(vinylidene fluoride) (PVDF), and in polysolfone (PSf) membranes] bio-Pd particles with that of commercially available powdered Pd–alumina in the presence of molecular hydrogen to dehalogenate diatrizoate [145]. The authors reported that first-order rate constants normalized with Pd concentration depending on pH were 2.5 to 30 times higher in the case of Pd–alumina than that of bio-Pd suspension. On the other hand, bio-Pd encapsulated in PSf membranes showed better efficiency than that of PVDF membranes at lower Pd loadings. Higher surface porosity of PSf membranes was considered to be responsible for better performance of PSf-encapsulated bio-Pd particles.

In a similar study, Hennebel et al. prepared biogenic Pd NPs supported on zeolite and biogenic Pd NPs encapsulated with alginate, polyacrylamide, polyurethane, and silica. These beads were utilized for dechlorination of trichloroethene in the presence of hydrogen gas. Under identical reaction conditions, biogenic Pd NPs achieved 5 to 46 times higher reaction rate constants than those achieved by encapsulated and supported biogenic Pd NPs [146]. Among encapsulated particles, biogenic Pd NPs encapsulated in silica beads exhibited the lowest rate constant, probably due to their smaller porosity and pore size. Also, the very low reactivity of silica support might have impaired the transfer of organic compound to the catalytic surface, resulting in lower results. Thus, the reactivity of the support toward the target compound, surface porosity, and pore size played an important role in determining the change in reactivity of biogenic Pd NPs in encapsulated form within various support materials.

Liu et al. studied the hydrodechlorination of monochlorobenzene (MCB) using Pt NPs stabilized with poly(vinyl-2-pyrrolidone) (PVP) in the presence of H_2 at atmospheric pressure. The authors demonstrated that the catalytic activity of PVP–Pt NPs was dependent on the PVP/Pt ratio of molar concentrations [138]. The highest removal and conversion rates of MCB were achieved at a PVP/Pt ratio of 20; lower results were obtained at PVP/Pt ratios of 10 and 40. At the lower ratio (PVP/Pt = 10), the stabilization of Pt nanoparticles was poor, resulting in agglomeration leading to less catalytic surface. On the other hand, at a higher ratio (PVP/Pt = 40), particles were well stabilized, but an excessive concentration of PVP hindered the hydrodechlorination of MCB. When a reaction was carried out in the presence of extra amounts of PVP, the extent of the MCB hydrodechlorination decreased and the selectivity of the product (benzene vs. cyclohexane) changed.

Baxter-Plant et al. compared biologically prepared Pd^0 (biogenic Pd), by three different sulfate-reducing bacterial strains, with chemically prepared Pd^0 (chemogenic Pd) for reductive dechlorination of 2-chlorophenol and polychlorinated biphenyls (PCBs) [144]. The dechlorination of target compounds was estimated from the release of chloride ions. The authors noted that the release of chloride from dechlorination of both target compounds was biphasic (initially slower, followed by rapid removal) for chemogenic Pd, whereas it was almost linear for biogenic Pd with respect to time. Also, the rates of release of chloride were much higher for biogenic Pd than for chemogenic Pd for 2-CP. The authors also studied the effect of the water solubility of

TABLE 3   Comparison of Cl Release by Bio-Pd and Chemical-Pd

		Rate of Chloride Release (μmol/min·mg Pd)	
Compound	Water Solubility (g/L at 20°C)	Bio-Pd (prepared by *Desulfiovibrio desulfuricans*)	Chemical-Pd
4-Chlorobiphenyl	1.46×10^{-3}	13.5 ± 1.2	1.55 ± 0.6
2,4,6-Trichlorobiphenyl	2.39×10^{-4}	8.3 ± 0.3	2.45 ± 0.2
2,3,4,5-Tetrachlorobiphenyl	1.39×10^{-5}	9.3 ± 0.2	0.3 ± 0.1

Source: [144].

PCBs on dechlorination by biogenic Pd and chemogenic Pd. It is well known that the water solubility of PCBs decreases with an increase in chlorine substitution. Rates of chloride release observed by Baxter-Plant et al. [144] from dechlorination of PCBs using biogenic Pd and chemogenic Pd are compared in Table 3.

It may be noted that the rate of chloride release decreased from 13.5 mol/minμmg Pd to 9.3 mol/min·mg Pd for biogenic Pd and from 1.55 mol/min·mg Pd to 0.3 mol/min·mg Pd for chemogenic Pd as the solubility of PCB decreased. The authors indicated that for biogenic Pd, a biomatrix facilitated access of non-aqueous fraction of higher-substituted PCB to catalytic surfaces. Moreover, among the bacterial strains used, biogenic Pd prepared from *Desulfiovibrio desulfuricans* demonstrated the highest activity. It seems that a change in morphology of similar supports (e.g., bacterial cells) also influences the catalytic activity. From the discussion above it may be concluded that the reactivity of a supported nanocatalyst is affected by (1) surface characteristics (porosity, pore size, etc.) of the support, (2) reactivity of the support with the target compound and hydrogen donor, and (3) loading of catalyst on the support.

4.7   Competition Between Target Compound and Hydrogen Source

Hydrodechlorination reactions are also found to be influenced by competition between the hydrogen donors and target compounds for sorption on catalyst surface. Jiao et al. prepared a thin film of Pd NPs on a glass carbon substrate by cyclic voltammetric deposition [141]. In this process, potential repeatedly swept from -0.35 to 0.8 V in a solution of H_2PdCl_4 and H_2SO_4 for 15 cycles. In contrast to potentiostatic deposition (-0.25 V for 10 min), the authors found that Pd NPs formed were smaller in the case of cyclic voltammetric deposition ($\sim$20 nm). Authors provide insight into the reaction mechanism by performing electrochemical dechlorination of carbon tetrachloride (CT) at three different potentials at the cathode (vs. SCE): -0.05 V (where adsorption of hydrogen begins on Pd), -0.15 V (where adsorption of hydrogen is over but evolution of hydrogen gas has not yet begun), and -0.2 V (where hydrogen gas evolution begins). CT removal efficiencies after 15 min of electrolysis were 83.7% at -0.05 V, 19.4% at -0.15 V, and 89.3% at -0.2 V. At -0.05 V, adsorption of hydrogen and CT on Pd surface was concomitant followed by dechlorination. On the other hand, at -0.15 V the monolayer of adsorbed hydrogen inhibited adsorption of CT, resulting in lower removal efficiency. At -0.2 V, hydrogen evolution frees the reactive surface for adsorption of CT while molecular hydrogen is absorbed, which subsequently dechlorinates CT. Thus, competitive sorption of target compound and hydrogen source plays an important role in determining the rate and extent of dechlorination.

Aramendia et al. [151] demonstrated competitive interaction of target compound and hydrogen donor. They studied dehalogenation of halobenzenes using potassium formate as a hydrogen donor. The authors reported that the rate of dehalogenation followed the order chlorobenzene $\gg$ bromobenzene $\gg$ fluorobenzene. Iodobenzene, which possesses the weakest bond strength (C—I bond, 234 kJ/mol), remained unaffected because it is more readily sorbed on the catalytic surface and thus inhibited sorption of hydrogen donors. On the other hand, fluorobenzene, which has the highest bond strength (C—F bond, 451 kJ/mol), was not adsorbed; reactive sites were occupied mainly by formic acid. Thus, irrespective of bond strength, excessive or inadequate sorption of target compounds on reactive surface can lead to incomplete dechlorination.

4.8   Effect of Hydrogen Donors

Hydrogen donors influence dehalogenation reactions by reacting with the catalyst and catalyst support. Lowry and Reinhard indicated that formation of undesired, partially dehalogenated intermediates may increase in H_2-limiting conditions because of the fact that cleaving of halogen from target compound is a reversible reaction [152]. Hennebel et al. studied the effect of various hydrogen donors, such as hydrogen gas, formic acid, formate, ethanol, and isopropanol, on dechlorination of trichloroethylene using biogenic Pd NPs [146,147]. Ethanol and isopropanol could not activate the biogenic Pd NPs for dechlorination. Among hydrogen gas, formate, and formic acid, the highest reaction rate constant was achieved in the presence of hydrogen gas, while the reaction kinetics with formic acid was slower and with formate the slowest. The authors indicated that with formate a slower, hydride mechanism dominated, while with formic acid a faster, radical (H-atoms) mechanism was favored. In the case of formate, when the pH of the reaction medium was reduced to 3 using HCOOH (HCOOH/HCOO$^-$ = 5), the reaction kinetics increased and almost matched that in the presence of $H_2.$ The authors indicated that the difference in results achieved using formic acid and formate was attributed to the difference in the rate of decomposition of these donors at the catalyst surface.

Roy et al. studied the effect of hydrogen sources (sodium borohydride, H_2, ammonium acetate, and hydrogen saturated water) on the dechlorination of chlorophenols using palladium and nickel as catalysts under mild reaction conditions [153]. Sodium borohydride and H_2-saturated water were inefficient sources of hydrogen in the presence of Pd/γ-alumina; chlorophenols were partially removed by dechlorination to form phenol, but concomitant dearomatization was not observed. On the other hand, in the presence of H_2 and ammonium acetate, chlorphenols were completely removed and reduced all the way to cyclohexanol and cyclohexane. Reactions using Ni^0 powder in the presence of sodium borohydride were strongly influenced by temperature. While chlorophenol removal was almost complete with the formation of small concentrations of cyclohexanol at 338 K, partial removal was achieved at 296 K. H_2 was an inefficient source of hydrogen at 296 or 353 K with Ni^0 powder as a catalyst. However; Raney Ni could facilitate complete removal of chlorophenol, yielding cyclohexanol at 296 K in the presence of H_2 and sodium borohydride as hydrogen sources. These results reveal that various hydrogen sources interact with different catalysts in different ways.

Decontamination of sites contaminated with Cr^{6+} by biological reduction has been attempted using *Desulfovibrio vulgaris*, and *Desulfovibrio* sp. Oz-7. However, reduction of Cr^{6+} to Cr could occur only in the presence of an agent able to make a complex with Cr^{3+}. Field-scale application of biological Cr^{6+} reduction may be uneconomical, due to the high cost of providing a complexing agent [154]. On the other hand, biogenic Pd NPs can be used effectively in the presence of hydrogen donors such as H_2 or formate to reduce Cr^{6+}. Mabbett et al. [154] reported 94 and 96% reduction of 700 μM Cr^{6+} in the presence of formate and H_2, respectively. However, lactate could achieve only partial reduction of hexavalent chromium. The authors indicated that the smaller formate molecule is chemisorbed and decomposed on a Pd surface in a better way than the bulky lactate molecule.

A colloidal suspension of nano-sized supported catalysts can accomplish improved mass transfer and enhanced reaction rates due to the higher surface area compared to that of a fixed bed. However, nanoparticles may easily escape with the reaction mass, resulting in the loss of precious catalyst and violation of the discharge limit on metals, especially in the case of water treatment. A relatively recent trend is to produce Pd NPs supported on magnetic NPs [155–157]. Hildebrand et al. reported the highest reactivity of Pd NPs supported on magnetite (Pd, 0.15 wt%) for hydrodechlorination of TCE (6000 L/g·min), with the advantage that these particles can easily be separated magnetically [146,147]. Authors also conducted multi-spiking experiments using the same Pd–magnetite NPs in a buffered reaction medium. The reactivity of Pd–magnetite (0.15 wt%) was reduced from 6700 L/g·min in the first cycle to 1600 L/g·min in the sixth cycle, indicating superior reusability. A similar decrease in catalytic activity was observed in multispiking experiments conducted for hydrodechlorination of monochlorobenzene. The authors predicted that the loss in catalytic activity may be a result of some type of "aging" of reactive surfaces and not necessarily due to accumulation of chloride ions. The authors also observed that an increase in Pd loading on magnetite from 0.15% to 5% resulted in a decrease in catalytic activity for TCE dechlorination from 6000 L/g·min to 520 L/g·min. Carbon monoxide chemisorption study indicated that Pd was dispersed better in the form of a monolayer at 0.15% loading, which resulted in better activity. On the other hand, three-dimensional Pd clusters were identified in electron microscopy at 5% Pd on magnetite, resulting in a lower exposed surface.

He et al. investigated the effect of temperature on the synthesis of Pd NPs by a simple one-step addition of ascorbic acid to a $Na_2PdCl_4\cdot3H_2O$ solution with sodium carboxymethyl cellulose (CMC) as a dispersing agent [158]. Authors reported that the mean size and polydispersivity of Pd NPs decreased with an increase in temperature from 22° to 95°C. For Pd NPs synthesized at 22°C, the observed reaction rate constant for the hydrodechlorination of trichloroethene was 112 L/(g Pd)·min, which increased almost six-fold when Pd NPs synthesized at 95°C were employed under identical reaction conditions [158].

Nutt et al. employed Pd–Au NPs (~20 nm) prepared using ascorbic acid as a reducing agent and studied the effect of coverage of Au NPs by Pd NPs on TCE dechlorination [159]. Authors prepared 1.9, 3.8, 5.7, and 11.4 wt% Pd on Au NPs and observed the change in the resonance peak of native Au NPs sol (pinkish tint), due to the addition of different amounts of Pd. It was observed that at lower Pd loadings (1.9 and 3.8 wt%), the mixture retained a pinkish hue. However, at higher Pd loadings the suspension displayed a gray–violet tint, indicating a submonolayer coverage of the Au surface by Pd NPs at a lower Pd loading. Dynamic light-scattering experiments also supported the observations above, showing an increase in hydrodynamic diameter of Pd–Au NPs from 20 nm at 1.9 wt% to about 150 nm at 5.7 wt%. It was reported that the observed reaction rate constant (L/(g Pd)·min) decreased with increased Pd loading. Pd–Au NPs (1.9 wt% Pd) exhibited about 15-fold faster reaction kinetics than that of pure Pd NPs (or fully covered Pd–Au NPs). It was speculated that enhanced activity of partially covered Pd–Au NPs may be attributed to an electronic (by way of donating electron density from Au to the Pd atoms) mixed metal site effect (formation of active Pd–Au surface species) or geometric (formation of Pd clusters on the Au surface) effects produced by the Au surface [160]. Pd coverage also affected the product distribution. For example, more than 90% of the TCE dechlorination product was ethane for Pd–Au NPs with ≥ 1 monolayer coverage, while about 70% of the TCE dechlorination product was ethene for Pd–Au NPs with <1 monolayer coverage. In a continuation of this study, Nutt et al. prepared smaller Au NPs (~4 nm) using a tannic acid/trisodium citrate reduction method and deposited on them metallic Pd by its reduction with hydrogen [161]. The authors investigated dechlorination of TCE using Pd–Au NPs with varying Pd loading and reported that at lower and higher Pd loadings (2.8 and 41.4 wt%) TCE dechlorination rate constants were significantly lower (498 and 419 L/(g Pd)·min, respectively) [161] than at intermediate Pd loadings (>1900 L/(g Pd)·min at 12.7 and 14.9 wt%), indicating a volcano-shaped dechlorination kinetics-Pd loading relationship. It was concluded from comparison with the earlier study [159] that optimum Pd loading for 20 and 4 nm Pd–Au NPs was dissimilar. Pd–Au NPs supported on magnesia and silica also accomplished dechlorination of TCE but at a much slower rate than that of unsupported Pd–Au NPs. Four-millimeter Pd–Au NPs supported on ion-exchange resins used as a packed bed could achieve almost complete dechlorination of TCE for up to four days [160].

4.9 Palladium Catalysts with Other Metals

Addition of a second metal to palladium catalysts is often found to improve catalytic activity. Enhancement of catalytic activity may be caused by changes in geometric and electronic properties, formation of mixed sites, and disappearance of the β Pd-H phase in the modified catalyst [162]. Golubina et al. (2006) studied the hydrodechlorination of chlorobenzenes using Pd–C modified with varying concentrations of Fe [163]. Authors reported 16% conversion of 1,4-dichlorobenzene (DCB) by 2 wt% Pd–C. On the other hand, using modified catalyst with a total of 5 wt% metal, conversion of DCB increased from 26% to 34% at Pd/Fe molar ratios of 1-0.5 to 1-7.7. When total metal content was increased to 11 to 12 wt% on C, authors noted that conversion of DCB was dependent on the molar Pd/Fe ratio. For example, DCB conversion was 34% at the Pd/Fe ratio of 1-0.4 (total metal 12 wt%) and complete conversion of DCB to benzene was observed at a Pd/Fe ratio of 1-1 (total metal 12 wt%). Similar results were noted for HDC of hexachlorobenzene (HCB), where authors observed that addition of Fe in amounts not exceeding the Pd content improved the catalytic performance of Pd–C. The authors describe the effect of Fe on the catalytic activity of palladium as follows:

1. *Stabilization of small metal particles.* Addition of Fe dilutes Pd. This results in either the formation of smaller particles that are more active and resistant to agglomeration or in the geometry of the active sites.

2. *Alloy formation.* Pd and Fe may form an alloy during reduction by hydrogen. Alloy formation may either change the electronic state of metals or change the geometry of active sites. These changes may alter the heat of adsorption of the target compound and hydrogen, thereby directly affecting the performance of catalyst.

3. *Mixed site formation.* Although the catalyst was prepared by impregnation of Pd and Fe salts on carbon followed by reduction by hydrogen at an elevated temperature ($\sim$773 K), the presence of Fe_2O_3 was observed in the surface layer. Thus, the ultimate form of the catalyst could be Pd–Fe_xOy–C, which may promote metal oxide interaction and change the heat of adsorption of the substrate and hydrogen.

4. *Prevention of Pd poisoning.* The hydrodechlorination of chlorinated organic compounds results in the formation of HCl. The presence of Fe may thermodynamically favor the formation of ferrous/ferric chloride over that of palladium chloride. Thus, poisoning of palladium due to HCl may be eliminated and catalytic activity may be improved. On the other hand, higher fractions of Fe may cover the active catalytic sites by ferric chloride, resulting in rapid deactivation of catalyst compared to that of Pd–C.

4.10 Deactivation and Poisoning of Palladium Catalysts

Palladium catalysts are very efficient in the removal of a variety of compounds of concern to the environment. Besides cost, deactivation and poisoning of Pd catalyst are important drawbacks to be considered from the view point of field application. Deactivation is the process by which activity of a catalyst is reduced. Deactivation of palladium catalysts can be caused by:

1. Blockage of the Pd surface by adsorption of minerals, natural organic matter, precipitation of salts, accumulation of debris, metal oxides, and the formation of biofilms.
2. Ostwald ripening in the case of nanoparticles where smaller NPs continue to shrink or dissolve and larger NPs continue to grow [156].
3. Dislodgement of Pd and formation of metal oxides on active catalyst sites in the case of Pd bimetallic systems.
4. Corrosion of support and leaching of Pd by hydrogen halide formed due to dehalogenation of halogenated organic compounds.
5. Formation of carbonaceous deposits (cocking) on the Pd surface, especially in gas-phase reactions.
6. Sintering of active Pd surfaces, especially at high-temperature gas-phase reactions.

Poisoning of Pd catalysts is due mainly to the presence of some inorganic and organic constituents in the reaction phase, such as reduced sulfur compounds, heavy metals, carbon monoxide [146,147,155,156], and HCl [150]. Poisoning may be avoided by pretreating the influent streams to remove such compounds. Poisoning may be reversible; that is, poisoned catalysts may be regenerated chemically using an oxidizing agent such as sodium hypochlorite or oxygen. Lowry and Reinhard have indicated the following possible reasons for the regeneration of catalyst by an oxidizing agent [152]:

1. Oxidation followed by desorption of reduced sulfur species or organic molecules strongly sorbed to the catalyst surface.
2. Oxidation and removal of coke deposits from the catalyst surface.
3. Redispersion of sintered Pd particles.
4. Oxidation of less active palladium hydride to more active metallic Pd or PdO.
5. Destruction of biofilm from catalyst surface.

Sometimes, Pd catalysts are selectively poisoned to fine tune selectivity towards a specific product. For example, use of lead compounds in preparation of Lindlar catalyst for hydrogenation of alkynes to alkenes ($-C\equiv C-$ to $-C=C-$) by blocking certain active sites [164]. Lowry and Reinhard studied the effect of common groundwater solutes—carbonate species (carbonate, bicarbonate, and carbon dioxide), sulfur species, and chloride—on TCE dechlorination by a bed of Pd/γ-alumina in the presence of hydrogen [165]. The authors observed rapid decline (from 45% to 33% TCE conversion) in first the first two to three days in the baseline TCE dechlorination study performed using deionized water. It was speculated that this drop in activity was due to microscale surface modification of catalyst. In a 60-day operation the TCE conversion dropped from 31% to 24%. However, the catalyst activity was restored almost completely by a regeneration run of 90 min using hypochlorite solution. It was found that none of the carbonate species had an adverse impact on TCE dechlorination. On the contrary, the presence of carbonate and carbonic acid appeared to reduce the rate of deactivation. On the other hand, formate was formed in the presence of HCO_3^- according to the following reaction:

$$HCO_{3-} + H_2 \overset{Pd}{\Rightarrow} HCOO^- + H_2O \qquad (16)$$

The presence of formate may stimulate biological growth.

The presence of sulfate or chloride did not show any adverse effect on dechlorination, whereas the presence of sulfite and HS^- resulted in rapid deactivation. Flushing with deionized (DI) water alone could not regenerate the columns, indicating chemisorption of these ions on catalyst and support surfaces. However, regeneration with a high dose (750 mg/L) of hypochlorite could reactivate the catalyst. In another report from the same research group [166], it was observed that the presence of chloride and sulfate lowered the dechlorination rate constant by almost 50% compared to that in the absence of these anions while studying hydrodechlorination of 1,2-dibromo-3-chloropropane (DBCP) by Pd/γ-alumina in the presence of hydrogen. Patel and Suresh observed that the rate of PCP dechlorination by Pd–Mg bimetal was reduced in the presence of chloride and remained unaltered in the presence of sulfate [134].

Hildebrand et al. employed Pd–magnetite NPs to study the effect of heavy metals and reduced the sulfur species on hydrodechlorination of TCE, chloroform, and monochlorobenzene [155]. Authors also compared the hydrodechlorination of TCE and chloroform in DI water and tap water and found that the reaction rate was biphasic (initially slow, then rapid) in DI water but monophasic and much slower in tap water. Authors concluded that constituents of tap water influenced not only the continuity but also the rate of reaction. Conducting hydrodechlorination studies in tap water treated by cation-exchange resins followed by granular activated carbon improved the catalytic activity to the level comparable to that of pure water. Considering that volatile organics from tap water are removed due to hydrogen purging prior to the experiment, authors indicated that the low activity of catalyst in tap water may be attributed to the presence of natural organic matters such as humic acid. Among the anions, HS^-, I^-, SO_3^{2-}, and MnO_4^- poisoned or reduced the catalytic activity. On the other hand, Cu^{2+}, Sn^{2+}, Pb^{2+}, and Hg^{2+} poisoned the catalyst.

Yuan and Keane studied the hydrodechlorination of 2,4-dichlorophenol (DCP) using Pd–C and Pd/γ-alumina in the presence of hydrogen [150]. Initial catalytic activity was identical for both types of catalysts; however, the reuse of catalysts revealed a significant reduction in catalytic activity. The rate of HDC using reused catalysts decreased to 10% and 66% of the initial rate (obtained using fresh catalysts), respectively, for Pd–C and Pd/γ-alumina. Although some Pd leaching was observed for both catalysts, the deactivation was the result of chemical poisoning due to HCl. Alumina could stabilize supported Pd better than carbon, and hence Pd/γ-alumina was less affected.

5   SUMMARY AND FUTURE PROSPECTIVE

INPs are one of the fastest-developing fields. INPs have a number of key physicochemical properties, such as high surface area, reactivity, optical and magnetic properties, and oxidation and reduction capacities, which make them attractive for water reatment. INPs are much more effective than that of conventional ZVI. The INP surface can be modified for groundwater treatment. The surface area of S-INPs is about three times higher than INPs. This led to increased removal of contaminants. Kanel et al. found that the surface-normalized rate constant of As(III) on INPs is about three orders of magnitude greater than ZVI. Similarly, the arsenic attached on nano iron oxides can be removed using a low-strength magnet [167]. The nano iron and INPs are not mobile and thus have limited use. Hence, surface-modified INPs have been developed and tested for removal of arsenic and TCE [3,109,110,168]. The S-INPs can be used as a CRB material to treat deep groundwater contaminants. Although there are some studies to test S-INPs [109,110,168] as a CRB material in column experiments, more research on the pilot and field scales is needed in the future. Kanel et al. studied S-INPs transport in a two-dimensional porous media–packed tank and found that an S-INP moves like a tracer without significant retardation and is driven by the density of the particles [117]. It is also found that the density effect cannot be distinguished by one-dimensional column experiments; hence, more studies are needed in two-dimensional systems. The applicability of INPs has been proven; further development is needed in the targeted synthesis of new materials with specific and selective properties for known applications.

Groundwater remediation through bimetallic catalysts is a more effective and desirable approach than monometallic catalysts For example, palladium-on-gold nanoparticles (Pd–Au NPs) have recently been shown to catalyze the hydrodechlorination of trichloroethene in water, at room temperature, and in the presence of hydrogen, with the most active Pd–Au material found to be ≥ 70 times more active than Pd supported on alumina on a per-Pd atom basis [36]. Laboratory-synthesized nanoscale bimetallic particles (Pd–Fe, Pd–Zn, Pt–Fe, Ni–Fe) have a larger surface area than those of commercially available microscale metal particles. Surface area normalized reactivity constants are about 100 times higher than those of microscale iron particles. Production of chlorinated by-products, frequently reported in studies with iron particles, is reduced notably, due to the presence of catalyst. The nanoparticle technology offers great opportunities for both fundamental research and technological applications in environmental engineering and science [33].

New INPs suitable for in situ and ex situ applications in water treatment will definitely attract further special attention in the forthcoming years. We envision that INPs will become critical components in industrial and public water purification systems as more progress is made toward the synthesis of new INPs that are cost-effective, efficient, and environmentally friendly for use.

Acknowledgments

This review was performed while the first author held a National Research Council Research Associateship Award at the Air Force Institute of Technology, Wright-Patterson Air Force Base, Ohio.

REFERENCES

1. Kanel, S. R., J. M. Greneche, and H. Choi, Arsenic(V) removal from groundwater using nano scale zero-valent iron as a colloidal reactive barrier material, *Environ. Sci. Technol.*, 2005;**40**(6):2045–2050.

2. Kanel, S. R., et al., Removal of Arsenic(III) from groundwater by nanoscale zero-valent iron, *Environ. Sci. Technol.*, 2005;**39**(5):1291–1298.

3. Li, L., et al., Synthesis, properties and environmental applications of nanoscale iron-based materials: a review, *Crit. Rev. Environ. Sci. Technol.*, 2006;**36**(5):405–431.

4. Zhang, W.-X., Nanoscale iron particles for environmental remediation: an overview, *J. Nanopart. Res.*, 2003;**5**: 323–332.

5. Huber, D. L., Synthesis, properties and applications of iron nanoparticles, *Small*, 2005;**1**(5):482–501.

6. Narr, J., T. Viraraghavan, and Y. C. Jin, Applications of nanotechnology in water/wastewater treatment: a review, *Fresenius Environ. Bull.*, 2007;**16**(4):320–329.

7. Tratnyek, P. G., and R. L. Johnson, Nanotechnologies for environmental cleanup, *Nano Today*, 2006;**1**(2):44–48.

8. Wiesner, M. R., Responsible development of nanotechnologies for water and wastewater treatment, *Water Sci. Technol.*, 2006;**53**(3):45–51.

9. Noubactep, C., A critical review on the process of contaminant removal in Fe^0–H_2O systems, *Environ. Technol.*, 2008;**29**(8):909–920.

10. Glavee, G. N., et al., Chemistry of borohydride reduction of iron(II) and iron(III) ions in aqueous and nonaqueous media: formation of nanoscale Fe, FeB, and Fe_2B powders, *Inorg. Chem.*, 1995;**34**(1):28–35.

11. Wang, C.-B., and W.-X. Zhang, Synthesizing nanoscale iron particles for rapid and complete dechlorination of TCE and PCBs, *Environ. Sci. Technol.*, 1997;**31**(7):2154–2156.

12. Giasuddin, A. B. M., S. R. Kanel, and H. Choi, Adsorption of humic acid onto nanoscale zerovalent iron and its effect on arsenic removal, *Environ. Sci. Technol.*, 2007;**41**(6):2022–2027.

13. Hoch, L. B., et al., Carbothermal synthesis of carbon-supported nanoscale zero-valent iron particles for the remediation of hexavalent chromium, *Environ. Sci. Technol.*, 2008;**42**(7):2600–2605.

14. de Caro, D., et al., Synthesis, characterization, and magnetic studies of nonagglomerated zerovalent iron particles: unexpected size dependence of the structure, *Chem. Mater.*, 1996;**8**(8):1987–1991.

15. Chen, S.-S., H.-D. Hsu, and C.-W. Li, A new method to produce nanoscale iron for nitrate removal, *J. Nanopart. Res.*, 2004;**6**: 639–647.

16. Hoag, G. E., et al., Degradation of bromothymol blue by "greener" nano-scale zero-valent iron synthesized using tea polyphenols, *J. Mater. Chem.*, 2009;**19**(45):8671–8677.

17. Zhang, W.-X., and D. W. Elliott, Applications of iron nanoparticles for groundwater remediation, *Remediat. J.*, 2006;**16**(2):7–21.

18. Wang, C. B., and W.-X. Zhang, Synthesizing nanoscale iron particles for rapid and complete dechlorination of TCE and PCBs, *Environ. Sci. Technol.*, 1997;**31**(7):2154–2156.

19. Zhang, W. X., Nanoscale iron particles for environmental remediation: an overview, *J. Nanopart. Res.*, 2003;**5**(3–4):323–332.

20. Li, X. Q., D. W. Elliott, and W.-X. Zhang, Zero-valent iron nanoparticles for abatement of environmental pollutants: materials and engineering aspects, *Crit. Rev. Solid State Materi. Sci.*, 2006;**31**(4):111–122.

21. Li, X.-Q., D. Brown, and W.-X. Zhang, Stabilization of biosolids with nanoscale zero-valent iron (nZVI), *J. Nanopart. Res.*, 2007;**9**(2):233–243.

22. Li, A., et al., Debromination of decabrominated diphenyl ether by resin-bound iron nanoparticles, *Environ. Sci. Technol.*, 2007;**41**(19):6841–6846.

23. Lien, H. L., and W.-X. Zhang, Hydrodechlorination of chlorinated ethanes by nanoscale Pd/Fe bimetallic particles, *J. Environ. Engi.*, 2005;**131**(1):4–10.

24. Song, H., and E. R. Carraway, Reduction of chlorinated ethanes by nanosized zero-valent iron: kinetics, pathways, and effects of reaction conditions, *Environ. Sci. Technol.*, 2005;**39**(16):6237–6245.

25. Sarathy, V., et al., Degradation of 1,2,3-trichloropropane (TCP): hydrolysis, elimination, and reduction by iron and zinc, *Environ. Sci. Technol.*, 2010;**44**(2):787–793.

26. Xu, Y., and W.-X. Zhang, Subcolloidal Fe/Ag particles for reductive dehalogenation of chlorinated benzenes, *Ind. Eng. Chem. Res.*, 2000;**39**(7):2238–2244.

27. Krause, R., et al., Fe–Ni nanoparticles supported on carbon nanotube-co-cyclodextrin polyurethanes for the removal of trichloroethylene in water, *J. Nanopart. Res.*, 2010;**12**(2):449–456.

28. Bezbaruah, A. N., J. M. Thompson, and B. J. Chisholm, Remediation of alachlor and atrazine contaminated water with zero-valent iron nanoparticles, *J. Environ. Sci. Health B*, 2009;**44**(6):518–524.

29. Satapanajaru, T., et al., Remediation of atrazine-contaminated soil and water by nano zerovalent iron, *Water Air Soil Pollut.*, 2008;**192**(1–4):349–359.

30. Naja, G., et al., Degradation of hexahydro-1,3,5-trinitro-1,15-triazine (RDX) using zerovalent iron nanoparticles, *Environ. Sci. Technol.*, 2008;**42**(12):4364–4370.

31. Feng, J., B. W. Zhu, and T. T. Lim, Reduction of chlorinated methanes with nano-scale Fe particles: effects of amphiphiles on the dechlorination reaction and two-parameter regression for kinetic prediction, *Chemosphere*, 2008;**73**(11):1817–1823.

32. Lin, Y., et al., Incorporation of hydroxypyridinone ligands into self-assembled monolayers on mesoporous supports for selective actinide sequestration, *Environ. Sci. Technol.*, 2005;**39**(5):1332–1337.

33. Zhang, W.-X., C. B. Wang, and H. L. Lien, Treatment of chlorinated organic contaminants with nanoscale bimetallic particles, *Catal. Today*, 1998;**40**(4):387–395.

34. He, F., D. Y. Zhao, and C. Paul, Field assessment of carboxymethyl cellulose stabilized iron nanoparticles for in situ destruction of chlorinated solvents in source zones, *Water Res.*, 2010;**44**(7):2360–2370.

35. Quinn, J., et al., Field demonstration of DNAPL dehalogenation using emulsified zero-valent iron, *Environ. Sci. Technol.*, 2005;**39**(5):1309–1318.

36. Nutt, M. O., et al., Improved Pd-on-Au bimetallic nanoparticle catalysts for aqueous-phase trichloroethene hydrodechlorination, *Appl. Catal. B*, 2006;**69**(1–2):115–125.

37. Shih, Y. H., et al., Dechlorination of hexachlorobenzene by using nanoscale Fe and nanoscale Pd/Fe bimetallic particles, *Colloids Surfaces A*, 2009;**332**(2–3):84–89.

38. Zhu, B. W., T. T. Lim, and J. Feng, Influences of amphiphiles on dechlorination of a trichlorobenzene by nanoscale Pd/Fe: adsorption, reaction kinetics, and interfacial interactions, *Environ. Sci. Technol.*, 2008;**42**(12):4513–4519.

39. Thompson, J. M., B. J. Chisholm, and A. N. Bezbaruah, Reductive dechlorination of chloroacetanilide herbicide (Alachlor) using zero-valent iron nanoparticles, *Environ. Eng. Sci.*, 2010;**27**(3):227–232.

40. Joo, S. H., A. J. Feitz, and T. D. Waite, Oxidative degradation of the carbothioate herbicide, molinate, using nanoscale zero-valent iron, *Environ. Sci. Technol.*, 2004;**38**(7):2242–2247.

41. Elliott, D. W., H. L. Lien, and W.-X. Zhang, Zerovalent iron nanoparticies for treatment of ground water contaminated by hexachlorocyclohexanes, *J. Environ. Qual.*, 2008;**37**(6):2192–2201.

42. Shu, H. Y., et al., Reduction of an azo dye Acid Black 24 solution using synthesized nanoscale zerovalent iron particles, *J. Colloid Interface Sci.*, 2007;**314**(1):89–97.

43. Trujillo-Reyes, J., et al., Removal of Indigo Blue in aqueous solution using Fe/Cu nanoparticles and C/Fe-Cu nanoalloy composites, *Water Air Soil Pollut.*, 2010;**207**(1–4):307–317.

44. Bokare, A. D., et al., Effect of surface chemistry of Fe–Ni nanoparticies on mechanistic pathways of azo dye degradation, *Environ. Sci. Technol.*, 2007;**41**(21):7437–7443.

45. Cheng, R., J. L. Wang, and W.-X. Zhang, Comparison of reductive dechlorination of p-chlorophenol using Fe^0 and nanosized Fe^0, *J. Hazard. Mater.*, 2007;**144**(1–2): 334–339.

46. Lee, J., J. Kim, and W. Choi, Oxidation on zerovalent iron promoted by polyoxometalate as an electron shuttle, *Environ. Sci. Technol.*, 2007;**41**(9):3335–3340.

47. Kang, S. H., and W. Choi, Oxidative degradation of organic compounds using zero-valent iron in the presence of natural organic matter serving as an electron shuttle, *Environ. Sci. Technol.*, 2009;**43**(3):878–883.

48. Varanasi, P., A. Fullana, and S. Sidhu, Remediation of PCB contaminated soils using iron nano-particles, *Chemosphere*, 2007;**66**(6):1031–1038.

49. Xu, J., and D. Bhattacharyya, Fe/Pd nanoparticle immobilization in microfiltration membrane pores: synthesis, characterization, and application in the dechlorination of polychlorinated biphenyls, *Ind. Eng. Chem. Res.*, 2007;**46**(8):2348–2359.

50. Choi, H., S. Agarwal, and S. R. Al-Abed, Adsorption and simultaneous dechlorination of PCBs on GAC/Fe/Pd: mechanistic aspects and reactive capping barrier concept, *Environ. Sci. Technol.*, 2009;**43**(2):488–493.

51. Agarwal, S., et al., Reactivity of substituted chlorines and ensuing dechlorination pathways of select PCB congeners with Pd/Mg bimetallics, *Environ. Sci. Technol.*, 2009;**43**(3):915–921.

52. Kim, J. H., P. G. Tratnyek, and Y. S. Chang, Rapid dechlorination of polychlorinated dibenzo-p-dioxins by bimetallic and nanosized zerovalent iron, *Environ. Sci. Technol.*, 2008;**42**(11):4106–4112.

53. Liao, C. J., et al., Treatment of pentachlorophenol-contaminated soil using nano-scale zero-valent iron with hydrogen peroxide, *J. Mol. Catal. A*, 2007;**265**(1–2):189–194.

54. Reddy, K. R., and M. R. Karri, Removal and degradation of pentachlorophenol in clayey soil using nanoscale iron particles, in *Geotechnics of Waste Management and Remediation*. 2008, ASCE Press, Reston, VA, 2008, pp. 463–469.

55. Schaefer, C. E., et al., Comparison of biotic and abiotic treatment approaches for co-mingled perchlorate, nitrate, and nitramine explosives in groundwater, *J. Contam. Hydrol.*, 2007;**89**(3–4):231–250.

56. Welch, R., and R. G. Riefler, Estimating treatment capacity of nanoscale zero-valent iron reducing 2,4,6-trinitrotoluene, *Environ. Eng. Sci.*, 2008;**25**(9):1255–1262.

57. Joo, S. H., et al., Quantification of the oxidizing capacity of nanoparticulate zero-valent iron, *Environ. Sci. Technol.*, 2005;**39**(5):1263–1268.

58. Lee, C., and D. L. Sedlak, Enhanced formation of oxidants from bimetallic nickel–iron nanoparticles in the presence of oxygen, *Environ. Sci. Technol.*, 2008;**42**(22):8528–8533.

59. Noradoun, C. E., and I. F. Cheng, EDTA degradation induced by oxygen activation in a zerovalent iron/air/water system, *Environ. Sci. Technol.*, 2005;**39**(18):7158–7163.

60. Lee, C., C. R. Keenan, and D. L. Sedlak, Polyoxometalate-enhanced oxidation of organic compounds by nanoparticulate zero-valent iron and ferrous ion in the presence of oxygen, *Environ. Sci. Technol.*, 2008;**42**(13):4921–4926.

61. Chang, M. C., et al., Remediation of soil contaminated with pyrene using ground nanoscale zero-valent iron, *J. Air Waste Manag. Assoc.*, 2007;**57**(2):221–227.

62. Elliott, D. W., and W.-X. Zhang, Field assessment of nanoscale bimetallic particles for groundwater treatment, *Environ. Sci. Technol.*, 2001;**35**(24):4922–4926.

63. Li, L., et al., Synthesis, properties, and environmental applications of nanoscale iron-based materials: a review, *Crit. Rev. Environ. Sci. Technol.*, 2006;**36**(5):405–431.

64. Liu, Y., T. Phenrat, and G. V. Lowry, Effect of TCE concentration and dissolved groundwater solutes on NUI-promoted TCE dechlorination and H-2 evolution, *Environ. Sci. Technol.*, 2007;**41**(22):7881–7887.

65. Liu, Y. Q., et al., Trichloroethene hydrodechlorination in water by highly disordered monometallic nanoiron, *Chem. Mater.*, 2005;**17**(21):5315–5322.

66. Geiger, C. L., et al., Nanoscale and microscale iron emulsions for treating DNAPL, in *Chlorinated Solvent and DNAPL Remediation: Innovative Strategies for Subsurface Cleanup*, S. M. Henry and S. D. Warner (Eds.), American Chemical Society: Oxford University Press, Washington, DC, 2003; pp. 132–140.

67. Hara, S. O., et al., Field and laboratory evaluation of the treatment of DNAPL source zones using emulsified zero-valent iron, *Remediat. J.*, 2006;**16**(2):35–56.

68. Quinn, J., et al., Field demonstration of DNAPL dehalogenation using emulsified zero-valent iron, *Environ. Sci. Technol.*, 2005;**39**(5):1309–1318.

69. Henn, K. W., and D. W. Waddill, Utilization of nanoscale zero-valent iron for source remediation: a case study, *Remediat. J.*, 2006;**16**(2):57–77.

70. Kanel, S. R., J. M. Grenèche, and H. Choi, Arsenic(V) removal from groundwater using nano scale zero-valent iron as a colloidal reactive barrier material, *Environ. Sci. Technol.*, 2006;**40**(6):2045–2050.

71. Liu, J. X., et al., Aqueous Cr(VI) reduction by electrodeposited zero-valent iron at neutral pH: acceleration by organic matters, *J. Hazard. Mater.*, 2009;**163**(1):370–375.

72. Manning, B. A., J. R. Kiser, and S. R. Kanel, Spectroscopic investigation of Cr(III)- and Cr(VI)-treated nanoscale zerovalent iron, *Environ. Sci. Technol.*, 2006;**41**(2):586–592.

73. Ponder, S. M., J. G. Darab, and T. E. Mallouk, Remediation of Cr(VI) and Pb(II) aqueous solutions using supported, nanoscale zero-valent iron, *Environ. Sci. Technol.*, 2000;**34**(12):2564–2569.

74. Rivero-Huguet, M., and W. D. Marshall, Reduction of hexavalent chromium mediated by micro- and nano-sized mixed metallic particles, *J. Hazard. Mater.*, 2009;**169**(1–3):1081–1087.

75. Wu, Y. J., et al., Chromium(VI) reduction in aqueous solutions by Fe_3O_4-stabilized Fe^0 nanoparticles, *J. Hazard. Mater.*, 2009;**172**(2–3):1640–1645.

76. Mondal, K., G. Jegadeesan, and S. B. Lalvani, Removal of selenate by Fe and NiFe nanosized particles, *Ind. Eng. Chem. Res.*, 2004;**43**(16):4922–4934.

77. Shokes, T. E., and G. Moller, Removal of dissolved heavy metals from acid rock drainage using iron metal, *Environ. Sci. Technol.*, 1999;**33**(2):282–287.

78. Uzum, C., et al., Application of zero-valent iron nanoparticles for the removal of aqueous Co^{2+} ions under various experimental conditions, *Chem. Eng. J.*, 2008;**144**(2):213–220.

79. Celebi, O., et al., A radiotracer study of the adsorption behavior of aqueous Ba^{2+} ions on nanoparticles of zero-valent iron, *J. Hazard. Mater.*, 2007;**148**(3):761–767.

80. Li, X. Q., and W.-X. Zhang, Iron nanoparticles: the core–shell structure and unique properties for Ni(II) sequestration, *Langmuir*, 2006;**22**(10):4638–4642.

81. Ponder, S. M., J. G. Darab, and T. E. Mallouk, Remediation of Cr(VI) and Pb(II) aqueous solutions using supported, nanoscale zero-valent iron, *Environ. Sci. Technol.*, 2000;**34**(12):2564–2569.

82. Gu, B., et al., Reductive precipitation of uranium(VI) by zero-valent iron, *Environ. Sci. Technol.*, 1998;**32**(21):3366–3373.

83. Cao, J., D. Elliott, and W.-X. Zhang, Perchlorate reduction by nanoscale iron particles *J. Nanopart. Res.*, 2005;**7**: 499–506.

84. Liou, Y. H., et al., Effect of precursor concentration on the characteristics of nanoscale zerovalent iron and its reactivity of nitrate, *Water Res.*, 2006;**40**(13):2485–2492.

85. Shin, K. H., and D. K. Cha, Microbial reduction of nitrate in the presence of nanoscale zero-valent iron, *Chemosphere*, 2008;**72**(2):257–262.

86. Sohn, K., et al., Fe(0) nanoparticles for nitrate reduction: stability, reactivity, and transformation, *Environ. Sci. Technol.*, 2006;**40**(17):5514–5519.

87. Li, S. J., et al., Reduction and immobilization of chromium(VI) by nano-scale Fe^0 particles supported on reproducible PAA/PVDF membrane, *J. Environ. Monitor.*, 2010;**12**(5):1153–1158.

88. Manning, B. A., et al., Spectroscopic investigation of Cr(III)- and Cr(VI)-treated nanoscale zerovalent iron, *Environ. Sci. Technol.*, 2007;**41**(2):586–592.

89. Rivero-Huguet, M., and W. D. Marshall, Reduction of hexavalent chromium mediated by micron- and nano-scale zero-valent metallic particles, *J. Environ. Monitor.*, 2009;**11**(5):1072–1079.

90. Shariatmadari, N., C. H. Weng, and H. Daryaee, Enhancement of hexavalent chromium Cr(VI) remediation from clayey soils by electrokinetics coupled with a nano-sized zero-valent iron barrier, *Environ. Eng. Sci.*, 2009;**26**(6):1071–1079.

91. Melitas, N., O. Chuffe-Moscoso, and J. Farrell, Kinetics of soluble chromium removal from contaminated water by zerovalent iron media: corrosion inhibition and passive oxide effects, *Environ. Sci. Technol.*, 2001;**35**(19):3948–3953.

92. Peterson, M. L., et al., Surface passivation of magnetite by reaction with aqueous Cr(VI): XAFS and TEM results, *Environ. Sci. Technol.*, 1997;**31**(5):1573–1576.

93. Manning, B. A., et al., Arsenic(III) and arsenic(V) reactions with zerovalent iron corrosion products, *Environ. Sci. Technol.*, 2002;**36**(24):5455–5461.

94. Cissoko, N., et al., Removal of Cr(VI) from simulative contaminated groundwater by iron metal, *Process Saf. Environ. Prot.*, 2009;**87**(6):395–400.

95. Su, C., and R. W. Puls, Arsenate and arsenite removal by zerovalent iron: effects of phosphate, silicate, carbonate, borate, sulfate, chromate, molybdate, and nitrate, relative to chloride, *Environ. Sci. Technol.*, 2001;**35**(22):4562–4568.

96. Su, C., and R. W. Puls, Arsenate and arsenite removal by zerovalent iron: kinetics, redox transformation, and implications for in situ groundwater remediation, *Environ. Sci. Technol.*, 2001;**35**(7):1487–1492.

97. Su, C., and R. W. Puls, In situ remediation of arsenic in simulated groundwater using zerovalent iron: laboratory column tests on combined effects of phosphate and silicate, *Environ. Sci. Technol.*, 2003;**37**(11):2582–2587.

98. Nurmi, J. T., et al., Characterization and properties of metallic iron nanoparticles: spectroscopy, electrochemistry, and kinetics, *Environ. Sci. Technol.*, 2005;**39**(5):1221–1230.

99. Joo, S. H., et al., Quantification of the oxidizing capacity of nanoparticulate zero-valent iron, 2005;**39**(5):1263–1268.

100. Zhu, H. J., et al., Removal of arsenic from water by supported nano zero-valent iron on activated carbon, *J. Hazard. Mater.*, 2009;**172**(2–3):1591–1596.

101. Myneni, S. C.B., T. K. Tokunaga, and G. E. Brown, Jr., Abiotic selenium redox transformations in the presence of Fe(II,III) oxides, *Science*, 1997;**278**(5340):1106–1109.

102. Lien, H. L., Y. S. Jhuo, and L. H. Chen, Effect of heavy metals on dechlorination of carbon tetrachloride by iron nanoparticles, *Environ. Eng. Sci.*, 2007;**24**(1):21–30.

103. Üzüm, Ç., et al., Application of zero-valent iron nanoparticles for the removal of aqueous Co^{2+} ions under various experimental conditions, *Chem. Eng. J.*, 2008;**144**(2): 213–220.

104. Li, X.-Q., and W.-X. Zhang, Iron nanoparticles: the core–shell structure and unique properties for Ni(II) sequestration, *Langmuir*, 2006;**22**(10):4638–4642.

105. Rumer, R. R., and M. E. Ryan, *Barrier Containment Technologies for Environmental Remediation Applications*, Wiley, New York, 1995.

106. Gillham, R. W., and S. F. O'Hannesin, Enhanced degradation of halogenated aliphatics by zero-valent iron, *Ground Water*, 1994;**32**: 958–967.

107. Giasuddin, A. B.M., S. R. Kanel, and H. Choi, Adsorption of humic acid onto nanoscale zerovalent iron and its effect on arsenic removal, *Environ. Sci. Technol.*, 2007;**41**(6):2022–2027.

108. He, F., et al., Stabilization of Fe–Pd nanoparticles with sodium carboxymethyl cellulose for enhanced transport and dechlorination of trichloroethylene in soil and groundwater, *Ind. Eng. Chem. Res.*, 2007;**46**(1):29–34.

109. Kanel, S. R., and H. Choi, Transport characteristics of surface-modified nanoscale zerovalent iron in porous media, *Water Sci. Technol.*, 2007;**55**(1–2):157–162.

110. Kanel, S. R., et al., Transport of surface-modified iron nanoparticle in porous media and application to arsenic(III) remediation *J. Nanopart. Res.*, 2007;**9**(5):725–735.

111. Lin, Y.-H., et al., Characteristics of two types of stabilized nano zero-valent iron and transport in porous media, *Sci. Total Environ.*, 2010;**408**(10):2260–2267.

112. Schrick, B., et al., Delivery vehicles for zerovalent metal nanoparticles in soil and groundwater, *Chem. Matter*, 2004;**16**(11):2187–2193.

113. He, F., and D. Zhao, Preparation and characterization of a new class of starch-stabilized bimetallic nanoparticles for degradation of chlorinated hydrocarbons in water, *Environ. Sci. Technol.*, 2005;**39**(9):3314–3320.

114. Quinn, J., et al., Field demonstration of DNAPL dehalogenation using emulsified zerovalent iron, *Environ. Sci. Technol.*, 2005;**39**(5):1309–1308.

115. Hydutsky, B. W., et al., Optimization of nano- and microiron transport through sand columns using polyelectrolyte mixtures, *Environ. Sci. Technol.*, 2007;**41**(18):6418–6424.

116. Saleh, N., et al., Surface modifications enhance nanoiron transport and NAPL targeting in saturated porous media, *Environ. Eng. Sci.*, 2007;**24**(1):45–57.

117. Kanel, S. R., et al., Two dimensional transport characteristics of surface stabilized zerovalent iron nanoparticles in porous media, *Environ. Sci. Technol.*, 2008;**42**(3):896–900.

118. CBEN, Center for Biological and Environmental Nanotechnology, *Annual Report*, 2007, Rice University, Houston, TX.

119. Matheson, L. J., and P. G. Tratnyek, Reductive dehalogenation of chlorinated methanes by iron metal, *Environ. Sci. Technol.*, 1994;**28**(12):2045–2053.

120. Lewis, F. A., *The Palladium Hydrogen Systems*, Academic Press, New York, 1967.

121. Wang, X. Y., et al., Dechlorination of chlorinated methanes by Pd/Fe bimetallic nanoparticles, *J. Hazard. Mater.*, 2009;**161**(2–3):815–823.

122. Patel, U., and S. Suresh, Dechlorination of chlorophenols by magnesium–silver bimetallic system, *J. Colloid Interface Sci.*, 2006;**299**(1):249–259.

123. Lien, H.-L. and W.-X. Zhang, Hydrodechlorination of chlorinated ethanes by nanoscale Pd/Fe bimetallic particles, *J. Environ. Eng.*, 2005;**131**(1):4–10.

124. Graham, L. J., and G. Jovanovic, Dechlorination of p-chlorophenol on a Pd–Fe catalyst in a magnetically stabilized fluidized bed: implications for sludge and liquid remediation, *Chem. Eng. Sci.*, 1999;**54**(15–16):3085–3093.

125. Patel, U. D., and S. Suresh, Dechlorination of chlorophenols using magnesium-palladium bimetallic system, *J. Hazard. Mater.*, 2007;**147**(1–2):431–438.

126. Lien, H. L., and W.-X. Zhang, Nanoscale Pd/Fe bimetallic particles: catalytic effects of palladium on hydrodechlorination, *Appl. Catal. B*, 2007;**77**(1–2):110–116.

127. Nagpal, V., et al., Reductive dechlorination of gamma-hexachlorocyclohexane using Fe–Pd bimetallic nanoparticles, *J. Hazard. Mater.*, 2010;**175**(1–3):680–687.

128. Liu, Y. H., et al., Catalytic dechlorination of chlorophenols in water by palladium/iron, *Water Res.*, 2001;**35**(8):1887–1890.

129. Zhu, B.-W., T.-T. Lim, and J. Feng, Reductive dechlorination of 1,2,4-trichlorobenzene with palladized nanoscale Fe^0 particles supported on chitosan and silica, *Chemosphere*, 2006;**65**(7):1137–1145.

130. Lin, C. J., Y. H. Liou, and S.-L. Lo, Supported Pd/Sn bimetallic nanoparticles for reductive dechlorination of aqueous trichloroethylene, *Chemosphere*, 2009;**74**(2):314–319.

131. Kung, K. H. S., and M. B. McBride, Bonding of chlorophenols on iron and aluminumoxides, *Environ. Sci. Technol.*, 1991;**25**(4):702–709.

132. Morales, J., R. Hutcheson, and I. F. Cheng, Dechlorination of chlorinated phenols by catalyzed and uncatalyzed Fe(0) and Mg(0) particles, *J. Hazard. Mater.*, 2002;**90**(1):97–108.

133. Kim, Y.-H., and E. R. Carraway, Dechlorination of pentachlorophenol by zero valent iron and modified zero valent irons, *Environ. Sci. Technol.*, 2000;**34**(10):2014–2017.

134. Patel, U. D., and S. Suresh, Effects of solvent, pH, salts and resin fatty acids on the dechlorination of pentachlorophenol using magnesium-silver and magnesium–palladium bimetallic systems, *J. Hazard. Mater.*, 2008;**156**(1–3):308–316.

135. Urbano, F. J., and J. M. Marinas, Hydrogenolysis of organohalogen compounds over palladium supported catalysts, *J. Mol. Catal. A*, 2001;**173**(1–2):329–345.

136. Zhu, J., et al., Carbon nanofiber-supported palladium nanoparticles as potential recyclable catalysts for the Heck reaction, *Appl. Catal. A*, 2009;**352**(1–2):243–250.

137. Choi, M., K. H. Shin, and J. Jang, Plasmonic photocatalytic system using silver chloride/silver nanostructures under visible light, *J. Colloid Interface Sci.*, 2010;**341**(1):83–87.

138. Liu, M. H., M. F. Han, and W. W. Yu, Hydrogenation of chlorobenzene to cyclohexane over colloidal Pt nanocatalysts under ambient conditions, *Environ. Sci. Technol.*, 2009;**43**(7):2519–2524.

139. Evans, B. R., et al., Palladium–bacterial cellulose membranes for fuel cells, *Biosens. Bioelectron.*, 2003;**18**(7):917–923.

140. Patel, U.D., S. Suresh, Complete dechlorination of pentachlorophenol using palladized bacterial cellulose in a rotating catalyst contact reactor, *J. Colloid. Interface Sci.*, 2008;**319**(2):462–469.

141. Jiao, Y. L., et al., Electrochemical reductive dechlorination of carbon tetrachloride on nanostructured Pd thin films, *Electrochem. Commun.*, 2008;**10**(10):1474–1477.

142. Iwakura, C., Y. Yoshida, H. Inoue, A new hydrogenation system of 4-methylstyrene using a palladized palladium sheet electrode, *J. Electroanal. Chem.*, 1997;**43**:43-45.

143. Baxter-Plant, V., I. P. Mikheenko, and L. E. Macaskie, Sulphate-reducing bacteria, palladium and the reductive dehalogenation of chlorinated aromatic compounds, *Biodegradation*, 2003;**14**(2):83–90.

144. Baxter-Plant, V. S., et al., Dehalogenation of chlorinated aromatic compounds using a hybrid bioinorganic catalyst on cells of *Desulfiovibrio desulfuricans*, *Biotechnol. Lett.*, 2004;**26**(24):1885–1890.

145. Hennebel, T., et al., Removal of diatrizoate with catalytically active membranes incorporating microbially produced palladium nanoparticles, *Water Res.*, 2010;**44**(5):1498–1506.

146. Hennebel, T., et al., Biocatalytic Dechlorination of trichloroethylene with bio-palladium in a pilot-scale membrane reactor, *Biotechnol. Bioeng.*, 2009;**102**(4):995–1002.

147. Hennebel, T., et al., Remediation of trichloroethylene by bio-precipitated and encapsulated palladium nanoparticles in a fixed bed reactor, *Chemosphere*, 2009;**76**(9):1221–1225.

148. Windt, W. D., P. Aelterman, and W. Verstraete, Bioreductive deposition of palladium (0) nanoparticles on Shewanella oneidensis with catalytic activity towards reductive dechlorination of polychlorinated biphenyls, *Environmen. Microbiol.*, 2005;**7**(3):314–325.

149. Lloyd, J. R., P. Yong, and L. E. Macaskie, Enzymatic recovery of elemental palladium by using sulfate-reducing bacteria, *Appl. Environ. Microbiol.*, 1998;**64**(11):4607–4609.

150. Yuan, G., and M. A. Keane, Catalyst deactivation during the liquid phase hydrodechlorination of 2,4-dichlorophenol over supported Pd: influence of the support, *Catal. Today*, 2003;**88**(1–2):27–36.

151. Aramendia, M.A., V. Borau, I.M. Garcia, C. Jimenez, J.M, Marinas, A. Marinas, F.J. Urbano, Hydrodehalogenation of aryl halides by hydrogen gas and hydrogen transfer in the presence of palladium catalysts, *Stud. Surf. Sci. Catal.*, 2000;**130**:2003–2008.

152. Lowry, G. V., and M. Reinhard, Pd-catalyzed TCE dechlorination in water: effect of [H2](aq) and H2-utilizing competitive solutes on the TCE dechlorination rate and product distribution, *Environ. Sci. Technol.*, 2001;**35**(4):696–702.

153. Roy, H. M., et al., Catalytic hyrodechlorination of chlorophenols in aqueous solution under mild conditions, *Appl. Catal. A*, 2004;**271**(1–2):137–143.

154. Mabbett, A.N., P. Yong, J. P. G. Farr, L. E. Macaskie, Reduction of Cr(Vl) by "Palldized" biomass of Desulfovibrio *desulfuricans* ATCC 29577, *Biotechnol. Bioengg.*, 2004;**87**(1):104–109.

155. Hildebrand, H., K. Mackenzie, and F. D. Kopinke, Highly active Pd-on-magnetite nanocatalysts for aqueous phase hydrodechlorination reactions, *Environ. Sci. Technol.*, 2009;**43**(9):3254–3259.

156. Hildebrand, H., K. Mackenzie, and F. D. Kopinke, Pd/Fe$_3$O$_4$ nano-catalysts for selective dehalogenation in wastewater treatment processes–influence of water constituents, *Appl. Catal. B*, 2009;**91**(1–2):389–396.

157. Laska, U., C.G. Frost, G. J. Price, P. K. Plucinski, Easy-separable magnetic nanoparticles-supported Pd catalysts: Kinetics, stability and catalyst reuse, *J. Catal.*, 2009;**268**:318–328.

158. He, F., et al., One-step "green" synthesis of Pd nanoparticles of controlled size and their catalytic activity for trichloroethene hydrodechlorination, *Ind. Eng. Chem. Res.*, 2009;**48**(14):6550–6557.

159. Nutt, M. O., J. B. Hughes, and M. S. Wong, Designing Pd-on-Au bimetallic nanoparticle catalysts for trichloroethene hydrodechlorination, *Environ. Sci. Technol.*, 2005;**39**(5):1346–1353.

160. Wong, M. S., et al., Cleaner water using bimetallic nanoparticle catalysts, *J. Chem. Technol. Biotechnol.*, 2009;**84**(2):158–166.

161. Nutt, M. O., et al., Improved Pd-on-Au bimetallic nanoparticle catalysts for aqueous-phase trichloroethene hydrodechlorination, *Appl. Catal. B*, 2006;**69**(1–2):115–125.

162. Coq, B., and F. Figueras, Bimetallic palladium catalysts: influence of the co-metal on the catalyst performance, *J. Mol. Catal. A*, 2001;**173**(1–2):117–134.

163. Golubina, E. V., et al., The role of Fe addition on the activity of Pd-containing catalysts in multiphase hydrodechlorination, *Appl. Catal. A*, 2006;**302**(1):32–41.

164. Albers, P., J. Pietsch, and S. F. Parker, Poisoning and deactivation of palladium catalysts, *J. Mol. Catal. A*, 2001;**173**(1–2):275–286.

165. Lowry, G. V., and M. Reinhard, Pd-catalyzed TCE dechlorination in groundwater: solute effects, biological control, and oxidative catalyst regeneration, *Environ. Sci. Technol.*, 2000;**34**(15):3217–3223.

166. Siantar, D. P., C.G. Schreier, C-S Chou, M. Rainhard, Treatment of 1,2-dibromo-3-chloropropane and nitratecontaminated water with zero-valent or hydrogen/palladium catalysts, *Wat. Res.*, 1995;**30**(10):2315–2322.

167. Yavuz, C. T., et al., Low-field magnetic separation of monodisperse Fe$_3$O$_4$ nanocrystals, *Science*, 2006;**314**(5801):964–967.

168. Schrick, B., et al., Delivery vehicles for zerovalent metal nanoparticles in soil and ground-water, *Chem. Matter*, 2004;**16**(11):2187–2193.

11

USE OF CARBON NANOTUBES IN WATER TREATMENT

VENKATA K. K. UPADHYAYULA

Oak Ridge Institute of Science and Education, Oak Ridge, Tennessee

JAYESH P. RUPARELIA

Institute of Technology, Nirma University, Ahmedabad, India

ABINASH AGRAWAL

Department of Earth and Environmental Sciences, Wright State University, Dayton, Ohio

Nanoscale Multifunctional Materials: Science and Applications, First Edition.
Edited by Sharmila M. Mukhopadhyay.
© 2012 John Wiley & Sons, Inc. Published 2012 by John Wiley & Sons, Inc.

1   INTRODUCTION TO CHANGING WATER TREATMENT NEEDS

The future of human beings on this planet may depend on our selfless commitment to protecting natural water resources from excessive exploitation and managing its use to prevent their exhaustion. Human society requires natural water for its survival, particularly in critical areas such as agriculture, energy production, and the manufacture of essential goods. Increasing industrialization, urbanization, and the emergence of global economy have led to significant improvement in the quality of life in the twentieth century; yet, human activities have also put immense pressure on the plentiful natural resources, including excessive release of greenhouse gases into the atmosphere, worldwide deforestation, plus overexploitation and contamination of freshwater systems. In particular, the problems related to freshwater quantity and quality have reached a critical level in many parts of the world. The key reasons for rising freshwater problems include: (1) depletion of freshwater resources on a global scale, due to excessive withdrawals to support food and energy production needs; (2) increased urbanization leading to depletion of fresh and groundwater reserves; (3) large-scale production of industrially manufactured goods; and (4) chemical farming techniques (i.e., application of fertilizers, pesticides, and insecticides) that are responsible for their increase in chemical and biological contaminant levels in freshwater systems. All such changes have heightened our concerns for drinking water contaminants and have influenced our efforts in the modern-day world in terms of identifying additional drinking water resources and addressing the growing challenges in water treatment. The large-scale application of water treatment technology by well-established techniques [particularly reverse osmosis (RO)] has made it feasible to provide a continuous supply of drinking water from sustainable raw water sources such as seawater. However, emerging techniques [e.g., nanofiltration (NF)] have shown excellent promise of a safe and potable supply of drinking water in a sustainable manner and can meet the rising demands since newer treatment methods are capable of removing almost all possible chemical and biological impurities from source waters.

Despite excellent water purification techniques that are currently available, millions of people worldwide are affected by severe shortages of freshwater supplies. Many also suffer from waterborne illnesses from drinking water that is laden with toxic contaminants and pathogens. Clearly, the main cause of the shortage of safe drinking water in many parts of the world is economical rather than technical. Many poor and rural areas in the under-developed and developing nations do not possess a centralized conventional water treatment facility, which can provide a partial solution in terms of treating drinking water contaminants released due to human activity. Obviously, it is beyond the financial means of poor communities to install and operate water treatment plants for providing a complete solution. Therefore, from a strategic standpoint, it is important that research and development should focus on creating technologies that bring about improved effectiveness in contaminant removal from drinking sources while remaining cost-effective and portable so that a safe supply of drinking water can be made available to financially backward regions of the world. Technologies supporting an alternative water treatment philosophy such as point of use (POU) and point of entry (POE) systems need to be developed at the research and policy formulation level [1]. This is because with POU systems, it is possible to obtain treated water qualities similar to that of a centralized water treatment facility, but at the same time, POU treatment systems have the advantage of portability, which makes them an attractive option in developing nations. On the other hand, implementation of POU/POE-based treatment methods can be of great advantage in developed nations as well, where they can serve as an additional measure to sustain the potablity of treated water that could possibly be compromised in the distribution systems. Among many POU-based treatment options available, treatment technologies based on adsorption and membrane filtration mechanisms are encouraging. Although membrane filtration techniques such as RO and NF can still be costly at the POU level, more attention is being given to adsorption-based POU systems, where efforts are under way to develop an adsorbent medium that can treat multiple contaminants in source water (both chemical and biological natured) and at the same time be cost-effective.

The rise of nanotechnology led to the discovery of nanomaterials of various kinds, which is considered to be a major breakthrough in many scientific applications, including water treatment research. Nanomaterials, particularly carbon nanotubes (CNTs), show promise as an adsorbent medium and, more importantly, as a POU treatment material for removal of chemical and biological contaminants from water systems. CNTs as adsorbent media are shown to express high capacities contaminant removal for a wide range of drinking water contaminants, superior to those of many other microporous sorbent media, such as activated carbon. Furthermore, the cost of CNTs that was earlier considered a significant major roadblock preventing large-scale applications such as water treatment is declining rapidly, thereby making the introduction of CNT filters as POU wayer treatment devices. Above attributes of CNTs can help realize POU-based water treatment because they can achieve significantly higher removal efficiencies for multiple contaminants in portable systems, similar to that achieved in a centralized water treatment facility. Application of CNTs as an adsorbent medium to concentrate chemical and biological impurities from water systems and the feasibility of CNT filters for POU water treatment forms the central theme of this chapter. The technical aspects of adsorption for various drinking water contaminants, including heavy metals, organics, and biological impurities such as microorganisms, natural organic matter (NOM), and biological toxins, are discussed in the chapter. The implications of implementing the CNT technology in water treatment and the challenges of its implementation are also discussed. Prior to introducing concepts of CNT adsorption technology for water treatment, details on major types of chemical and biological contaminants found in drinking water sources are presented in the next section.

2 TYPES OF CONTAMINANTS PRESENT IN INFLUENT WATERS

As explained above, a major drawback of extensive utilization of human-made products and our heavy dependence on chemical farming techniques is an increased

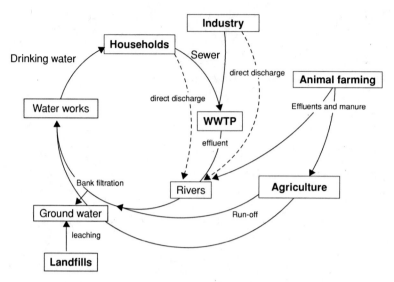

FIGURE 1 Possible avenues responsible for the introduction of chemical and biological contaminants into influent waters of a drinking water treatment plant. (From [2].)

toxic contaminant budget in freshwater systems. In Figure 1, possible anthropogenic sources responsible for the introduction of toxic contaminants into surface water and groundwater from the feedwater to a drinking water treatment plant are shown. As seen in the figure, anthropogenic activities such as agriculture, wastewater treatment plant, industries, and households can be major centers of contaminant transport into waterworks supplying drinking water. Most state-of-the-art wastewater treatment plants are not designed to treat most of emerging contaminants, and their metabolites can escape elimination in the treatment train and enter the aquatic environment via sewage effluents [2].

In Figure 2, major types of chemical and biological contaminants that might possibly enter a water treatment plant are shown. In addition, the figure represents classification based on the unintentional and intentional entry of contaminants. Unintentional contaminants have either originated from anthropogenic sources as shown in Figure 1 or occur naturally in the raw water system. Intentionally added contaminants are biothreat pathogens and biological toxins, which are technically not present in source waters but always pose a serious risk to drinking water treatment and distribution systems because these contaminants can get into the system by external means: by any event involving tampering with the treatment infrastructure. Such types of incidents, termed *bioterrorism attacks*, unfortunately are on the rise, especially in developed nations such as the United States [3].

Based on the entry route described in Figure 2, drinking water contaminants can be broadly classified into two types: unintentional addition and intentional addition. Details regarding these contaminants are discussed below.

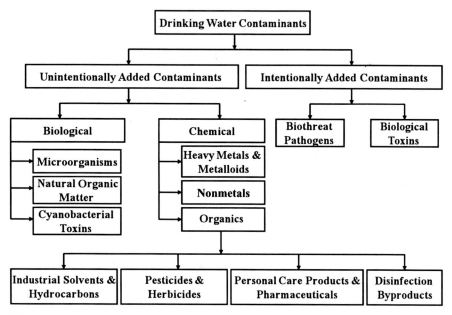

FIGURE 2   Types of unintentionally and intentionally added chemical and biological contaminants typically found in influent water of drinking water treatment plants.

2.1   Unintentional Addition of Drinking Water Contaminants

As shown in Figure 1 and described in Figure 2, there are many routes through which multiple water contaminants can enter a treatment plant. Here the word *unintentional* is used to suggest that as a consequence of increased industrial, agricultural, and urban activities, contaminants are discharged into freshwater systems and eventually enter a treatment plant when the influent water is drawn from these sources. Again, these contaminants can be broadly divided into three types: organic, inorganic, and biological contaminants.

Organic Contaminants   Organic contaminants are considered to be one a major threat to freshwater systems because of their potent nature to induce sickness or even chronic illness such as cancer in a healthy human being. Organic chemicals such as benzene, phenols [4], PCBs [5,6], chlorinated compounds [7], pharmaceuticals, personal care products, surfactants, and various industrial additives chemicals [8, 9] are not metabolized and discharged into sewers and reach wastewater treatment plants. These compounds may eventually enter the drinking water when untreated or partially treated wastewater is discharged into freshwater bodies. Many organic compounds are known to cause damage to the kidneys, stomach, and nervous system and they can be responsible for the development of cataracts, reproductive difficulties, and greater cancer risk in humans.

Heavy Metal Contaminants   Heavy metal contaminants are a major concern in drinking water treatment plants. Metals, usually in the form of ions, occur in water both as a result of natural processes and as a consequence of human activities. Heavy metals are natural components of the Earth's crust. They cannot be degraded or destroyed, but they can dissolve in water during runoff and industrial activities. Effluents from industries such as metal plating, chemical manufacturing, metallurgical industries, mining operations, and tanneries contain many toxic substances, especially heavy metals. Soils surrounding many military bases are also contaminated with metals and affect the quality of groundwater and surface water [10–13]. Some metals associated with these activities are cadmium (Cd), chromium (Cr), lead (Pb), nickel (Ni), zinc (Zn), and mercury (Hg). As trace elements, some heavy metals (e.g., copper, selenium, zinc) are essential for maintaining metabolism in the human body. However, at higher concentrations they can lead to poisoning [14]. Heavy metals are listed as priority pollutants because they are not only toxic but also possess high mobility and therefore can be transported easily in water systems. Heavy metals tend to bioaccumulate in living organisms, particularly in human beings, and are shown to induce mutagenic, teratogenic, and carcinogenic effects [14].

Biological Contaminants   The third type classified under the unintentional addition category is that of biological contaminants. Biological contaminants are broadly classified into three types:

1. Microorganisms of pathogenic and nonpathogenic forms, such as bacteria, viruses, protozoa, and other unicellular living organisms which seek entry into treatment plants through raw water sources or can grow intrinsically inside water supply and distribution systems due to the prevalence of poor engineering conditions, such as corrosion of pipes and longer residence times, resulting in stagnation of water in the pipelines [15,16].
2. The presence of natural organic matter (NOM) may contribute to extensive growth of microorganisms in water systems. NOM provides assimilable organic carbon (AOC), a major carbon and energy source supplier for heterotrophic microorganisms in drinking water systems. NOM contains a diverse group of organic compounds, all formed due to decomposition of animal and plant residues [17–24]. The presence of NOM is problematic because a portion of NOM gets oxidized in the treatment plant upon interacting with strong oxidizing agents such as chlorine and ozone disinfectants, and this oxidation process converts the high-molecular-weight compounds of NOM to simple organic acids which serve as a carbon source for heterotrophic bacteria [17,22].
3. Biological toxins such as microcystins, which are formed in a treatment plant as a result of raw water drawn into the plant from sources containing a large concentration of filamentous bacteria called cyanobacteria, commonly referred to as harmful algal blooms (HABs) [25]. The existence of HABs in natural waters is unsafe for many reasons, but what really matters for a water treatment plant is their scum-forming ability on the surface of the water. The vesicles

inside the vacuoles of cyanobacterial cells serve as a piston, where due to the constant production and utilization of high-molecular-weight carbohydrate compounds in the vacuoles, a sequential filling and emptying mechanism takes place inside the cells. This accounts for the buoyancy, which causes millions of cells to migrate to the surface and form a layer of scum which ultimately is transported into water treatment plants [25]. Once in the plants, due to the rigorous treatment conditions, the cells get lysed and subsequently release intracellular toxins into the water stream that are extremely difficult to treat and disinfect using conventional treatment practices [25,26]. Among cyanobacterial toxins, the most commonly occurring and notorious forms include derivatives of microcystin toxins [25].

Biological contaminants pose real challenges to drinking water treatment plants for many obvious reasons. Organic and inorganic contaminants may or may not be present in influent water, but biological contaminants are not only confirmed to be present but are present in largely diverse composition. In Table 1, types of microorganisms that can generally be found in water treatment plants are listed. Because the microbial contaminants entering a treatment plant can be of variable composition depending on the source, the treatment and removal efficiencies of microbes in a conventional treatment plant may vary significantly and with the probability that some species are not removed at all. For example, the average microbial removal efficiencies for the coagulation and sedimentation process varies from 27 to 74% for viruses, 32 to 87% for bacteria, and between 0 and 94% for algal species [27]. Similarly, variations in physical conditions, such as temperature, affects the removal rates of biological contaminants [28]. According to Zhang and DiGiano [28], the bacterial growth rate doubles for every $10°C$ increase in influent water temperature. On the other hand, despite treating raw water to the highest standards of potability, it is very difficult to maintain the same standards of potability in treated water distribution systems. The biostability of treated water can be influenced by a number of factors following the treatment, which cannot be controlled easily. For example, long hydraulic residence times of treated water [29–31], biofilm formation in distribution systems [15,27,32–39], and the corrosion of pipelines [37,40,41] can severely affect the biostability of treated water. In Table 2, problems due to the occurrence of microorganisms and possible reasons for their growth in conventional drinking water treatment and distribution systems are summarized.

2.2 Intentional Addition of Drinking Water Contaminants

In addition to unintentionally added chemical and biological contaminants, in some countries (e.g., the United States) the risk of intentional addition of bioterrorism agents into the treatment and distribution systems is alarmingly high. Bioterrorism is an act of terrorism where biothreat agents (e.g., dangerous disease-causing waterborne pathogens and biological toxins) are used to contaminate potable water supplies by tampering intentionally with infrastructure, particularly distribution systems of the treatment plant. These attacks raised serious security concerns, owing to the

TABLE 1 Types of Microorganisms Commonly Found in Drinking Water Treatment Plants and Their Sources of Entry/Existence

Microorganism	Type	Examples of Common Waterborne Pathogens	Possible Sources
Viruses	Enteric	Rota virus, norovirus, hepatitis A	Occur largely as a result of contamination with sewage discharge and human excreta
Bacteria	Excreted pathogens	*Salmonella* spp., *Shigella* spp., enterovirulent *E. coli*, *Vibrio cholera*, *Yersini enterocolitica*, *Campylobacter jejuni*, and *C. coli*	Sewage contamination of source waters
	Free living bacteria (natural habitat)	*Pseudomonas aeruginosa*, species of *Klebsiella* and *Aeromonas*, *Legionella pneumophila*	Potentially grow in source waters rich in organic matter and phosphorus-based nutrients
	Free-living bacteria (cyanobacteria)	*Microcystis aeruginosa*, Anabaena flos aquae, Planktotrix (Oscillatoria)	1. Potentially grow under high eutrophication conditions 2. Capable of releasing harmful toxins (hepatoxins and neurotoxins)
	Nuisance organisms	Iron- and surfur-reducing bacteria	Presence of these organisms indicates severe corrosion and promotes the growth of pathogens such as *Pseudomonas aeruginosa*
Protozoa	Enteric protozoa	*Cryptosporidium parvum*, *Giardia*, *Entamoeba histolytica*	Occur widely as parasites in the gut of humans and other animals
	Free-living protozoa (natural habitat)	*Naegleria fowleri*, *Acanthamoeba* species	Their occurrence in water systems is not via an enteric route and therefore their ecology is more complex than that of enteric protozoa
	Protozoa causing taste and odor issues	*Vannella*, *Saccamoeba*, *Ripidomyxa*	

Source: Data from [16].

TABLE 2 Problems Due to the Occurrence of Microorganisms and Possible Reasons for Their Growth in Conventional Drinking Water Treatment Plants and Distribution Systems

Treatment Stage	Problems	Possible Reasons	Ref.
Chemical pretreatment stage	There are, large variations in pathogen removal efficiencies.	1. Turbidity increases bacteria growth. 2. Pathogen removal efficiencies range from 0.05 log to 5.5 log removal. 3. Some bacteria can get enmeshed in floc formed by metal coagulants and escape the treatment. The enmeshment shields pathogens from disinfection. 4. Filamentous bacteria and some enteric viruses cannot be easily removed during the coagulation stage. 5. Metal coagulation process results in lysis of cyanobacteria cells, thereby releasing intracellular toxins into treatment systems that cannot be removed easily.	[25,27, 145–149]
Filtration stage	Filtration is not the ultimate barrier for removal of pathogens.	1. Pathogen removal efficiencies of conventional filters are highly dependent on chemical pretreatment. 2. Seasonal effects such as change in temperature affect pathogen removal efficiencies. Low water temperatures cause reduction in removal efficiencies. 3. Short perturbation events such as change in peak loading and change in chemical pretreatment conditions can cause the passage of certain filamentous bacteria (e.g., cyanobacteria).	[150,151]

(Continued)

TABLE 2 *(Continued)*

Treatment Stage	Problems	Possible Reasons	Ref.
Disinfection stage	Many disinfectant-resistant microbes escape the treatment	1. Many bacterial species (e.g., *Bacillus anthracis*, *Clostridinum botulinum*, Norwalk virus, and oocysts) are resistant to the disinfection process. 2. NOM reacts with disinfection by-products to form assimilable organic carbon (AOC) which serves as food and energy for microbial growth. 3. Some cyanobacterial toxins (e.g., anatoxin) are highly persistent and not readily destroyed, even by powerful disinfectants such as chlorine dioxide and ozone.	[42,152–159]
Distribution systems infrastructure	Longer hydraulic residence times of treated water may lead to microbial growth.	1. Long residence times result in improper water velocity distribution in the system, depleted disinfectant residuals, and rise in temperature. As a result, sedimentation can occur, and favorable habitat for bacterial growth is created. 2. Long residence times also promote the formation of DBPs and AOC, which in turn promote bacterial growth.	[28–29, 31,160]
	Growth of microbes in pipelines depends on materials used.	1. Biofilm activity is on the order iron pipes > plastic pipes > copper pipes. 2. Iron pipes also support diversified microbial biota. 3. Smaller pipes, due to their large surface/volume ratio, run out of chlorine residual faster than do bigger pipes and create favorable conditions for biofilm growth. 4. Packing materials and rubber gaskets used in pipe systems leach nutrients to support bacterial growth.	[160]

Prevailing conditions in distribution systems	Biofilm formation is a serious problems in distribution pipelines.	1. Biofilms have the ability to metabolize recalcitrant organic compounds. 2. Biofilms are resistant to disinfection. 3. Biofilms are surrounded by dense layers of extracellular polysaccharides that protect them from disinfection. 4. Their formation and sustenance depends on temperature, disinfectant residual, and the presence of nutrients such as AOC/MAP.	[15,27,32–39]
	Corrosion promotes microbial growth.	1. Iron and steel surfaces supply all the necessary macronutrients, such as carbon, fixed nitrogen, and fixed phosphorus, that are sufficient to promote the bacterial growth to problematic levels. 2. Corroded iron surfaces can easily adsorb substances that are rich in organic carbon (e.g., humic acid) that favor biofilm formation. 3. Iron pipes contain 0.02–1 wt% phosphorus, which gets oxidized with corrosion process and simulates microbial growth in the pipelines.	[37,40,41]

destructive potential of biothreat agents used in such attacks, which could leave long-lasting and damaging impacts on the public health and economic wellness of a society. The incidents of bioterrorism attacks, particularly on water distribution systems where treated water flows, have increased significantly in the past five years [3]. Once a biothreat pathogen enters a treated water distribution system, it may be difficult to detect them since the majority of biothreat agents are tasteless, colorless, and odorless. Furthermore, it is even more difficult to deactivate the pathogen in distribution systems because most pathogens have a biofilm-forming tendency, and some biothreat and weaponized agents, such as *Bacillus anthracis* (anthrax spores), are highly resistant to disinfection [42]. A list of weaponized (probably weaponized) biothreat agents (microorganisms and biological toxins) that have been confirmed as potential water threats (or probable water threats) is provided in Table 3. Information on infective inhalation dosages, their corresponding critical water dosages, and the stability rates of each of these agents is included in the table.

3 CARBON NANOTUBES AS ADSORBENT MEDIA IN WATER TREATMENT

CNTs represent prominent structures in the family of carbon nanomaterials and have been investigated extensively for their potential application as adsorbent media in water treatments, particularly for the removal of organic, inorganic, and biological contaminants. CNTs comprise a thin layer of graphene sheets rolled into concentric cylinders with both ends capped. If only one layer of a graphene sheet is rolled, it is called as a single-walled CNT (SWCNT), and if multiple, yet separate walls or layers of graphene are rolled into concentric cylinders, they form multiwalled CNTs (MWCNTs). The diameter of individual SWCNTs is about 1 to 2 nm, yet in some cases it may extend up to 3 nm [45]. The diameter of MWCNTs lies between 2 and 50 nm [46]. Further, the length of CNTs can vary from several micrometers to a few millimeters [47], and recently to centimeters [48,49]. Both SWCNTs and MWCNTs can potentially be utilized as adsorbent media to remove chemical and biological contaminants from water media.

3.1 Importance of Structural Properties of CNTs in the Adsorption Process

In their bulk state, CNTs have unique structural properties which are responsible for their potential as adsorbent media in water treatment. In bulk form CNTs possess *aggregated pores* [50], formed through the aggregation of tens and hundreds of individual CNTs due to van der Waals forces. Figure 3a shows the formation of CNT bundles [51]. It has been suggested that such aggregated pores of CNTs possess four distinct categories of sites suitable for adsorption that can capture various contaminants from water systems [50,52,53]. These categories can be identified qualitatively as follows (see Figure 3b): (1) type 1 sites, located in the hollow spaces within individual CNTs; (2) type 2 sites, located on the outer surface of individual CNTs, yet

TABLE 3 Biothreat Pathogens and Biological Toxins That Can Be Weaponized and That Are a Confirmed Threats or Potential Threats) to Water Systems[a]

Biothreat Agent	Weaponized (Yes/No)	Water Threat (Yes/No)	Stability in Water Measured in Time	Median Infective Dose[b]		Chlorine Resistance Information
				A	B	
Bacterial Pathogens						
Bacillus anthracis	Yes	Yes	2 years (spores)	57	171	Resistant
Brucella spp.	Yes	Probable	20–72 days	100	300	Not known
Vibria cholarae	Not known	Yes	6 weeks	10	30	Easily killed
Clostridium perfiringens	Probable	Unlikely	No data	1×10^6	1×10^7	Resistant
Francisella tulerensis	Yes	Yes	90 days	1×10^6	3×10^6	Resistant
Salmonella spp.	Not known	Yes		100	300	Easily killed
Shigella spp.	Not known	Yes	8 days	100	300	Easily killed
Yerisinia pestis (plague)	Probable	Yes	16 days	1	2	Easily killed
Rickettsia						
Coxiella burnetti (Q fever)	Yes	Possible	160 days	1	1	Not known
Viruses						
Hepatitis A	Not known	Yes	No data	1	1	Not known
Variola	Possible	Possible	Days to weeks	No data	Not known	
Protozoa						
Cryptosporidium oocysts	Not known	Yes	Days to weeks	1	3	Resistant

(Continued)

333

TABLE 3 (*Continued*)

Biothreat Agent	Weaponized (Yes/No)	Water Threat (Yes/No)	Stability in Water Measured in Time	Median Infective Dose[b]		Chlorine Resistance Information
				A	B	
Biological Toxins						
Aflatoxin	Yes	Yes	Probably stable	10		Probably tolerant
Anatoxin A	Not known	Probable	Days	No data		Probably tolerant
Botulinum toxins	Yes	Yes	Stable	6×10^{-6}		Medium resistance
Microcystins	Possible	Yes	Probably stable	10		Resistant at 100 ppm
Ricin	Yes	Yes	Stable	2		Resistant at 10 ppm
Saxitoxin	Possible	Yes	Stable	0.1		Resistant at 10 ppm
Staphylococcal enterotoxins	Probable	Yes	Probably Stable	0.01		Not known
T2 mycotoxins	Probable	Yes	Stable	26	8.7	Resistant
Tetradotoxins	Possible	Yes	Probably stable	0.1		Inactivated at 50 ppm

Source: Based on data from [42–44].

[a] ID_{50} for spores, spores/liter; all other microorganisms, organisms/liter; toxins = μg/L.

[b] Median infective dose A = 15 L/day; B = 5 L/day for 7 days.

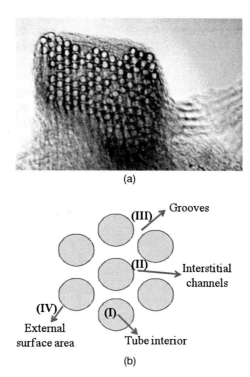

FIGURE 3   (a) TEM of a bundle of SWCNTs (from [51]); (b) hypothetical adsorption site regions in SWCNTs (from [53]).

in the interstitial channels between three or more neighboring CNTs; (3) type 3 sites located on the grooves present on the periphery of CNT bundles, and (4) type 4 sites, located on the exterior surface of the outermost CNTs [51,54–56]. It is suggested that adsorption to type 1 sites can be selective for the adsorption of contaminant molecules of smaller size (e.g., gas molecules), particularly if the tubes have open ends [50,53,57]. Type 2 sites, located in the interstitial spaces between CNTs can also be effective in the adsorption of contaminants, provided that the spacing is wide enough to accommodate adsorbate molecules [53]. However, the majority of the adsorption occurs in CNTs on types 3 and 4 sites. Type 4 sites, and to some extent type 3 sites, are easily accessible to biological contaminants of larger dimensions (greater than 10 nm) and possess varying degrees of symmetry [58]. The ability of CNTs to effectively capture biological contaminants of relatively larger dimensions (e.g., pathogens) is not observed in other microporous adsorbent media, including activated carbon, which is why CNTs have received special attention as an adsorbent medium for water treatment applications.

The diameter of CNTs can significantly influence the adsorption of contaminants, particularly those of large dimensions such as microorganisms, and the capture of biological contaminants from water is more effective by CNTs in their semidispersible state [112]. For example, the capture/removal of *Streptococcus mutans* by both single- and multiwalled CNTs has revealed in recent investigations that the precipitation of *S. mutans* on multiwalled CNTs of 30 nm diameter is greater than that of single- and multiwalled CNTs of 200 nm diameter, which suggested that the precipitation efficiency of *S. mutans* may depend on the CNT diameter and that multiwalled CNTs of intermediate diameter (30 nm) may offer a larger external surface area and thus the probability of greater contact with the bacteria. Structural defects of CNTs may also contribute to favorable adsorption of certain contaminants; for example, CNTs of poor quality and with more defects possess a higher surface area and exhibit greater sorption capacity toward aqueous Pb than do those of aligned CNTs [59].

3.2 Importance of Functional Properties of CNTs in the Adsorption Process

The functional properties of CNTs include its specific surface area, ratio of micropore to mesopore volume, and presence of functional moieties on the CNT surface. Surface modification of CNTs can have a profound influence on the removal of aqueous-phase contaminants via adsorption; surface modification can significantly alter the structural properties of CNTs: increase their Brunauer–Emmett–Teller (BET) surface area, and change the micropore/mesopore volume ratio. Further, surface modification techniques can also induce changes in the functional groups present on edges and sidewalls of CNTs that will affect its interactions with sorbate species in a positive (i.e., promote the adsorption of sorbate species) or a negative (i.e., reduce the adsorption of the sorbate species) manner.

The average BET specific surface area and mesopore volume of pristine and unmodified CNTs is reported to be 250 m^2/g and 0.85 cm^3/g, respectively [60]. However, the BET surface area can be increased and the mesopore volume of CNTs can be altered by its purification through a variety of techniques, including (1) chemical purification, where the CNTs are treated with strong acids (e.g., HNO_3) or bases (e.g., KOH and NH_3); (2) treatment with gaseous compounds such as CO_2, air, and ozone [61,62]; and (3) thermal treatment, where purification of CNTs can be achieved by heating them to specific temperatures [63]. A summary of various CNT purification techniques to increase BET surface area and to alter the micropore/mesopore volume ratio of CNTs is given in Table 4. The purification process of CNTs is aimed primarily at the removal of impurities, such as amorphous carbon associated with freshly prepared CNTs, the decomposition of impurities blocking CNT pore entrances, and the incorporation of additional functional groups on the CNT surface [63]. Although all three purification protocols noted above can enhance the BET surface area of CNTs, they produce functionally nonidentical surfaces that can interact differently with various adsorbate species. As a result, the adsorption capacities of the same contaminant on CNTs modified via two different purification routes may be quite different.

For example, the oxygen-containing functional groups on CNTs enhance the adsorption of polar species [64–66]. However, the presence of oxygen-containing functional groups (e.g., $-C{=}O$, $-OH$, and $-COOH$) on CNTs is not beneficial for the adsorption of biological contaminants. They can create a large number of

TABLE 4   CNT Purification Techniques and Their Typical Characteristics

CNT Purification Technique	Characteristics	Refs.
	Treatment with Acids	
HNO_3 modified	1. Acid treatment produces oxygen-containing carboxyl and hydroxyl groups. 2. Since acid treatment induced charged groups on a CNT surface, it is beneficial when considering CNTs for metal adsorption.	[71,161]
	Treatment with Bases	
NH_3 modified	1. NH_3 treatment removes carboxyl and hydroxyl groups. This method of treatment is suitable for sorption of biological contaminants since it increases the mesopore volume of CNTs.	[161]
KOH modified	1. KOH treatment increases micropores and mesopores. 2. The greater the KOH/CNT ratio, the higher the pore volume.	[62]
	Treatment with Gases	
Air activated	1. Has a much smaller micropore volume. 2. Air activation removes catalyst metals and amorphous carbon.	[62, 162–164]
CO_2 activated	1. CO_2-activated CNTs have a large micropore volume. Suitable for adsorption of smaller contaminants.	[62]
Ozone treated	1. Opens end caps and introduces holes in sidewalls. 2. Ozonolysis oxidizes carbon atoms and forms oxygen-containing groups.	[61]
	Treatment with Heat	
Heat treatment	1. Functional groups blocking pore entrances are thermally decomposed, thereby enhancing the external surface area. Best purification method to use to prepare material for adsorption of biological contaminants.	[21]

Source: [63].

negative charges on the CNT surface, thereby reducing the sorption capacities of negatively charged contaminants such as NOM [21]. Thus, CNT surface modification and purification techniques must be chosen according to the adsorbate species of interest because the extent of its adsorption on a CNT surface depends on structural and functional properties of CNTs, which in turn depend on the modification route adapted. Similarly, for biological contaminants, purification techniques that can increase mesopore content and reduce the micropore content of CNTs should be chosen with due consideration. The mesopore volume and BET surface area of CNTs can be enhanced by heat treatment for the capture of large biological contaminants, whereas other techniques of purification that enhance micropore volume of CNTs (e.g., air activation, acid treatment) will not be suitable for the removal of biological contaminants [21,61,62].

4 ADSORPTION OF CONTAMINANTS FROM WATER SYSTEMS ON CARBON NANOTUBES

Among the group of nanosorbent media envisioned to have a promising future in water treatment applications, CNTs has gained substantial attention, where the removal capacity of CNTs for heavy metals, organics, and biological contaminants in aqueous systems has shown excellent promise. In this section, we discuss the technical aspects of the adsorption of aqueous-phase chemical (heavy metals and organics) and biological contaminants by CNTs.

4.1 Adsorption of Chemical Contaminants on Carbon Nanotubes

CNTs have shown exceptional adsorption capability and high adsorption efficiency for various organic pollutants such as benzene [67], 1,2-dichlorobenzene [68], trihalomethanes [69], polycyclic aromatic hydrocarbons (PAHs) [70], and heavy metals [65,66,71–76]. The maximum sorption capacities of raw and surface-oxidized CNTs for selected heavy metal and organic contaminants, as calculated by the Langmuir equation, are given in Table 5. The experimental data for the majority of heavy metal and organic contaminants is consistent with the Langmurian isotherm model, which suggests that the micromolecular pollutants may form a monolayer on the surface of CNTs.

The maximum sorption capacity estimate for heavy metals on CNTs based on Langmuir model varies for different metals. Their adsorption uptake of heavy metals follows the order $Pb^{2+} > Ni^{2+} > Zn^{2+} > Cu^{2+} > Cd^{2+}$ (Table 5). The maximum sorption capacity for aqueous-phase metal ions are reported to be in the range 10 to 11 mg/g for Cd [74], 49 to 97 mg/g for Pb [59,71,73], 9 to 47 mg/g for Ni [65,76], and 11 to 43 mg/g for Zn [66,72].

The sorption capacities of metal ions by raw CNTs are much lower but increase significantly after oxidization, which can modify the exteriors of pristine CNTs as better sorbents; that is, this process can open the tips and remove the blockage surrounding the pores or tube openings, and it also attaches oxygen-containing polar functional groups to the CNT surface. The attachment of polar functional groups on CNT surface can enhance their dispersal in water and increase the exchange of electrons with metal ions, thus facilitating greater adsorption. As a result, the sorption capacities of CNTs modified with polar functional groups are significantly greater for removal of metal ions than are conventional sorbents employed for water treatment [77]. However, the adsorption behavior of CNTs modified with polar (oxygen-containing)

TABLE 5   Comparison of Surface Area and Maximum Sorption Capacities for Sorption of Metal Ions and Organic Contaminants on Different Types of CNTs

Sorbent	BET Surface Area (SA) (m^2/g)	Sorbate	q_m (mg/g)	q_m/SA (mg/m^2)	Ref.
SWCNTs	577	Ni^{2+}	9.22	0.0160	[65]
SWCNTs(NaOCl)	397	Ni^{2+}	47.85	0.1205	
MWCNTs	448	Ni^{2+}	7.53	0.0168	
MWCNTs(NaOCl)	307	Ni^{2+}	38.46	0.1253	
SWCNTs	590	Zn^{2+}	11.23	0.0190	[72]
SWCNTs(NaOCl)	423	Zn^{2+}	43.66	0.1032	
MWCNTs	435	Zn^{2+}	10.21	0.0235	
MWCNTs(NaOCl)	297	Zn^{2+}	32.68	0.1100	
MWCNTs15	181	Cu^{2+}	2.35	0.0130	[165]
MWCNTs15a	279	Cu^{2+}	5.26	0.0189	
MWCNTs15	181	TCP	0.03	0.0002	
MWCNTs15a	279	TCP	0.03	0.0001	
MWCNTs	159.7	1,3-Ddinitrobenzene	41.4	0.2592	[78]
MWCNTs(HNO_3)	274	1,3-Dinitrobenzene	48.7	0.1777	
MWCNTs	159.7	m-Nitrotoluene	41.4	0.2592	
MWCNTs(HNO_3)	274	m-Nitrotoluene	47.8	0.1745	
MWCNTs	159.7	p-Nitrophenol	31.7	0.1985	
MWCNTs(HNO_3)	274	p-Nitrophenol	41.1	0.1500	
MWCNTs	159.7	Nitrobenzene	27.8	0.1741	
MWCNTs(HNO_3)	274	Nitrobenzene	37.5	0.1369	

functional groups toward organic contaminants is not as promising as much as heavy metal ions. Recent investigations indicate that impurities such as amorphous carbon and metal particles are removed from raw CNTs (particularly MWCNTs) during oxidization treatment, which often results in modest increases in their surface area and pore volumes (Table 5). While the sorption capacity of organic contaminants on MWCNTs (q_m) modified with oxygen-containing functional groups is greater than that of their raw counterparts (see Table 5), their sorption capacity normalized on the basis of surface area (q_m/SA) however shows that the presence of polar, oxygen-containing functional groups on CNTs may actually reduce the adsorption of organic contaminants on CNTs [78].

Adsorption Mechanism of Chemical Contaminants on CNTs   The interspecies interactions at the solid–water interface play a key role in the adsorption of chemical contaminants on CNTs. The adsorption of heavy metals may be controlled largely by the chemical interactions occurring between metal species in the aqueous phase and functional groups on the CNT surface, and the structural properties of CNTs (i.e., the surface area and pore volume) may express less influence. Aqueous metal ions have different affinities for various functional groups, such as the carboxylic and phenolic groups present on CNT surfaces [64–66]. Although the mechanisms of

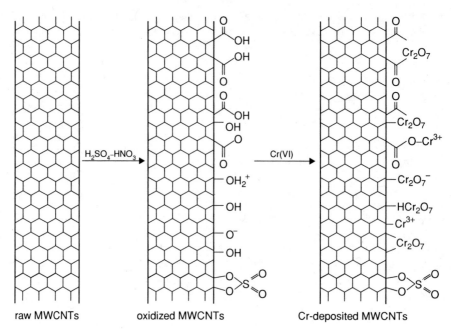

raw MWCNTs   oxidized MWCNTs   Cr-deposited MWCNTs

FIGURE 4   Major mechanism for the adsorption of Cr on oxidized MWCNT surfaces (from [80]).

interaction between heavy metal ions and CNT surfaces are complex, ion exchange may be the dominant process in removal. The carboxylic and phenolic functional groups on the CNT surface can undergo ion exchange with aqueous metal ions, in place of protons [79], and surface-oxidized CNTs can express better uptake of metal ions, due to their superior ion-exchange capacity than that of pristine CNTs [74]. The schematic of enhanced sorption of Cr(VI) on CNT surface after acid treatment [80] is shown in Figure 4, where the interactions of a CNT surface with Cr(VI) became favorable following the attachment of ionizable functional groups, leading to increased adsorption.

The adsorption of organic contaminants on CNT surfaces is controlled by electrostatic and hydrophobic interactions, yet the surface chemistry of adsorbent media (CNTs) pose a significant influence on the adsorption process. Typically, the aromatic molecules are physisorbed on the carbon surface by interactions between the π electrons of the aromatic ring and graphene layers [81], but the overall physisorption capacity may be reduced in the case of oxidized CNTs [82]. The polar functional groups on CNT surfaces can give rise to hydrogen bonding with water that can adversely affect the adsorption for aromatic compounds. Recent studies have indicated that a CNT surface functionalized with oxygen-containing groups can decrease the sorption efficiencies of organic contaminants for the following reasons: (1) the adsorption of water molecules on a CNT surface can displace organic contaminants, (2) dispersive and repulsive interactions between functional groups of a CNT surface and organic molecules, and (3) the presence of hydrogen bonding [78,83]. Further, another study has revealed that incorporated surface oxides can create polar regions and thus reduce the surface area available for naphthalene sorption; for example, a 10% increase in surface oxygen concentration resulted in a decrease of naphthalene sorption by as much as 70% [84].

Factors Affecting the Adsorption of Chemical Contaminants on CNTs   In addition to the structural and functional properties, the adsorption efficiencies of chemical contaminants on CNTs are also affected by the physiochemical properties of an aqueous medium, such as its pH and ionic strength. Removal of heavy metals from aqueous medium by CNTs may depend on the solution pH, which affects the surface charge of the CNTs and the degree of ionization and speciation of the adsorbates. The maximum M^{2+} (M = metal) removal was observed at pH 8 to 11, and it decreases at pH < 8, which may be attributed to the following factors: (1) the CNT surface is more negatively charged at a higher pH, which causes greater electrostatic attraction of positively charged metal species: for example, M^{2+} [77]; (2) greater competition between H^+ and M^{2+} species for the sorption sites on CNT surface at a lower pH; and (3) in acidic conditions (pH < 8), the hydrolysis of M^{2+} can create species, such as $[M(OH)^+$ and $M(OH)_2{}^0]$, that express reduced electrostatic interactions with the CNT surface.

Changes in pH can also affect the adsorption of certain organic pollutants on CNTs, as pH controls speciation of the ionizable adsorbate in solution and the polar functional groups on a CNT surface. For example, 4-nitrophenol can exist in ionized and neutral forms at circumneutral pH ($pK_a \sim 7.1$); at solution pH 4 to 7 (i.e., pH < pK_a), the dominant form of 4-nitrophenol is neutral molecule, which shows greater affinity toward SWCNTs. At pH > 7, 4-nitrophenol exists in ionized forms that have a lower affinity toward SWCNTs. Further, at pH < 4.0, the functional groups on an SWCNT surface become negatively charged due to deprotonation, and therefore its ability to adsorb decreases. Therefore, optimum pH for 4-nitrophenol adsorption on SWCNTs is typically about 7.0 [85]. A change in pH can also affect the solubility of an organic compound, which in turn influences its sorption capacity on CNTs. For example, the uptake of resorcinol on MWCNTs increased with a decrease in pH value [86], since the solubility of resorcinol is pH dependent and decreases with a decrease in solution pH.

Recent studies have further revealed that the desorption of heavy metal ions from CNT surface is also pH dependent, and it is quite efficient at pH $\sim$ 1 [66,76]. Li et al. reported that desorption of Pb^{2+} increased with decreasing pH, and complete removal of Pb^{2+} from a CNT surface was achieved at pH $\sim$ 2 [75]. In a recent study with Cr(IV), the desorption from unfunctionalized and functionalized (hydroxyl and carboxyl groups) MWCNTs occurs rapidly at a higher pH, particularly at pH 14 [87]. This may be attributed to deprotonation of dichromate ions ($Cr_2O_7{}^{2-}$) at high pH which are then easily repelled and outcompeted by excess hydroxyl ions (OH^-). The desorption techniques of heavy metals from CNTs are significant from the standpoint of regeneration and reuse of CNT filters for metal removal and the cost of operation. The ionic strength of the aqueous medium also has a modest influence on metal ion sorption onto CNTs. In studies of Cd^{2+}, Cu^{2+}, and Pb^{2+} sorption on CNTs, a decrease in sorption capacity with an increase in ionic strength has been reported [74].

4.2   Adsorption of Biological Contaminants on Carbon Nanotubes

CNTs has also shown potential in other environmental applications, particularly in water treatment. The removal of various chemical and biological contaminants from contaminated water systems with CNTs as adsorbent media has been a subject of study in recent years. While the adsorption of chemical contaminants is discussed elsewhere in the chapter, this section focuses on the application of CNTs as adsorbent media for the removal of biological contaminants in source waters before it enters the drinking water treatment plant.

Adsorption Removal Efficiencies of Biological Contaminants by CNTs   Studies related to the removal of biological contaminants (i.e., microorganisms, NOM, and cyanobacterial toxins) from contaminated water systems using CNTs as an adsorbent medium is reported by various groups. In particular, data pertaining to the adsorption of different bacteria types on single- and multiwalled CNTs reported by various authors suggest that CNT-based adsorption filters can serve as an excellent means of capturing the pathogens from raw water sources entering a drinking water treatment plant. In a study involving *B. subtilis* spores conducted by Upadhyayula et al. [88], the adsorption affinity of the pathogen on single-walled CNTs, is 27 to 37 times greater than its affinity on activated carbon and NanoCeram. Such high microbial immobilization capacity of CNTs can be attributed to the accessible external surface area provided by CNTs in comparison to other sorbents due to restricted openings in their microporous structure. Figure 5 shows SEM images of *B. subtilis* spores adsorbed on powdered activated carbon, NanoCeram, an alumina-based (AlOOH) adsorbent medium, and on single-walled CNTs obtained [88], which shows that the spores do not show much adherence on the surface of activated carbon and NanoCeram, whereas single-walled CNTs appear to capture spore population effectively because of their fibrous structure and the accessible external surface area for bacterial sorption.

According to the batch adsorption studies with *B. subtilis*, *E. coli*, and *S. aureus* on CNT and other adsorbent media conducted by Upadhyayula et al. [88] and Deng et al. [60], their adsorption affinity of the three bacteria on single-walled CNTs is exceptionally high. The adsorption affinities of above bacteria on SWCNTs is decrease in the following order: *S. aureus* > *E. coli* > *B. subtilis* spores. Single-walled CNTs show greater adsorption of *S. aureus* because of the smaller size of these spores (~0.5 nm), which contributes to faster diffusion and greater occupancy of bacteria on CNTs. On the other hand, the *B. subtilis* spores have lower adsorption affinity among the three, which might be due to their biological composition, which affects the adherence behavior. With respect to the three bacterial species, the Freundlich isotherm correlates the experimental values better than does the Langmuir isotherm.

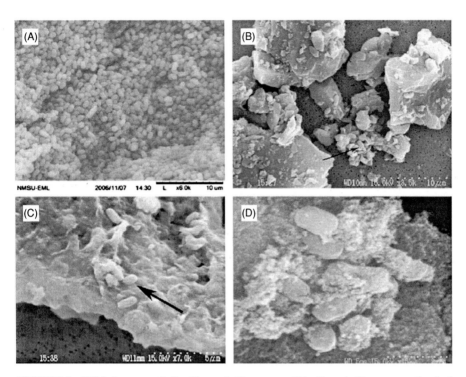

FIGURE 5   SEM images of *Bacillus subtilis* spores (A); *B. subtilis* spores adsorbed on powdered activated carbon (B); on NanoCeram™ (C); on single-walled CNTs (D). (From [88].)

The bacterial adsorption on porous media is multilayered and thus better characterized by the Freundlich isotherm rather than the Langmuir isotherm, which assumes a monolayer adsorption of the adsorbate [60,88].

Adsorption of bacteria on SWCNTs is characterized not only by high sorption capacities but also rapid adsorption kinetics [60,88]. In Figure 6, the adsorption kinetics of the three bacteria species mentioned above at different initial concentrations, corresponding to equilibrium concentration, and time taken to reach equilibrium are provided. Figure 6 is a plot between the ratio of the amount of bacteria adsorbed [colony-forming units (CFU)] per gram of CNTs at a given time (M_t) in minutes to the maximum amount adsorbed (M_{max}) versus time in minutes. At equilibrium concentration C_e (CFU/mL), the adsorbed bacterial mass approaches a maximum value, expressed as $M_t = M_{max}$, and the time taken for the bacteria to reach equilibrium [i.e., the time taken by the bacteria to get transported from the adsorbing phase in liquid to the adsorbed phase on a solid (CNTs)] was 30 min for *B. subtilis* spores, and only 5 min for *S. aureus* cells. This suggests that adsorption of bacteria on CNTs occur almost instantaneously as characterized by kinetics data. Moreover,

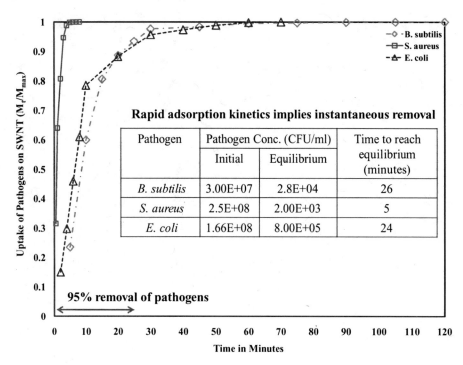

FIGURE 6 Adsorption kinetics data of *B. subtilis* spores, *E. coli*, and *S. aureus* on single-walled CNTs. The corresponding initial concentration C_0 (CFU/mL), equilibrium concentration C_e (CFU/mL) and time taken by the bacterial cells to reach equilibrium are shown in the inset. (From [88], with permission.)

the concentrations of bacteria (CFU/mL) used in batch adsorption studies conducted by Upadhyayula et al. [88] and Deng et al. [60] are high. In real-world situations, the maximum concentration of fecal coliforms present in raw drinking water drawn from a surface water source such as rivers will be approximately 1000 CFU/100 mL [89]. Thus, due to the high microbial adsorption capacity and rapid adsorption kinetics of CNTs, the CNT-based adsorption filters operating in a real-time environment can probably be used indefinitely with longer service times and minimal regeneration times.

While quantification of adsorption capacities and adsorption kinetics of bacteria on single-walled CNTs has been determined by Upadhyayula et al. [88] and Deng et al. [60], others [90] have reported similar bacterial adsorption behavior on multiwalled CNTs. Moon and Kim [90] unveiled the potential of utilizing multi-walled CNTs as a bacterial concentrating agent based on principles of adsorption and magnetic separation. In their study, they used clusters of multiwalled CNTs initially to adsorb bacterial cells, followed by applying a magnetic field to completely separate the mixture of multiwalled CNTs and the adsorbed bacteria. Four independent studies were conducted using four different bacterial strains—*E. coli*,

B. subtilis, P. putida, and *A. globiformis*—all approximately at a concentration of 2×10^7 CFU per liter of solution. Moon and Kim [90] suggest that multiwalled CNT clusters are able to nonselectively serve as a universal adsorbent medium, and the adsorption capacities of all four bacteria were similar. Biological composition and surface characteristics (e.g., hydrophobicity and surface charge) of the cells did not influence their adsorption behavior on multiwalled CNT clusters. This is a very promising finding regarding MWCNTs that can serve as a universal adsorbent for the removal of pathogens. However, unlike single-walled CNTs, the Langmuir isotherm model curves correlated the experimental data better than the Freundlich isotherm. The concentration of bacteria used by Moon and Kim [90] was lower, i.e., 2×10^4 CFU/mL, whereas the concentrations used by Upadhyayula et al. [88] and Deng et al. [60] were higher than 5×10^5 CFU/mL. Given the large accessible external surface area of CNTs, it is possible that all bacterial cells are accommodated on a single monolayer that is in contact with the adsorbent, and therefore the data fit the Langmuir isotherm characterized by monolayer adsorption.

Among the other biological contaminants, NOM adsorption on CNTs received significant research attention, as many scientists examined the potential of NOM removal by CNT adsorbent media, instead of PAC and GAC. The removal efficiency of NOM by commercial-grade microporous adsorbent media, such as GAC, is limited by its pore size and is affected by the presence of specific functional groups on the surface. NOM composition is highly polydispersed, with sizes ranging between 0.5 and 5 nm, whereas the size of the pore opening in GAC is only 1 nm [18]. Thus, the fraction of NOM smaller than 1 nm can get sorbed by GAC efficiently. Size exclusion is a principal mechanism governing the adsorption of NOM to the microporous media, such as GAC. The presence of functional groups on the surface also exerts an influence on NOM adsorption, where low-molecular-weight fractions get preferentially sorbed. Since NOM molecules always express a net negative surface charge, their adsorption on the GAC surface bearing cationic functional groups is greater than GAC without surface charge [17,18,91].

However, the adsorption mechanism of NOM on CNTs is entirely different. In CNTs, the adsorption and removal efficiency of NOM depends on its accessibility to the surface area and aggregated pores of CNTs. The surface areas and pore sizes of CNT are capable of concentrating NOM molecules that are much smaller in size than microorganisms. This is evident from the results obtained by Lu and Su [21], where the adsorption of all forms of NOM [i.e., the dissolved organic carbon (DOC) and assimilable organic carbon (AOC)] is higher on CNTs than on GAC. Although adsorption of both DOC and AOC fractions of NOM on CNTs is higher than on GAC; one fundamental difference between the two is that the low-molecular-weight fraction of NOM is prefentially sorbed into GAC, whereas the high-molecular-weight NOM has greater affinity towards CNTs [21,92]. It appears that the high mesoporous fractions in CNTs can accommodate high-molecular-weight fractions of NOM. It can also be seen from the results of Su and Lu [21] that the heat treatment of CNTs enhanced the adsorption of NOM presumably because heat-treated CNTs can have reduced, pore blockages and net surface charge, which may contribute to the higher adsorption capacity compared to pristine CNTs [24]. According to Hyung and Kim

[19], the adsorption of NOM on multiwalled CNTs depends greatly on NOM type, meaning that it depends on NOM's carbon functionality, since NOM express various forms, such as aliphatic, heteroaliphatic, aromatic, carbonyl, carboxyl, and acetal carbons [19]. However, regardless of the source (i.e., river, lake, soil, etc.) or the type (i.e., humic, fulvic, or the bulk fractions), the NOM rich in aromatic carbon content exhibited a stronger affinity toward multiwalled CNTs. This is because of the strong $\pi-\pi$ electron interaction between the outer surface of CNTs and the aromatic moieties of NOM compounds [19]. Finally, the adsorption of NOM on CNTs is an exothermic process, and the adsorption capacity decreases with an increase in temperature from 5°C to 45°C. This implies that CNT adsorbent media function effectively at all water temperatures normally seen in a drinking water treatment plant.

The prospect of CNT application to remove NOM from water systems may offer greater benefits in comparison to GAC from a practical standpoint, for the following reasons: (1) studies conducted by Su and Lu [21] indicate that desorption of NOM from a CNT surface is higher than from GAC. In CNTs, the adsorbed NOM molecule faces less resistance (due to the presence of wide-open aggregate pores) in comparison to GAC, where the molecules get desorbed from the internal pore regions and therefore face greater resistance. The DOC recovery from CNTs after 15 min of reactivation time is 96% against the recovery of only 57% in GAC [24]. (2) for a given temperature, the DOC recovery from CNTs is higher than from GAC, suggesting that operation of CNT filters might require lower energy consumption than operation of GAC filters. Also, CNTs registered higher removal efficiencies than GAC registered, even after periodic cycles of reactivation; the weight loss of CNT adsorbent media is less (6%) than the weight loss of GAC media (18%). This suggests that due to their high adsorption and desorption capacities, lower reactivation temperature, and lower weight loss, CNTs can probably function more economically than GAC filters in the removal of NOM from water systems.

Finally, the adsorption of another biological contaminant, the cyanobacterial toxins on CNT media, is studied by a few groups [93,94]. As explained earlier, the cyanobacterial toxins are formed in water treatment plants, due to the presence of cyanobacteria species in raw waters, which will get lysed during the treatment process and release harmful internal toxins. The most common form of cyanobacterial toxin is the microcystins (MCs). While other forms of treatment (e.g., coagulation) and disinfection techniques fail miserably to remove MCs from water systems and adsorption is the preferred technique for their removal, it was well documented that adsorption of MCs on wood-based activated carbon expressed higher MC removal capacities than those of PAC because of their slightly mesoporous content. Because of their large mesoporous structural dominance, CNTs showed sorption capacities superior to those of PAC [93,95,96]. The adsorption capacity result obtained by Yan et al. [93] indicate that two factors, the specific surface area and the external diameter of CNTs, determine the amount of adsorption of MCs on CNTs. While the former is responsible for enhanced adsorption of MCs, the size of the latter critically sets the upper limit of adsorption of MCs on CNTs. For example, a CNT with an outside diameter of 10 nm adsorbs only 5.9 mg/g of adsorbent, whereas a CNT with an outside diameter of 2 nm adsorbs 14.8 mg/g of adsorbent, suggesting that CNTs with pore dimensions closer to molecular dimensions of MCs are better candidates for adsorptive removal of the material [93]. Thus, the adsorption of MCs on CNTs is significant due to following factors: (1) MCs diffuses faster from the adsorbing phase to the adsorbed phase, as evidenced by rapid adsorption kinetics, reaching the equilibrium concentrations in less than 10 min; in comparison, with activated carbon, MCs took more than an hour to reach the equilibrium condition [93]; (2) the interaction between MCs and CNTs is strong because the adsorption affinity of every adsorption site on CNTs expresses the same affinity toward MCs; and (3) MCs forms a monolayer on CNT surface with data better correlated with the Langmuir isotherm [93]. Further, the desorption of MCs from smaller-diameter CNTs is less than that from larger ones indicating that structural properties, especially the size of CNTs, play an important role in determining their removal of MCs from water systems [93].

Factors Influencing Removal Efficiencies of Biological Contaminants by CNTs. Unlike chemical contaminants, where adsorption is influenced primarily by the structural properties and to some extent by the physical properties of the solution, adsorption in biological contaminants is greatly affected by the physiochemical and biological properties of the adsorbate species. Adsorption of microorganisms on porous media depends on multiple factors, such as pH, ionic strength, temperature, and the NOM content of the solution, and on the composition of the microorganisms themselves, which is characterized by properties such as hydrophobicity, surface charge, presence of pili, and flagella [97]. Bacterial attachment on a porous medium is a four-step process [98]. In the first stage, bacteria from the bulk can be transported to the surface of the porous media in a variety of ways. At this stage, bacteria motility might play a crucial role in the transportation process. For example, bacteria actively transport themselves to the surface by using their motile machinery, such as flagella and pili, where they get the opportunity to launch on a solid surface. However, it is not only bacteria having higher motility that can easily get transported to the surface—for two reasons. First, the movement of motile bacteria is erratic and might not always be in the direction of the surface, and second, bacterial cells can get mobilized via other forms of motion, such as diffusion and convection, where the flow of fluid is higher than the motion of the bacteria, and therefore the nonmotile bacteria can get transported to the surface equally well. Bacteria from the bulk solution can also be transported to an adhering surface such as a porous medium through electrophoretic motion induced by their surface electrical charge. Once the bacteria reach porous media, the next step in the process is initial attachment, where the cells get either reversibly or irreversibly attached to the surface of porous media. At this stage of the process, biological characteristics of adhering cells (e.g., surface hydrophobicity, net surface charge, motility, surface texture) greatly affect the attachment phenomena. In fact, the influence of biological characteristics (particularly the surface hydrophobicity and surface charge) is so profound that on the same adsorbent media, attachment of certain bacterial species with typical surface properties can be compared to other types of bacteria with differing surface characteristics [97,98]. As a rule of thumb, hydrophobic bacteria can get attached to hydrophobic surfaces and hydrophilic bacteria to hydrophilic surfaces [97]. Similarly, since a majority of bacteria have a net negative surface charge, they may get attached to adsorbent surfaces with positively charged functional groups [97,99]. However, there are a number of exceptions to this rule, as bacterial adhesion on porous media is a complex function of many parameters. Initial stages of bacterial attachment to adsorbent media is also characterized by a series of reversible adhesion stages, where the attached bacteria get dislodged from the adsorbent surface as a result of mild shear or by bacteria's own mobility [98]. Reversible adhesion of bacteria on porous media is also a function of various physiological conditions, such as pH, ionic strength, presence of mineral oxides, and NOM in the solution [97,99–103]. For example, the adsorption of certain types of bacteria on porous media is higher at low pH than high pH [104]. Thus, any slight change in the pH conditions of the solution can lead to detachment of cells from the surface. Similarly, the changes in the ionic strength of the solution (depends on the varying composition of mineral oxides) can cause reversible attachment of bacterial cells from adsorbent surface [97,103,105]. In the third stage of the process, that is, after establishing the initial contact with porous media, they try to make the contact more permanent by releasing surface-active compounds and special surface structures, such as fibrils, which assist in strengthening the link between cells and a solid surface, thereby leading to their firm irreversible attachment [98]. The last stage, followed by firm initial attachment, is bacterial colonization, where adhered bacteria grow and multiply into new colonies, supplemented by release of a slimy polymeric substance called extracellular polysaccharides (EPSs). The newly formed colonies of microorganisms get encapsulated in the EPS material, thus resulting ultimately in the formation of a matured biofilm [98].

The effect of the biological composition of microbes and the physiochemical properties of the solution on adsorption of microorganisms on CNTs needs to be established. However, based on bacterial adsorption studies conducted by Upadhyayula et al. [88] and Deng et al. [60] using single-walled CNTs and Moon and Kim [90] on multiwalled CNTs, it appears that the biological composition of cells may have less influence when being adsorbed onto a fibrous material such as CNTs, which are characterized by having a vast mesoporous regime and a large external surface that is easily accessible by microorganisms. Particularly the studies conducted by Moon and Kim [90] showed no significant difference in the adsorption capacities of four types of bacterial species inclusive of gram-negative, gram-positive, and spore-forming bacterial strains, all having different biological compositions. On the other hand, the adsorption capacities of three bacteria reported by Upadhyayula et al. [88] and Deng et al. [60] tend to be different. For example, the adsorption of *S. aureus* cells showed an adsorption affinity 100-fold higher than *E. coli* cells at concentrations tested by Deng et al. However, without further investigation, it is difficult to judge whether the difference in adsorption capacities of the two bacteria reported by Deng et al. is due to the biological composition of the cells or is just due their physical dimensions given the fact that the size of *S. aureus* cell is almost half the size of an *E. coli* cell. The

BET surface area of single-walled CNTs used by Deng et al. is 250 m^2/g, and since they have used 0.1 g in their study, the BET surface area of the sorbent is thus 25 m^2. In a hypothetical example, assuming that only 10% (i.e., 2.5 m^2 of surface area) is available and accessible by the microbes, the number of S. aureus cells with a typical surface area of 0.25 $\times$ 10^{-12} m^2 (assuming that the size of each cell is 0.5 μm) that can be adsorbed on single-walled CNTs is 4 $\times$ 10^{13}, whereas the number of E. coli cells, with a typical surface area of 2 $\times$ 10^{-12} m^2 (assuming that the shape of an E. coli is cylindrical and the size is 1 μm $\times$ 2 μm), is 1.25 $\times$ 10^{12}. This corresponds to a 100-fold magnitude difference in adsorption affinity of S. aureus and E. coli on single-walled CNTs, which can be due to their size difference.

Deng et al. [60] also conducted batch adsorption studies using mixed cultures, using a mixed population of cells with varying initial concentrations in ratios of 1 : 3, 1 : 1, and 3 : 1, respectively. The adsorption capacity data of these mixed cultures can be well correlated with the pure component Freundlich isotherm equation, which indicates strongly that the adsorption proceeds in a noncompetitive mode (i.e., single-walled CNTs are able to offer separate sites of adsorption for both S. aureus and E. coli bacteria, and bacteria do not compete for adsorption sites) [60]. On the other hand, it would be interesting to explore the adsorption capacity of various bacteria on single-walled CNTs at more realistic concentrations, similar to those used by Moon and Kim [90]. In such a case, unlike at high concentrations of adsorbing bacteria, where the difference in sorption capacities is clearly evident [60], at low concentrations the difference in the sorption capacities on CNTs may be insignificant. Nevertheless, it is tempting to conclude that in the case of fibrous and mesoporous materials and those providing large accessible external surface areas, such as CNTs, the biological nature of adsorbed cells may have less influence on the extent of adsorption. However, this claim needs to be validated by rigorous scientific data.

Although factors influencing the adsorption of biological contaminants on CNTs are yet to be determined, the effects of physiochemical conditions such as pH, temperature, and ionic strength on the adsorption of other biological contaminants (e.g., NOM) are reasonably well established [19,21]. In the case of NOM adsorption on CNTs:

1. pH values below 5 favor the adsorption process because having acidic pH values results in an increase in H^+ ions, which is advantageous for the adsorbtion of negatively charged NOM molecules. On the other hand, an increase in pH values beyond the isoelectric point results in the addition of OH^- ions, which leads to competition between NOM molecules and OH^- groups for adsorption sites that result in reduced adsorption.

2. A change in a solution's ionic strength affects the stability of CNTs in the solution, thereby affecting the sorption capacity [19]. CNTs form a stable suspension in aqueous media as a result of adsorption of NOM on its surface. The stability of CNTs increases with increased sorption of NOM on their surface [19]. However, at higher ionic strengths, the suspension of

CNTs in the solution is reduced, but the amount of NOM adsorbed remains the same.

3. An increase in temperature decreases the time required by DOC and AOC fractions of NOM to reach adsorption equilibrium; that is, the adsorption kinetics increases as the temperature increases. The rise in temperature causes the solution viscosity to decrease, due to which the diffusion rate of DOC and AOC within the spores of CNTs increases at a faster rate [24].

Finally, adsorption of biological toxins such as MCs on CNTs also depends on the water quality parameters, such as the pH of the solution. Similar to activated carbon removal of MCs, CNT adsorbent is higher at lower pH, for two reasons: (1) the solubility of MCs, mainly MC-LR, decreases with an increase in pH, and (2) at pH values above 7, the specific surface available for adsorption is decreased. It was stated previously that the adsorption capacity of MCs is a function of CNT outside diameter and decreases with an increase in diameter [93]. This is because the specific surface area of CNTs decreased with an increase in CNT outside diameter, and at a higher pH value of the solution, due to the greater aggregation of CNTs, the specific surface area of CNTs will decrease further, thus causing a decrease in the adsorption of MCs. Furthermore, at high pH, the negative charge potentials displayed by CNTs with larger outer diameters (10 to 30 nm) is higher than CNTs with smaller diameters (2 to 10 nm), which is another reason contributing to a decrease in sorption capacities of MCs by CNTs with larger diameters at high pH values [93].

Beneficial Aspects of Utilizing CNTs for Adsorption of Biological Contaminants
Utilizing CNTs as adsorbents appears to offer multiple benefits in terms of their use to capture and remove biological contaminants from drinking water systems. CNT adsorbent media can potentially have four principal advantages when used for the adsorption of biological contaminants, discussed below.

Possible Absence of Microbial Clogging of Filters The operational efficiency of most filters utilizing carbon-based adsorbent media such as GACs and PACs is affected by the growth and persistence of pathogenic microorganisms that clog the pore surface of the adsorbent media [106,107]. Enteric pathogens such as *Yersinia enterocolitica, Salmonella typhimurium*, and enterotoxigenic pathogens such as *E. coli* are shown to colonize on the surface of GAC filter beds and accumulate to populations of 10^5 to 10^7 CFU/g in a 14-day period, even after complete sterilization of filter beds [106]. This suggests that the use of either GAC or PAC filters requires close monitoring of their breakthrough profiles to prevent release of any carbon particles entrained with pathogens downstream of the process. Although the majority of carbon-based porous media such as GAC allow entrainment of bacterial pathogens on their surfaces, certain materials, such as Aqualen (activated carbon fibers) [98] and some forms of CNTs (e.g., single-walled CNTs) [108–110], may not allow the accumulation of pathogenic bacterial cells. This may be attributed to the material's cytotoxic nature, which causes the metabolic activity of cells to cease after they get

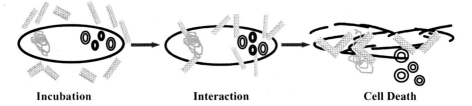

| Incubation | Interaction | Cell Death |

FIGURE 7   Atimicrobial effect of CNTs on bacterial cells.

attached to the surface. For example, the microorganisms (*E. coli*) adsorbed onto Aqualen fibers are shown to produce enhanced levels of EPS material to promote their adhesion and colonization on the material. But subsequently, the cells get deactivated as the Aqualen fibers squeeze out and eventually adsorb away the EPS material from microbes, resulting in cell starvation, loss of membrane integrity, and ultimately, cell death [98]. It was observed that the metabolic activity of microbes such as *E. coli*, *A. tumefaciens*, and *S. cerevisiae* adsorbed onto the fibers of Aqualen ceased in less than 24 h, which resulted in complete deactivation of adsorbed cell population [98]. In a few other studies reported so far, CNTs were shown to exhibit a cytotoxic nature, leading to the death of a wide range of microorganisms, such as bacteria (e.g., *Micrococcus lysodeikticus* [111], *S. mutans* [112], *E. coli* [108–110], *Salmonella* spp. [113], and bacteria endospores (*B. cereus*) [114]), protozoan species (e.g., *Tetrahymena pyriformis* [115]), and viruses (e.g., MS2 bacteriophage [116]). The antimicrobial effect of CNTs on bacteria is contact based, where the extent of damage to a bacterial cell depends on the probability of contact between CNTs and the bacterial cell surface. Figure 7 is, a schematic explaining the antimicrobial action of CNTs on bacterial cells.

As shown in Figure 7, the needlelike thin fibers of CNTs impinge: that is, punch holes in the outer cell structure of a bacterial cell, disrupt the intracellular metabolic pathways of the cell, and subsequently, by weakening the membrane integrity of the cell, the intracellular contents of the cell are released, resulting in cell death. It is believed that the cells undergo severe oxidative stress when subjected to the impingement process. However, the antimicrobial effect of CNTs on microorganisms depends on many factors corresponding to its physical dimensions, such as size, shape, number of walls (i.e., whether single- or multiwalled CNTs), and on its chemical properties, such as the presence of surface functional groups [109]. In general, loosely packed (i.e., debundled) highly dispersed short-length tubes exhibit more cytotoxicity toward bacterial cells than do other types [109]. This is because dispersivity increases the probability of contact between the tubes with bacterial cell surface, whereas short tubes easily penetrate the cell surface of bacteria and can induce greater damage. It was thought initially that due to their short length, single-walled CNTs can be more toxic to cells because of their ability to penetrate the cell structure. Further investigations conducted by Kang et al. [109] concluded that the dispersivity of CNTs is more important than their length because a higher cell lysis effect was observed with highly dispersed multiwalled CNTs than with short single-walled

CNTs, suggesting that dispersivity of CNTs is the most important property of all those that influence the cytotoxic efficiency of CNTs. Both single- and multiwalled CNTs are shown to have toxic effects on *E. coli* cells in experiments conducted by Arias and Young [113] using different cell types, *Salmonella typherium* (gram-negative) and *B. subtilis* (gram-positive, spore-forming). The cells were exposed to single- and multiwalled CNTs containing various functional groups (e.g., —COOH, —OH, and —NH$_2$). Their results indicate the following:

1. Single-walled CNTs exhibited a strong antimicrobial effect contributing to a 7 log reduction in cell populations, particularly *S. typhirium*, but the antimicrobial effect is dependent neither on the charge and functional groups associated with the CNT surface nor on cell shape, structure, and biological composition. On the other hand, the antimicrobial action of single-walled CNTs containing these functional groups depends on concentration and the solution media in which the cells are suspended [113]. The antimicrobial effect of both single-walled CNT-OH and CNT-COOH modified tubes is evident at concentrations of 50 μg/mL on a bacterial population of 10^7 CFU/mL *S. typhirium* cells. The bacteria were completely inactivated by increasing the concentrations to 250 μg/mL in 15 min [113]. The antimicrobial effect is also dependent on the solution in which the cells are suspended. A rapid reduction in cell population is achieved in deionized water, whereas the decrease in cell numbers was not significant while the cells are under the influence of buffering media such as phosphate buffer saline.

2. The multiwalled CNTs modified with functional groups such as —OH, —COOH, and —NH$_2$ failed to decrease the cell counts of the bacteria, suggesting a lack of antimicrobial effect of multiwalled CNTs on these cell types [113]. The antimicrobial effect of acid-modified multiwalled CNTs on *S. typhirium* cells obtained by Arias and Young [113] was different (expressed no antimicrobial action) from those obtained from Kang et al. [109], who have used *E. coli* cells (expressed antimicrobial action). The difference may be attributed to different cell composition [given that both groups have used similar, i.e., 5 × 10^6 CFU/mL and 10^7 CFU/mL cell concentrations in their experiments, and have used the same type of acid treatment, i.e., using a H$_2$SO$_4$–HNO$_3$ mixture (3 : 1 v/v)] or, more important, due to the extent of van der Waals forces of attraction between the molecules on a CNT surface and a cell surface that can be different for different cells [108,113]. Another interesting point of difference between the studies conducted by these two groups is that for almost a similar concentration of cells used, the concentration of acid-modified multiwalled CNTs used by Kang et al. [109] was 20 μg/mL (for 5 × 10^6 CFU/mL of *E. coli* cells), whereas the concentration of acid-modified multiwalled CNTs used by Arias and Young [113] is in the range 100 to 500 μg/mL (for 10^7 CFU/mL *S. typhirium* cells). This leads again to a hypothesis of whether the antimicrobial effect by any means is dependent on the upper limit of concentration of multiwalled CNTs used. However, this hypothesis needs to be validated.

Nevertheless, the antimicrobial action of CNTs over a wide range of bacterial species may prove to be advantageous when the medium used as an adsorbent removes biological contaminants, because unlike other porous media, such as GAC and PAC, bacterial entrainment on the surface of the adsorbent may be prevented.

Possible Absence of Competitive Adsorption   Biological contaminants such as NOM are known for their notorious ability to hinder the removal of many micropollutants using adsorption mechanism. Adsorption of micropollutants such as atrazine [58] and microcystins [93] on microporous media is affected severely by the presence of NOM because it restricts the access of micropollutants to the micropore regions of the adsorbent media [117–119]. NOM can prevent access of micropollutants either by blocking the pore regions of adsorbent media or by entering into direct competition with these adsorbents, displacing them and thereby preventing them from getting sorbed on the pores. Li et al. [118] observed that NOM, whose molecular weight ranges from 200 to 700 Da, is mainly responsible for the pore blockage effect of PAC, which reduced the diffusion rate constant and sorption capacities of atrazine on PAC. However, the pore blockage effect by NOM is shown to decrease even if the medium contains a smaller fraction of mesopores [118]. Similarly, adsorption of nonbiodegradable organics by GAC filters is reduced from 50% to less than 10% over time, due to the presence of NOM as a competitive adsorbate molecule that eventually displaces a micropollutant by occupying a greater portion of adsorption sites [119]. Another disadvantage of the presence of NOM as competitive adsorbate is that it promotes fouling of sorption sites by favoring the growth of microorganisms, therefore resulting not only in the premature saturation of GAC filter beds but also in leakage of pathogenic microorganisms on the downstream side into potable waters [119]. Due to their large mesopore content, CNTs might not pose the problem of pore blockage or a displacement effect when NOM is present as coadsorbate. Although this claim needs to be validated, the possibility of microbial fouling of CNT filters as a result of the presence of NOM is less likely for several other reasons. First, since the accessible external surface area of CNTs is significantly higher, it will require large numbers of microbial colonies to reach saturation. Second, the antimicrobial nature of CNTs may prevent the fouling caused by microorganisms. Thus, using CNT filters for the removal of biological contaminants can have a better service life even in the presence of competitive adsorbate molecules.

Possible Utilization of CNTs for Point-of-Use Treatment Applications   Although access to a safe and continuous supply of potable water is the fundamental right of every citizen of every country on the planet, many people (especially the residents of poor and severely water-stressed countries) are not fortunate enough to exercise this basic right. Ironically, even in this modern world and despite having advanced water treatment infrastructure, such as reverse osmosis and nanofiltration systems, a population greater than 1 billion worldwide are living without access to a safe and adequate supply of drinking water [120]. According to the World Health Organization (WHO), it is estimated that more than 2 to 3 million deaths occur per year due to the consumption of nonpotable water laden with waterborne pathogens responsible for the spread of diseases such as diarrhea and cholera [120]. Lack of a potable water supply, due either to physical or economic water scarcity, is a major issue with many developing nations, but protecting treated water supplies from bioterrorism attacks has emerged as a major national security issue in developed countries such as the United States.

Point-of-use (POU) and point-of-entry (POE) treatment systems can provide an effective solution to water treatment problems in both developed and developing nations. In POU water treatment the water is treated at the point of consumption, such as a filter placed near the tap faucet in a house. In POE treatment water is treated at a designated entry point such as where water enters an office building or other facility. In other words, POE is considered a reduced version of a centralized water treatment facility. POU and POE treatment systems can serve as effective solutions in both developing and developed nations because in areas lacking centralized water treatment infrastructure, and where water treatment becomes an individual choice, POU and POE systems can be implemented that can efficiently treat incoming water to a desired level of potability. POU treatment systems implemented in various developing countries showed promising results in the treatment of both biological contaminants, mainly pathogens [120–122], and chemical contaminants such as arsenic [120]. According to the results obtained by Souter et al. [120], the POU treatment system is able to remove bacterial pathogens (e.g., *E. coli*, *S. typhirium*, *V. cholerae*), cysts (e.g., *Cryptosporidium parvum*, and *Giardia*), and viruses (polio and rota viruses) from a concentration of 10^8 per liter to undetected levels. In another study, implementation of POU systems treated raw water effectively, which reduced diarrhea incidence in the region by 93% and prevalence by 85% [122].

While a number of treatment methods, such as flocculent–disinfection filter aids and membrane techniques such as reverse osmosis, ion exchange, and adsorption, are available to be used as POU and POE treatment systems, removal of contaminants via adsorption is considered attractive because of its simplicity and low cost. POU systems based on the utilization of adsorbent media such as GAC can effectively remove a wide range of organic contaminants, natural metal impurities such as arsenic compounds, and offer moderate removal capacities of certain heavy metals and metals bound to organic contaminants of concern [123]. However, micropore-dominant media such as GAC and PAC are unable to remove biological contaminants such as pathogens [123,124]. According to Snyder et al., although PAC filter beds are able to reduce the number of coliforms in product waters, indicator organisms are still being detected in these waters, which strongly indicates the limitation of GAC and PAC filters for the removal of biological contaminants from water systems [124]. In addition, GAC and PAC filters provide a favorable atmosphere for bacterial growth, which can be aggravated in the presence of contaminants such as NOM in water systems [124]. With the advent of novel adsorbent media such as CNTs, the prospects of adsorption-based POU and POE treatment systems for the removal of biological contaminants, particularly waterborne pathogens, appear to be encouraging. CNT media can confer six advantages when used as POU treatment systems for treating biological contaminants. First, due to their large microbial and NOM concentration capacities, they are expected to function more efficiently than activated carbon filters. Second, as opposed to GAC adsorbent media, where filter clogging and premature saturation of filters are a major problem, in CNT filters the cytotoxic nature of the material to most commonly found microorganisms would prevent the likely possibility of bacterial growth on the filter surface. Third, since adsorption of microorganisms on CNTs is observed to be nonselective, the media can likely be used universally for removing multiple pathogens, including biothreat agents [90]. Fourth, compared to other systems, such as reverse osmosis, CNT filters can operate at reasonably low pressures [125]. For example, Mostafavi et al. [125] were able to concentrate viruses and bacteria from water systems using CNT filters at pressures of less than 11 bar, indicating that CNT filters may incur low energy-related costs compared to those of membrane-based processes. Fifth, the reusability rate of CNT filters is also expected to be high. Polymeric filters (e.g., cellulose acetate) undergo irreversible surface property changes as a result of bacterial adsorption [116], and regeneration of these filter media is not possible by simple thermal means. Due to their excellent mechanical properties, CNT filters do not undergo such irreversible deformational changes and at the same time can be regenerated and reused by simple thermal treatment methods [116]. Finally, the weight loss of adsorbent after repeated use is comparatively less than the weight loss of media in GAC filters, suggesting that the release of CNTs into the environment may be minimal and that the filter replacement costs may be lower [24].

Thus, CNT filters can serve as potential candidates to be deployed as POU and POE treatment systems in the near future. CNT-based POU systems can be used as a treatment of choice in developing nations, where the lack of financial affordability to invest in costly water treatment infrastructure prevents people from having access to safe and continuous water supplies. On the other hand, CNT POU and POE adsorption filters can also be used in developed nations as a precautionary measure to ease the burden of maintaining potability of treated water in distribution systems in the event of a bioterrorism attack.

Possible Removal of Biological Toxins Though the Adsorption–Biodegradation Mechanism Using CNTs as Adsorbent Media As noted earlier, microcystin derivatives (MCs) pose a serious problem in drinking water treatment plants and can be removed effectively via adsorption as a principal mechanism. Adsorbent media such as GAC are known for their ability to remove MC derivatives from water systems [95,96,126]. However, it is also evident from the studies of Huang et al. [95] that the adsorption capacity of MC derivatives on GAC is decreased if NOM is present as a coadsorbate. However, this disadvantage can be converted to an advantage if the mechanism of biodegradation is combined with adsorption. This means that by growing bacterial biofilms capable of degrading the MCs on a GAC surface, it is possible initially to adsorb the contaminant from water systems and subsequently achieve its biodegradation via the biofilm bacteria [126]. This can be done by utilizing MCs as a secondary substrate by bacteria capable of degrading the toxin, where NOM is the primary substrate utilized by the bacteria [126]. Various authors have demonstrated that the MC derivatives from contaminated water systems can be removed through a combination adsorption–biodegradation mechanism [95,96,126]. It has been determined that although adsorption plays a vital role, biodegradation becomes an efficient mechanism in the removal of MCs [126]. According to a study conducted by Wang et al. [126], adsorption breakthrough of MCs occurred after several days of use of a nonsterile GAC filter, whereas MC breakthrough was not evident for one month in a sterile GAC column. This suggests that adsorption of MCs is reduced by the presence of an active biofilm on the GAC surface, as the microbial matter (both cells and extracellular polymeric substances) hinders the transfer of MCs into the internal adsorption sites of GAC. However, once biodegradation begins, it will be the dominant mechanism in removing MCs. This is evident from studies conducted by Wang et al. [126], who observed a sudden increase in MC removal after a month of operation by GAC filter beds containing active biofilms. Although the formation of mature biofilms on GAC media is important for commencement of the biodegradation process, the efficiency of such a process depends heavily on factors such as temperature, initial bacterial density, type of bacteria population, and the ability of the bacteria to produce degrading enzymes under various environmental conditions [126].

Many scientists proposed that production of biodegrading enzymes by biofilm bacteria is the most critical step in determining the efficiency of the biodegradation process [127]. This depends on the availability of toxin to biofilm-forming microbes, which in turn depends on the ability of the porous material to host such biofilms on their surfaces. This means that a porous medium which is able to provide a large accessible surface area for bacterial immobilization, and which can provide a favorable environment for subsequent colonization and biofilm formation, is a good candidate for use as a substrate to host microbial biofilms applicable in the bioremediation of toxins such as MC derivatives. This same inquisitiveness led to the idea of applying CNT adsorbent media to grow useful biofilms that can bioremediate toxic substances. Although not many studies have been conducted in this area of research, the prospects for such studies utilizing CNTs as a substrate to harvest bioremediating biofilms appear encouraging. Although the intrinsic nature of CNTs to prepared is to be cytotoxic, the extent of the cytotoxicity depends on the type of bacteria and the physical characteristics of the CNTs [109]. Furthermore, the cytotoxic nature of CNTs can be varied and can even be eliminated completely through a surface modification process by adding certain functional groups to the surface of CNTs.

A study conducted by Yan et al. [94] to remove MC derivatives (MC-LR and MC-RR) from solution using *Ralstonia solanacearum* bacterial biofilms grown on CNTs showed that the removal rate of MC derivatives via adsorption–biodegradation mechanisms is five to six times higher in CNTs than in other media. The combined adsorption–biodegradation mechanism increased the MC removal capacity by 20% over removal on CNTs by sorption alone and biodegradation alone by *R. solanacearum* bacteria [94]. CNTs can serve as perfect media for the removal of MCs via adsorption and biodegradation because of their exceptional ability to provide a modulating effect by adsorbing high concentrations of the toxin from the bulk and regulating its availability by releasing it slowly to the immobilized microbes. Sorption capacities of MCs on CNTs are observed to be higher than in other media, including GAC, and because CNT can offer a better surface area of immobilization for microbes, the biodegradability of MCs on CNTs is also higher than that of other media because of enhanced bioavailability (i.e., toxin is made available more readily to a large population of bacteria). Thus, compared to other micropore media, such as GAC and PAC, CNTs are expected to perform well not only in terms of adsorbing MCs but in biodegrading them as well.

5   CHALLENGES ASSOCIATED WITH THE USE OF CNT TECHNOLOGY IN WATER TREATMENT PLANTS

Implementation of CNT adsorption technology for the removal of biological contaminants in drinking water treatment plants is associated with a number of challenges that need to be resolved. The three most pressing problems that may hinder the progress of commercialization of CNT technologies applicable in water treatment are cost, safety and environmental impact, and lack of availability of scale-up data. Until recently, the cost of CNTs was one of the dominant factors that offset all its technical advantages. With the cost of CNTs as high as $75 to 100 per gram, they lose heavily to commercially available products such as powdered and granular activated carbon, which probably cost only a few dollars per kilogram. However, due to the rapid improvements made in CNT production technology in recent years, with batch synthesis techniques being replaced by continuous production techniques, consumption of costly carbon precursors such as acetylene is being replaced by that of cheaper and cleaner sources such as natural gas, and because fixed catalytic metal catalyst beds requiring high energy are being replaced by floating and fluidized catalyst beds, the cost of CNT synthesis is gradually being reduced [128–135]. The average cost of CNTs produced using catalytic chemical vapor deposition (CCVD) is projected to be $25 to 38 per kilogram in the near future, which may be reduced further to $10 per kilogram [132]. Thus, the prospective utilization of CNTs for large-scale environmental applications, particularly in water treatment, can be promising provided that there is a rapid reduction in cost. On the other hand, since CNTs' adsorption capacity for all types of biological contaminants is seen to be higher than that of microporous media, consumption of the material in water treatment plants may not be higher than that of activated carbon.

The second major challenge involved in using CNTs as adsorption filters for the capture of biological contaminants is the need to mitigate the health and ecological risks associated with the use of material on a large scale. Some portion of CNTs may eventually be lost from filter beds due to repeated cycles of water purification, mainly during the backwashing stage. The lost CNT media can possibly enter wastewater treatment plants or be discharged directly into freshwater systems such as rivers and lakes. In either case, they tend to pose serious side effects. For example, CNTs can impose respiration inhibition–related toxicity on the activated sludge microbial consortium of a wastewater treatment plant, suggesting that the presence of CNTs in a wastewater treatment plant may lead to destabilization of activated sludge floc

[136]. In another study by Rodrigues and Elimelech the single-walled CNT-coated substratum has 10 times less biofilm colonization than that of a substratum without them, suggesting that the presence of CNTs can have a toxic impact even on attached growth wastewater treatment processes, such as trickling filters [137]. On the other hand, if CNTs are released directly into freshwater systems, they can cause a variety of problems. For example, the natural hydrophobic state of CNTs can induce toxic effects in benthic organisms such as amphipods [138]. The metal impurities associated with CNTs, which can easily be mobilized and made bioavailable at nearly neutral pH conditions (i.e., between 5.5 and 7.4) are seen to produce toxic effects in fish and daphnids [139]. CNTs that have a more hydrophilic nature tend to induce a hemolytic response in human red blood cells based on the inflammatory response observed in mice [140,141]. Similarly, soluble CNTs induce a genotoxic effect in amphibian life forms such as *Xenopus* larvae [142]. Furthermore, the toxic effects of CNTs can be increased due to the presence of certain impurities, such as NOM, in water systems [143]. However, the relatively good news with respect to the toxicity potential of CNTs is that their bioaccumulation in aquatic species is much lesser than that of compounds such as PAH and pyrene, where it is observed that CNTs do not readily adsorb into organisms' tissues [144]. However, further investigation is needed in this area to reach confirmation.

Finally, critical data (e.g., pilot-plant data representing the breakthrough of various biological contaminants, particularly pathogens) are missing for CNTs, which makes it difficult to make judgments on scale-up of the process to a commercial scale. Data established on adsorption isotherms and adsorption kinetics may be good for characterizing the potential of CNTs to remove biological contaminants, but it is the breakthrough data that could answer many practical questions that are needed. Establishing practical operational data on CNT adsorbent media is not easy because CNTs come in different forms (i.e., single- and multiwalled), in different diameters and lengths, in different chemical compositions and surface chemistries (depending on the functional groups), with different interaction abilities with the environment (i.e., some are hydrophobic, some are hydrophilic) and with varying levels of cytotoxicity. The largely diversified material nature of CNTs is reflected in their performance in removing contaminants from water systems (i.e., some forms of CNTs are better adsorbent media than others). For example, variations in size, length, and dispersivity of tubes can have a marked influence on the adsorption of bacteria by CNTs [112]. The large material variability of CNTs not only influences their performance in terms of removal capacities of biological contaminants, but is also reflected in their antimicrobial nature. Some forms (e.g., short length and dispersible CNTs) are relatively more toxic to cells than are other forms [109], and the toxicity can be altered due to the presence of contaminants such as NOM in water systems [143]. Thus, the establishment of data validating the relationship between the performance of CNTs as adsorbent media and their functionality is of utmost importance before considering the media for commercial applications. If these challenges are mitigated, CNT adsorbent media are going to provide solutions for many tough drinking water treatment problems, such as providing safe and continuous water supplies in developing nations and enhanced water security from bioterrorism attacks in developed nations.

6 CONCLUSIONS

Technical details on the utilization of CNTs as adsorbent media for the removal of chemical and biological contaminants in drinking water, the practical advantages of the implementation of CNT adsorption technology, and the challenges associated with its commercial realization have been the focus of this chapter. These details provide conclusive evidence that CNT-based adsorption filters have the ability to capture multiple pollutants from a contaminated water system. The use of CNT adsorption filters to remove contaminants will be a major breakthrough in water treatment technology because they can possibly achieve at a portable level standards of treatment similar to those seen in centralized water treatment facilities. Thus, CNTs are expected to solve problems in many developing nations where the infrastructure to construct an advanced central water treatment facility is either lacking or unaffordable. At the same time, CNT sorption filters can also be used as POU and POE treatment devices in developed nations as an additional protective measure to enhance the safety and quality of treated drinking water, especially when the risk of bioterrorism attacks on drinking water distribution systems are alarming high. A major issue of concern prohibiting the entry of CNT adsorbent in the water industry is its toxic nature, which must be addressed carefully. Interaction with CNTs in aqueous media is toxic to human beings and aquatic species. The presence of CNTs at certain concentrations is shown to affect operational efficiencies of the biological activated sludge process because the material is toxic to the microbial consortium residing in the wastewater treatment plant. It is therefore important to address these issues by performing rigorous research and to establish release (i.e., weight loss from adsorption filters), fate, and transport information on CNTs. By having these issues solved in the near future, it can be expected that CNT adsorption technology is going to play a dominant role in changing the course of the water treatment industry in coming years.

REFERENCES

1. Adrianes, P., et al., Intelligent infrastructure for sustainable potable water: a roundtable for emerging transnational research and technology development needs, *Biotechnol. Adv.*, 2003;**22**(1–2):119–134.

2. Petrovic, M., S. Gonzalez, and D. Barcelo, Analysis and removal of emerging contaminants in wastewater and drinking water, *Trends Anal. Chem.*, 2003;**22**(10):245–255.

3. Nuzzo, J. B., The biological threat to the U.S. water supplies: toward a national water security policy, *Biosecurity Bioterrorism Biodefense Strategy Pract. Sci.*, 2006;**4**(2):147–159.

4. Busca, G., et al., Technologies for the removal of phenol from fluid streams: a short review of recent developments, *J. Hazard. Mater.*, 2008;**160**(2–3):265–288.

5. Haritash, A. K., and C. P. Kaushik, Biodegradation aspects of polycyclic aromatic hydrocarbons (PAHs): a review, *J. Hazard. Mater.*, 2009;**169**(1–3):1–15.

6. Lopez, M. C. C., Determination of potentially bioaccumulating complex mixtures or organochlorine compounds in wastewater: a review, *Environ. Int.*, 2003;**28**(8):751–759.

7. Pokhrel, D., and T. Virarahavan, Treatment of pulp and paper mill wastewater: a review, *Sci. Total Environ.*, 2004;**333**(1–3):37–58.

8. Snyder, A. A., et al., Role of membranes and activated carbon in the removal of endocrine disruptors and pharmaceuticals, *Desalination*, 2007;**202**(1–3):156–181.

9. Bolong, N., et al., A review of the effects of emerging contaminants in wastewater and options for their removal, *Desalination*, 2009;**239**(1–3):229–246.

10. Bailey, S. E., et al., A review of potentially low cost sorbents for heavy metals, *Water Res.*, 1999;**33**(11):2469–2479.

11. Kurniawan, T. A., et al., Physico-chemical treatment techniques for wastewater laden with heavy metals, *Chemi. Eng. J.*, 2006;**118**(1–2):83–98.

12. Ngah, W. S. W., and M. A. K. M. Hanafiah, Removal of heavy metal ions from wastewater by chemically modified plant wastes as adsorbents: a review, *Bioresource Technol.*, 2008;**99**:33935–3948.

13. Kurniawan, T. A., et al., Comparisons of low cost adsorbents for treating wastewaters laden with heavy metals, *Sci. Total Environ.*, 2006;**366**(2–3):409–426.

14. Kakkar, P., and F. N. Jaffery, Biological markers for metal toxicity, *Environ. Toxicol. Pharmacol.*, 2005;**19**(2):335–349.

15. Momba, M. N. B., et al., An overview of biofilm formation and its impact on the deterioration of water quality, *Water SA*, 2000;**26**(1):59–66.

16. NHMRC and NRMMC, *Australian Drinking Water Quality Guidelines*, National Health and Medical Research Council, Canberra, Australia, 2004.

17. Cheng, W., S. A. Dastgheib, and T. Karanfil, Adsorption of dissolved organic natural matter by modified activated carbons, *Water Res.*, 2005;**39**:2281–2290.

18. Dastgheib, S. A., T. Karanfil, and W. Cheng, Tailoring activated carbons for enhanced removal of natural organic matter from natural waters, *Carbon*, 2004;**42**:547–557.

19. Hyung, H., and J. H. Kim, Natural organic matter (NOM) adsorption to multiwalled carbon nanotubes: effect on NOM characteristics and water quality parameters, *Environ. Sci. Technol.*, 2008;**42**(12):4416–4421.

20. Li, F., et al., Factors affecting the adsorption capacity of dissolved organic matter onto activated carbon: modified isotherm analysis, *Water Res.*, 2002;**36**:4592–4604.

21. Lu, C., and F. Su, Adsorption of natural organic matter by carbon nanotubes, *Sep. Purif. Technol.*, 2007;**58**:113–121.

22. Matilainen, A., N. Vieno, and T. Tuhkanen, Efficiency of activated carbon filtration in the natural organic matter removal, *Environ. Int.*, 2006;**32**:324–331.

23. Schreiber, B., et al., Adsorption of dissolved organic matter onto activated carbon: the influence of temperature, adsorption, wavelength, and molecular size, *Water Res.*, 2005;**39**:3449–3456.

24. Su, F., and C. Lu, Adsorption kinetics, thermodynamics, and desorption of natural dissolved organic matter by multiwalled carbon nanotubes, *J. Environ. Sci. Health A*, 2007;**42**:1543–1552.

25. Hoehn, R. C., and B. W. Long, Toxic cyanobacteria (bule green algae): an emerging concern, *Nat. Water Toxins,* 1999.

26. Keijola, A. M., et al., Removal of cyanobacterial toxins in water treatment processes: laboratory and pilot scale experiments, *Toxicol. Assess. Int. J.,* 1988;**3**:643–656.

27. Lechevallier, M. W., and K. Au Keung, *Water treatment and pathogen control, in Imapact of Treatment on Microbial Quality: A Review Document on Treatment Efficiency to Remove Pathogens,* International Water Association, London, 2004.

28. Zhang, W., and F. A. DiGiano, Comparison of bacterial regrowth in distribution systems using free chlorine and chloramine: a statistical study of causative factors, *Water Res.,* 2002;**36**(6):1469–1482.

29. Kerneis, A., et al., The effects of residence time on the biological quality in a distribution network, *Water Res.,* 1995;**29**(7):1719–1727.

30. USEPA, *Guidance Manual for Compliance with Interim Enhanced Surface Water Treatment Rule: Turbidity Provisions,* U.S. Environmental Protection Agency; Washington, DC, Chapter 7, 1999.

31. GAO, *Effect of Water Age on Distribution Systems Water Quality.* Office of Groundwater and Drinking Water, Washington, DC, 2002.

32. Berry, D., C. Xi, and L. Raskin, Microbial ecology of drinking water distribution systems, *Curr. Opin. Biotechnol.,* 2006;**17**:297–302.

33. Bretter, I., and M. G. Hofle, Molecular assessment of bacterial pathogens: a contribution to drinking water safety, *Curr. Opin. Biotechnol.,* 2008;**19**:274–280.

34. Codony, F., et al., Effect of chlorine, biodegradable dissolved organic carbon and suspended bacteria on biofilm development in drinking water systems, *J. Basic Microbiol.,* 2002;**42**(5):311–319.

35. Gagnon, G. A., et al., Disinfectant efficacy of chlorite and chlorine dioxide in drinking water biofilms, *Water Res.,* 2005;**39**:1809–1817.

36. Hallam, N. B., et al., The potential for biofilm growth in water distribution systems, *Water Res.,* 2001;**35**(17):4063–4071.

37. Szabo, J. G., Persistence of microbiological agents on corroding biofilm in a model drinking water system following intentional contamination, *in Civil and Environmental Engineering, University of Cincinatti, Cincinatti,* OH, 2006.

38. Pozos, N., et al., UV disinfection in a model distribution system: biofilm growth and microbial community, *Water Res.,* 2004;**38**:3083–3091.

39. Martiny, A. C., et al., Long term succession of structure and diversity of a biofilm formed in model drinking water distribution systems, *Appl. Environ. Microbiol.,* 2003;**69**(11):6899–6907.

40. Lechevallier, M. W., et al., Examining the relationship between iron corrosion and the disinfection of biofilm bacteria, *J. Am. Water Works Assoc.,* 1993;**85**(7):111–123.

41. Volk, C., et al., Practical evaluation of iron corrosion control in a drinking water distribution system, *Water Res.,* 2000;**34**(6):1967–1974.

42. Horman, A., Assessment of microbial safety of drinking water produced from surface water under field conditions, *in Food and Environmental Hygiene,* University of Helsinki, Helsinki, Finland, 2005.

43. Burrows, W. D., and S. E. Renner, Biological warfare agents as threats to potable water, *Environ. Health Perspect.,* 1999;**107**(12):975–984.

44. Hickman, D. C., *A Chemical and Biological Warfare Threat: USAF Water Systems at Risk*, Counterproliferation Paper 3, 1999, Air War College, Montgomery, AL, 1999.

45. Javey, A., M. Shim, and H. Dai, Electrical properties and devices of large diameter single-walled carbon nanotubes, *Appl. Phys. Lett.*, 2002;**80**(6):1064–1066.

46. Shaffer, M. S. P., and J. K. W. Sandler, Carbon nanotubes/nanofibre composites, in *Processing and Properties of Nanocomposites*, S. G. Advani, (Ed.), World Scientific, Hackensack, NJ, 2006.

47. Huang, S., X. Cai, and J. Liu, Growth of millimeter long horizontally aligned single walled carbon nanotubes on flat substrates, *J. Am. Chem. Soc.*, 2003;**125**:5636–5637.

48. Zheng, L. X., et al., Ultra-long single wall carbon nanotubes, *Nat. Mater.*, 2004;**3**:673–676.

49. Wang, X., et al., Fabrication of ultralong and electrically uniform single-walled carbon nanotubes on clean substrates, *Nano Lett.*, 2009;**9**(9):3137–3141.

50. Agnihotri, S., et al., Structural characterization of single walled carbon nanotube bundles by experiment and molecular simulation, *Langmuir*, 2005;**21**(3):896–904.

51. Thess, A., et al., Crystalline ropes of metallic carbon nanotubes, *Science*, 1996;**273**:483–487.

52. Zhao, J., et al., Gas molecule adsorption in carbon nanotubes and nanotubes bundles, *Nanotechnology*, 2002;**13**(2):195–200.

53. Agnihotri, S., M. Rostam-Abadi, and M. J. Rood, Temporal changes in nitrogen adsorption properties of single walled carbon nanotubes, *Carbon*, 2004;**42**(12–13):2699–2710.

54. Rao, C. N. R., et al., Fullerenes, nanotubes, onions and related carbon structures, *Mater. Sci. Eng. Rep.*, 1995;**15**(6):209–262.

55. Dai, H., Carbon nanotubes: opportunities and challenges, *Surf. Sci.*, 2002;**500**(1–3):218–241.

56. Paradize, M., and T. Goswami, Carbon nanotubes: production and industrial applications, *Mater. Des.*, 2007;**28**(5):1477–1489.

57. Talapatra, S., et al., Gases do not adsorb on the interstitial channels of close-ended single walled carbon nanotube bundles, *Physi. Revi. Lett.*, 2000;**85**(1):138–141.

58. Yan, X. M., et al., Adsorption and desorption of atrazine on carbon nanotubes, *J. Colloid Interface Sci.*, 2008;**321**:30–38.

59. Li, Y. H., et al., Different morphologies of carbon nanotubes effect on the lead removal from aqueous solution, *Diamond Relat. Mater.*, 2006;**15**:90–94.

60. Deng, S., et al., Adsorption equilibrium and kinetics of microorganisms on single walled carbon nanotubes, *IEEE Sensors*, 2008;**8**(6):954–962.

61. Benny, T. H., T. J. Bandosz, and S. S. Wong, Effect of ozonolysis on the pore structure, surface chemistry, and bundling of single walled carbon nanotubes, *J. Colloid Interface Sci.*, 2008;**317**:375–382.

62. Chen, Y., et al., Pore structures of multiwalled carbon nanotubes activated by air, CO_2 and KOH, *J Porous Mater.*, 2006;**13**:141–146.

63. Upadhyayula, V. K. K., et al., Application of carbon nanotube technology for removal of contaminants in drinking water: a review, *Sci. Total Environ.*, 2009;**408**:1–13.

64. Bandosz, T. J., J. Jagiello, and A. Schwarz, Effect of surface chemical groups on energetic heterogenity of activated carbons, *Langmuir*, 1993;**9**:2518–2522.

65. Lu, C., and C. Liu, Removal of nickel(II) from aqueous solution by carbon nanotubes, *J. Chemi. Technol. Biotechnol.*, 2006;**81**:1932–1940.

66. Lu, C., H. Chiu, and C. Liu, Removal of zinc(II) from aqueous solution by purified carbon nanotubes: kinetics and equilibrium studies, *Ind. Eng. Chem. Res.*, 2006;**45**:2850–2855.

67. Gauden, P. A., et al., Thermodynamic properties of benzene adsorbed in activated carbons through silica templates and their applications to the adsorption of bulky dyes, *Chemi. Mater.*, 2006;**12**:3337–3341.

68. Peng, X., et al., Adsorption of 1,2-dichlorobenzene from water to carbon nanotubes, *Chem. Phys. Lett.*, 2003;**379**(1–2):154–158.

69. Lu, C., Y. Chung, and K. Chang, Adsorption of trihalomethanes from water with carbon nanotubes, *Water Res.*, 2005;**39**(6):1183–1189.

70. Yang, K., L. Zhu, and B. Xing, Adsorption of polycyclic aromatic hydrocarbons by carbon nanomaterials, *Environ. Sci. Technol.*, 2006;**40**:1855–1861.

71. Li, Y., et al., Adsorption of cadmium(II) from aqueous solution by surface oxidized carbon nanotubes, *Carbon*, 2003;**41**(5):1057–1062.

72. Lu, C., and H. Chiu, Adsorption of zinc(II) from water with purified carbon nanotubes, *Chem. Eng. Sci.*, 2006;**61**:1138–1145.

73. Li, Y., et al., Lead adsorption on carbon nanotubes, *Chem. Physi. Lett.*, 2002;**357**(3–4):263–266.

74. Li, Y., et al., Competitive adsorption of Pb^{2+}, Cu^{2+}, and Cd^{2+} on carbon nanotubes, *Water Res.*, 2003;**39**(4):605–609.

75. Li, Y., et al., Adsorption thermodynamic, kinetic and desorption studies of Pb^{2+} on carbon nanotubes, *Water Res.*, 2005;**39**(4):605–609.

76. Chen, C., and X. Wang, Adsorption of Ni(II) from aqueous solution using oxidized multi wall carbon nanotubes, *Ind. Eng. Chem. Res.*, 2006;**45**(26):9144–9149.

77. Rao, G. P., C. Lu, and F. Su, Sorption of divalent metal ions from aqueous solution by carbon nanotubes: a review, *Sep. Purif. Technol.*, 2007;**58**(1):224–231.

78. Shen, X. E., et al., Kinetics and thermodynamics of sorption of nitroaromatic compounds to as grown and oxidized multiwalled carbon nanotubes, *J. Colloids Interface Sci.*, 2009;**330**(1):1–8.

79. Gao, Z., et al., Investigation of factors affecting adsorption of transition metals on oxidized carbon nanotubes, *J. Hazard. Mater.*, 2009;**167**(1–3):357–365.

80. Hu, J., et al., Removal of chromium from aqueous solution by using oxidized multiwalled carbon nanotubes, *J. Hazard. Mater.*, 2009;**162**(2–3):1542–1550.

81. Moreno-Castilla, C., Adsorption of organic molecules from aqueous solutions on carbon materials, *Carbon*, 2004;**42**(1):83–94.

82. Jankowska, H., et al., Adsorption from benzene–ethanol binary solutions on activated carbons with different contents of oxygen surface complexes, *Carbon*, 1983;**21**(2):117–120.

83. Pan, B., and B. S. Xing, Adsorption mechanisms or organic chemicals on carbon nanotubes, *Environ. Sci. Technol.*, 2008;**42**(24):9005–9013.

84. Cho, H. H., et al., Influence of surface oxides on the adsorption of naphthalene onto multi walled-carbon nanotubes, *Environ. Sci. Technol.*, 2008;**42**(8):2899–2905.

85. Zhu, S., et al., Single walled carbon nanohorn as new solid-phase extraction adsorbent for determination of 4-nitrophenol in water sample, *Talanta*, 2009;**79**(5):1441–1445.

86. Liao, Q., J. Sun, and L. Gao, The adsorption of resorcinol from water using multiwalled carbon nanotubes, *Colloids Surf. A*, 2008;**312**(2–3):160–165.

87. Pillay, K., E. M. Cukrowska, and N. J. Coville, Multi walled carbon nanotubes as adsorbents for removal of parts per billion levels of hexavalent chromium from aqueous solution, *J. Hazard. Mater.*, 2009;**166**(2–3):1067–1075.

88. Upadhyayula, V. K. K., et al., Adsorption of *Bacillus subtilis* on single walled carbon nanotube aggregates, activated carbon and nanoceram™. *Water Res.*, 2009 in press.

89. Eleria, A., and R. M. Vogel, Predicting fecal coliform bacteria levels in the Charles River, Massachusets, USA, *J. Am. Water Resources Assoc.*, 2005;**03111**:1195–1209.

90. Moon, H. M., and J. W. Kim, Carbon nanotube clusters as universal bacterial adsorbents and magnetic separation agents, *Biotechnol. Prog.*, 2009;**26**(1):179–185.

91. Newcombe, G., Charge vs. porosity: some influences on the adsorption of natural organic matter (NOM) by activated carbon, *Water Sci. Technol.*, 1999;**40**(9):191–198.

92. Fairey, J. L., G. E. Speitel, Jr., and L. E. Katz, Impact of natural organic matter on monochloramine reduction by granular activated carbon: the role of porosity and electrostatic surface properties, *Environ. Sci. Technol.*, 2006;**40**:4268–4273.

93. Yan, H., et al., Adsorption of microcystins by carbon nanotubes, *Chemosphere*, 2006;**62**:142–148.

94. Yan, H., et al., Effective removal of microcystins using carbon nanotubes embedded with bacteria, *Chin. Sci. Bull.*, 2004;**49**:1694–1698.

95. Huang, W. J., B. L. Cheng, and Y. L. Cheng, Adsorption of microcystin-LR by three types of activated carbon, *J. Hazard. Mater.*, 2007;**141**:115–122.

96. Pendleton, P., R. Schumann, and S. H. Wong, Microcystin-LR adsorption by activated carbon, *J. Colloid Interface Sci.*, 2001;**240**:1–8.

97. Stevik, T. K., et al., Retention and removal of pathogenic bacteria in wastewater percolating through porous media: a review, *Water Res.*, 2004;**38**:1355–1367.

98. Pimenov, A. V., et al., The adsorption and deactivation of microorganisms by activated carbon fiber, *Sep. Sci. Technol.*, 2001;**36**(15):3385–3394.

99. Pereira, M. A., et al., Influence of physico-chemical properties of porous microcarriers on the adhesion of an anaerobic consortium, *J. Ind. Microbiol. Biotechnol.*, 2000;**24**:181–186.

100. Chenu, C., and G. Stotzky, Fundamentals of soil particle microorganisms interactions, in *Interactions Between Soil Particles and Microorganisms*, P. M. Huang, J. M. Bollag, and N. Sensi (Eds.), Vol. 8, Wiley, Chichester, UK, 2002.

101. Samonin, V. V., and E. E. Elikova, A study of the adsorption of bacterial cells on porous materials, *Microbiology*, 2001;**73**(6):696–701.

102. Scholl, M. A., and R. W. Harvey, Laboratory investigations on the role of sediment surface and groundwater chemistry in transport of bacteria through a contaminated sandy aquifier, *Environ. Sci. Technol.*, 1992;**26**(7):1410–1417.

103. Yee, N., J. B. Fein, and C. J. Daughney, Experimental study of pH, ionic strength and reversibility behavior of bacteria–mineral deposition, *Geochim. Cosmochim. Acta*, 2000;**64**(4):609–617.

104. Gerba, C. P., et al., Adsorption of polio virus onto activated carbon in wastewater, *Environ. Sci. Technol.*, 1975;**9**(8):727–731.

105. Abu Lail, L. I., An atomic force microscopy study of bacterial adhesion to natural organic matter–coated surfaces in the environment, in *Civil and Environmental Engineering*, Worchester Polytechnic Institute, Worchester, MA, 2006.

106. Camper, A. K., et al., Growth and persistence of pathogens on granular activated carbon filters, *Appl. Environ. Microbiol.*, 1985;**50**(6):1378–1382.

107. Momani, F. L., D. W. Smith, and M. G. El-din, Degradation of cyanobacteria toxin by advanced oxidation process, *J. Hazard. Mater.*, 2008;**150**:238–249.

108. Kang, S., et al., Antibacterial effects of carbon nanotubes: size does matter, *Langmuir*, 2008;**24**(13):6409–6413.

109. Kang, S., S. M. Mauter, and M. Elimelech, Physiochemical determinants of multiwalled carbon nanotube bacterial cytotoxicity, *Environ. Sci. Technol.*, 2008;**42**(19):7528–7534.

110. Kang, S., et al., Single walled carbon nanotubes exhibit strong antimicrobial activity, *Langmuir*, 2007;**23**(17):8670–8673.

111. Nepal, D., et al., String antimicrobial coatings: single walled carbon nanotubes armored with biopolymers, *Nano Lett.*, 2008;**8**(7):1896–1902.

112. Akasaka, T., and F. Watari, Capture of bacteria by flexible carbon nanotubes, *Acta Biomater.*, 2009;**5**:607–612.

113. Arias, L. R., and L. Yang, Inactivation of bacterial pathogens by carbon nanotubes in suspensions, *Langmuir*, 2009;**25**(5):3003–3012.

114. Krishna, V., et al., Photocatalytic disinfection with titanium dioxide coated multiwalled carbon nanotubes, *Trans. IChemE B*, 2005;**83**(B4):393–397.

115. Zhu, Y., et al., Dependence of cytotoxicity of multiwalled carbon nanotubes on the culture medium *Nanotechnology*, 2006;**17**:4668–4674.

116. Brady-Estevez, A. S., S. Kang, and M. Elimelech, A single walled carbon nanotube filter for removal of viral and bacterial pathogens, *Small*, 2008;**4**(4):481–484.

117. Matsui, Y., et al., Effect of natural organic matter on powdered activated carbon adsorption of trace contaminants: characteristics and mechanism of competitive adsorption, *Water Res.*, 2003;**37**:4413–4424.

118. Li, Q., et al., Pore blockage effect of NOM on atrazine adsorption kinetics of PAC: the roles of PAC pore size distribution and NOM molecular weight, *Water Res.*, 2003;**37**:4863–4872.

119. Simpson, D. R., Biofilm processes in biological active carbon water purification, *Water Res.*, 2008;**42**:2839–2848.

120. Souter, P. F., et al., Evaluation of a new water treatment for point of use household applications to remove microorganisms and arsenic from drinking water, *J. Water Health*, 2003;**01.2**:73–84.

121. Sagara, J., Study of filtration for point of use drinking water treatment in Nepal, in *Civil Engineering and Applied Mechanics*, McGill University; Montreal, Quebec, Canada, 1999.

122. Doocy, S., and G. Burnham, Point of use water treatment and diarrhoea reduction in the emergency context: an effectiveness trial in Liberia, *Trop. Med. Int. Health*, 2006;**2**(10):1542–1552.

123. Silverstein, I., *Investigation of the Capability of Point of Use/Point of Entry Treatment Devices as a Means of Providing Water Security*. National Homeland Security Research Center, USEPA; Cincinnati, OH, 2006.

124. Snyder, J. W., et al., Effect of point of use activated carbon filters on the bacteriological quality of rural groundwater supplies, *Appl. Environ. Microbiol.*, 1995;**61**(12):4291–4295.

125. Mostafavi, S. T., M. R. Mehrnia, and A. M. Rashidi, Preparation of nanofilter from carbon nanotubes for application in virus removal from water, *Desalination*, 2009;**238**(1–3):271–280.

126. Wang, H., et al., Discriminating and assessing adsorption and biodegradation removal mechanisms during granular activated carbon filtration of microcystin toxins, *Water Res.*, 2007;**41**:4262–4270.

127. Singh, R., D. Paul, and R. K. Jain, Biofilms: implications in bioremediation, *Trends Microbiol.*, 2006;**14**(6):389–397.

128. Agboola, A. E., et al., Conceptual design of carbon nanotube processes, *Clean Technol. Environ. Policy*, 2007;**9**:289–311.

129. Andrews, R. C., et al., Continuous production of aligned carbon nanotubes: a step closer to commercialization, *Chem. Phys. Lett.*, 1999;**303**:467–474.

130. Cheng, H. M., et al., Large-scale and low-cost synthesis of single-walled carbon nanotubes by the catalytic pyrolysis of hydrocarbons, *Appl. Phys. Lett.*, 1998;**72**(25):3282–3284.

131. Colomer, J. F., et al., Large-scale synthesis of single-wall carbon nanotubes by catalytic chemical vapor deposition (CCVD) method, *Chem. Phys. Lett.*, 2000;**317**:83–89.

132. Dalton, A. B., et al., Continuous carbon nanotube composite fibers: properties, potential applications, and problems, *J. Mater. Chem.*, 2003;**14**:1–3.

133. Endo, M., T. Hayashi, and Y. Ahm Kim, Large-scale production of carbon nanotubes and their applications, *Pure Appl. Chem.*, 2006;**78**(9):1703–1713.

134. Qiang, Z., et al., Large scale production of carbon nanotube arrays on the sphere surface from liquified petroleum gas at low cost. *Chin. Sci. Bull.*, 2007;**52**(21):2896–2902.

135. Wang, Y., et al., The large-scale production of carbon nanotubes in a nano-agglomerate fluidized-bed reactor, *Chem. Phys. Lett.*, 2002;**364**:568–572.

136. Luongo, L. A., and X. Zhang, Toxicity of carbon nanotubes to the activated sludge process, *J. Hazard. Mater.*, 2010;**178**:356–362.

137. Rodrigues, D. F., and M. Elimelech, Toxic effects of single walled carbon nanotubes in the development of *E. coli* biofilm, *Environ. Sci. Technol.*, 2010. (in press).

138. Kennedy, A. J., et al., Factors influencing the partitioning and toxicity of nanotubes in the aquatic environment, *Environ. Toxicol. Chem.*, 2008;**27**(9):1932–1941.

139. Hull, M. S., et al., Release of metal impurities from carbon nanomaterials influences aquatic toxicity, *Environ. Sci. Technol.*, 2009;**43**:4169–4174.

140. Lin, Y., et al., Characterization of functionalized single-walled carbon nanotubes at individual nanotube-thin bundle level, *J. Phys. Chem. B*, 2003;**107**(38):10453–10457.

141. Saxena, R. K., et al., Enhanced in vitro and in vivo toxicity of poly-dispersed acid-functionalized single walled carbon nanotubes, *Nanotoxicology*, 2007;**1**(4):291–300.

142. Mouchet, F., et al., Characterization and in vivo ecotoxicity evaluation of double wall carbon nanotubes in larvae of the amphibian Xenopus laevis, *Aquat. Toxicol.*, 2008;**87**:127–137.

143. Kang, S., S. M. Mauter, and M. Elimelech, Microbial cytotoxicity of carbon based nanomaterials: implications for river water and wastewater effluent, *Environ. Sci. Technol.*, 2009;**43**(7):2648–2653.

144. Petersen, E. J., Q. Huang, and W. J. Weber Jr, Ecological uptake and depuration of carbon nanotubes by Lumbriculus variegatus, *Environ. Health Perspect.*, 2008;**116**(4):496–500.

145. Assavasilavasukul, P., et al., Effect of pathogen concentrations on removal of *Cryptosporidium* and *Giardia* by conventional drinking water treatment, *Water Res.*, 2008;**42**:2678–2690.

146. Kfir, R., et al., Studies evaluating the applicability of utilizing the same concentration techniques for the detection of protozoan parasites and viruses in water, *Water Sci. Technol.*, 1995;**31**(5–6):417–423.

147. Warhurst, A. M., et al., Adsorption of cyanobacterial hepatoxin microcystin-LR by a low cost activated carbon from seed husks of the pan tropical tree *Moringa oleifera*, *Sci.Total Environ.*, 1997;**207**:207–211.

148. Montgomery, J. M., *Water Treatment Principles and Design*. Wiley, New York; 1985.

149. Liu, G., An investigation of UV disinfection performance under the influence of turbidity and particulates for drinking water applications, *in Civil Engineering*, University of Waterloo; Waterloo, Ontario, Canada, 2005.

150. Brazos, B. J. and J. T. O'Conner, Relative contributions of regrowth and aftergrowth to the number of bacteria in a drinking water distribution system, presented at the Water Quality Technology Conference, AWWA, Baltimore, MD, 1987.

151. Dugan, N. R. and D. J. Williams, Cyanobacteria passage through drinking water filters during pertubation episodes as a function of cell morphology, coagulant and initial filter loading rate, *Harmful Algae*, 2006;**5**:26–35.

152. Charnock, C., and O. Kjonno, Assimilable organic carbon and biodegradable dissolved organic carbon in Norwegian raw and drinking waters, *Water Res.*, 2000;**34**(10):2629–2642.

153. Clark, R. M., et al., Development of C_t equation for the inactivation of Cryptosporidium oocysts with chlorine dioxide, *Water Res.*, 2003;**37**:2773–2783.

154. Escobar, I. C., A. A. Randall, and J. S. Taylor, Bacterial growth in distribution systems: effect of assimilable organic carbon and biodegradable dissolved organic carbon, *Environ. Sci. Technol.*, 2001;**35**:3442–3447.

155. Frias, J., F. Ribas, and F. Lucena, Effects of different nutrients on bacterial growth in pilot distribution system, *Antonie van Leeuwenhoek*, 2001;**80**:129–138.

156. Lehtola, M. J., et al., Formation of biofilms in drinking water distribution networks: a case study in two cities in Finland and Lativa, *J. Ind. Microbiol. Technol.*, 2004;**31**:489–494.

157. Puil, M. L., *Biostability in drinking water distribution systems study at pilot scale,* in *Civil and Environmental Engineering*, University of Central Florida, Orlando, FL, 2004.

158. Rose, L. J., et al., Monochloramine inactivation of bacterial select agents, *Appl. Environ. Microbiol.*, 2007;**73**(10):3437–3439.

159. Rose, L. J., et al., Chlorine inactivation of bacterial bioterrorism agents, *Appl. Environ. Microbiol.*, 2005;**71**(1):566–568.

160. Lechevallier, M. W., Conditions favouring coliform growth and HPC bacteria growth in drinking water and on water contact surfaces, in *Heterotrophic Plate Counts and Drinking Water Safety*, J. Bartram, et al. (Eds.), International Water Association; London, 2003.

161. Liao, Q., J. Sun, and L. Gao, Adsorption of chlorophenols by multiwalled carbon nanotubes treated with HNO_3 and NH_3, *Carbon*, 2008;**46**:544–561.

162. Lee, S.M., et al., Pore characterization of multiwalled carbon nanotubes modified by KOH, *Chem. Phys. Lett.*, 2005;**416**:251–255.

163. Liu, Y., Z. Shen, and K. Yokogawa, Investigation of preparation and structures of activated carbon nanotubes, *Mater. Res. Bull.*, 2006;**41**:1503–1512.

164. Niu, J. J., et al., An approach to carbon nanotubes with high surface area and large pore volume, *Microporous Mesoporous Mater.*, 2007;**100**: 1–5.

165. Chen, C., et al., Adsorption kinetic, thermodynamic and desorption studies of Th(IV) on oxidized multiwall carbon nanotubes, *Colloids Surf. A,* 2007;**302**(1–3):449–454.

12

PEPTIDE NANOTUBES IN BIOMEDICAL AND ENVIRONMENTAL APPLICATIONS

BYUNG-WOOK PARK AND DONG-SHIK KIM

Department of Chemical and Environmental Engineering, University of Toledo, Toledo, Ohio

1 BACKGROUND

Self-assembly plays a major role in building all biological systems. Even from a very simple bacterium of only a few hundred types of proteins to complex mammals, their

Nanoscale Multifunctional Materials: Science and Applications, First Edition.
Edited by Sharmila M. Mukhopadhyay.

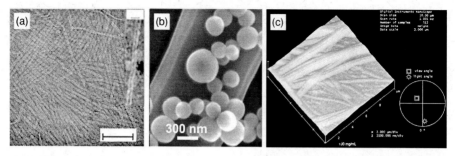

FIGURE 1 (a) TEM image of a fibrillar peptide–dendron hybrid (PDH) (scale bar: 100 nm; inset, high-magnification TEM image of a single nanotube scale bar: 10 nm) (from [7]); (b) SEM image of spherical polypeptides of lyophilized peptide (D-Phe-D-Phe) (from [8]); (c) AFM image of tubular peptide nanotubes (from unpublished data of the authors).

various nanoscale machines, neural networks, and cellular structures are built from assemblies of building blocks through the self-assembly mechanism. Peptide-based assemblies have attracted considerable attention, as they are biocompatible, easy to produce in large amounts, and are composed of diverse structural and functional properties.

Since the pioneering work of Ghadiri et al. [1] in the early 1990s, peptide nanotubes (PNTs) have been recognized as important structural elements in nanotechnology. Just as in carbon nanotubes (CNTs), the tubular structure of PNTs offers a variety of applications in chemistry, biochemistry, materials science, medicine, and in biomedical and environmental areas. The most appealing characteristic of PNTs is their biocompatibility. As short polymers of amino acids linked by peptide bonds, peptides resemble proteins, and the self-assembled form of peptides such as PNTs are compatible with many biosystems and biomaterials. Unlike typical protein, however, peptides have remarkable thermal and chemical stability [2], which possibly expands their applications to microelectronic and microelectromechanics (MEMS) processes and fabrication.

Under certain conditions, peptides form a precise spherical (i.e., vesicle) [3], fibrillar [4,5], or tubular structure [6] through various self-assembly mechanisms. Figure 1 shows examples of these structures. Here we discuss only the tubular structure of peptide assemblies in terms of its basic chemistry and applications. Applications are divided into two areas: biomedical and environmental. Existing and suggested applications in these areas are summarized and possible future applications are proposed.

First, some basic information on PNTs is reviewed briefly. Understanding PNTs' unique characteristics helps to provide better insight into apparently obvious applications for them and to connect their unique chemical properties with corresponding novel applications. It will also enable the application to be expanded to wider fields.

2 SYNTHESIS AND CHARACTERISTICS OF PEPTIDE NANOTUBES

Unlike human-made structures, well-ordered three-dimensional structures of PNTs can be formed spontaneously into complex structures upon the association of simple building blocks. This type of bottom-up approach has been chosen as a natural means to build organic systems for many organisms, while the top-down approach is the conventional method used in microtechniques. Peptide nanotubes are synthesized through molecular self-assembly, which is mediated by weak noncovalent bonding such as that in hydrogen bonds, ionic bonds, hydrophobic interaction, and van der Waals interaction. As weak as they may be in isolation, when combined they provide the structural stability of self-assembled nanomaterials. Structural conformation and interaction with other molecules are influenced by these bonds as well.

All biomolecules, including peptides and proteins, interact with each other and self-organize to form well-defined structures that render them specific functionalities. The structure of PNTs is determined by the existing molecules and synthesis conditions. To use them properly, the structure–property relationship of PNTs should be well understood. Three different types of PNTs are reported based on the types of peptide building blocks: dipeptide, bolaamphiphiles, and surfactant-like. Dipeptides are the simplest peptide building blocks, derived from the diphenylalanine motif of the Alzheimer's β-amyloid peptide. As its name implies, diphenylalanine consists of two alanine ($\overset{H_3C}{\underset{H_2N}{>}}C-\overset{O}{\overset{\parallel}{C}}-OH$) molecules and phenyl (⬡-) groups. The structure of dipeptide is unique, and a model was suggested by Görbiz [9]. The formation of β-amyloid fibrils was first found in association with a large number of major diseases, including Alzheimer's, type 2 diabetes, prion diseases, and Parkinson's disease [10–13]. The deposition of amyloid fibrils in various tissues and organs is characteristic of the diseases. Although it is well established that the accumulation of insoluble protein deposits in the form of amyloids in tissues and organs is associated with these diseases, it has been less than a decade since prefibrillar assembles were first seen to possibly represent the cytotoxic elements that are responsible for cell death in infected tissues. The self assembly mechanism not only provides a bottom-up approach to the synthesis of novel materials, but also helps us better understand the onset of these diseases.

Diphenylalanine [Figure 2(a)] can be relatively easily assembled into a dipeptide PNT [Figure 2(b)] [14]. Two different methods are used to build the self-assembled PNT. First, lyophilized diphenylalanine is dissolved in 1,1,1,3,3,3-hexafluoro-2-propanol at a concentration of 100 mg/mL. Then, 20 μL of 100-mg/mL PNT solution is dropped into 980 μL of deionized water. Second, lyophilized dipehnylalanine is dissolved in deionized water and sonicated for 30 min, following which the solution is stirred with a magnetic bar at 70 °C, 800 rpm until the vial turns completely transparent. The vial is cooled slowly at room temperature, and nanostructures are formed in the vial.

Surfactant-like peptides such as glycine and aspartic acid have been investigated for self-assembled peptides [3,15]. Amphiphilic molecules are well known to form micelles and bilayers through the interaction between hydrophilic polar heads and

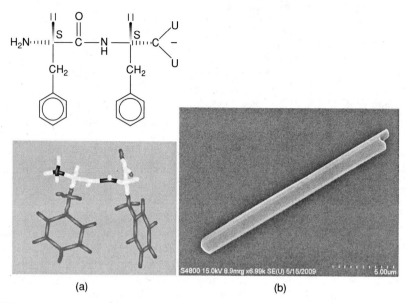

FIGURE 2   (a) Molecular structure of diphenylalanine (L-Phe-L-Phe) (from [9]); (b) SEM image of a diphenylalanine PNT (from unpublished data of the authors).

hydrophobic tails. These peptides also have a polar hydrophilic head and a hydrophobic tail. Glycine and aspartic acid are reported to build stable proteins in many biomaterials. Glycine [Figure 3(a)] is found in all types of collagen, silkworm silk, and bioadhesives of marine mussels. Therefore, PNTs made of surfactant-like peptides are not only advantageous in biomedical, industrial, and environmental applications but are also beneficial for investigating natural synthesis mechanisms of biomaterials and their effects on human health and the environment. Aspartic acid [Figure 3(b)] has two hydroxyl groups on the sides, which carry two negative charges at the hydrophilic head. The carboxylate anion of aspartic acid is known as aspartate, and the L-isomer of aspartate is one of the 20 proteinogenic amino acids (i.e., the building blocks of protein).

A surfactant peptide is composed of a number of glycine molecules (4 to 10) in the tail group, with two aspartic acid groups in the head (Figure 4). A peptide consisting of 6 units of glycine and 1 unit of aspartic acid forms nanotubes 30–50 nm

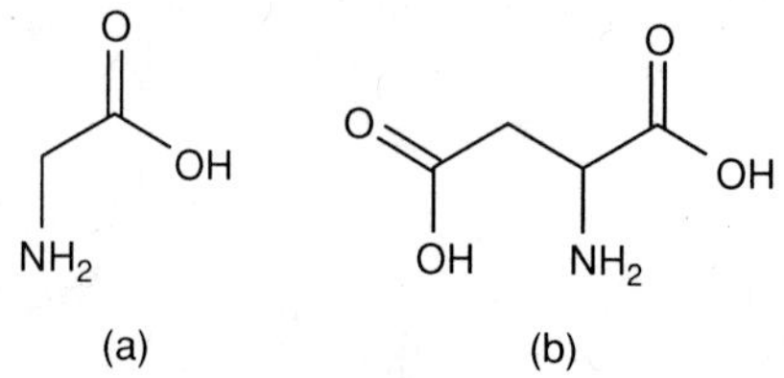

FIGURE 3   (a) Glycine; (b) aspartic acid.

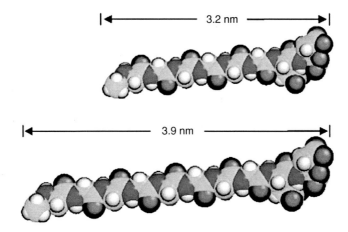

FIGURE 4   Molecular structure of glycine–aspartic acid peptides. (From [3].)

in diameter. Figure 5 shows an example of PNT formation and the application of surfactant peptide.

In addition to these linear amino acid–based PNTs, cyclic peptides can form very stable PNTs. Pioneered by Ghadiri [1], eight alternating D- and L-amino acids, cyclo [-(D-Ala-Glu-D-Ala-Gln)$_2$-], form PNTs that are highly mechanically, chemically, and thermally stable. The PNTs were reported to persist for long periods of time in common organic polar and nonpolar solvents, including DMF and DMSO. They can withstand repeated centrifugation and vortex mixing. This type of PNTs was reported to be stable to highly acidic (pH 1) and highly basic solutions (pH 14) [16].

These amino acids link each other to form a cyclic peptide through amide bonding of the two terminal groups. In a controlled acidic solution, the cyclic peptides spontaneously stack up on top of each other and are held together by hydrogen bonding. This assembly results in the peptide nanotube shown in Figure 6. It is interesting and worth noting that cyclic peptides with proper side-chain modification can penetrate cellular lipid bilayers to form a highly effective transmembrane ion channel that can pump in glucose. Therefore, they can be used as tools to study transmembrane transport mechanisms. Some peptides are investigated for alternative antimicrobial

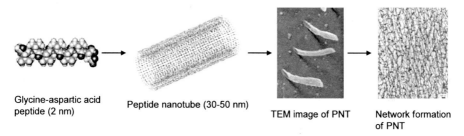

Glycine-aspartic acid peptide (2 nm)

Peptide nanotube (30-50 nm)

TEM image of PNT

Network formation of PNT

FIGURE 5   Formation of PNT and a PNT network. (From [15].)

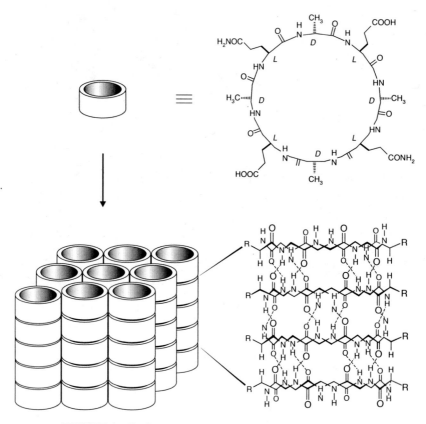

FIGURE 6   Cyclic peptide and its PNT formation. (From [15].)

development, as they kill the cells by destroying their membrane without causing antimicrobial resistance [17].

Bolaamphiphilic compounds consist of a hydrophobic chain with two polar head groups at its extremities. Due to their chemical structure, these compounds can form monolamellar membranes, vesicles, or fibrous organizations in aqueous solution. These synthetic surfactants are, in fact, a copy of the lipids that form the cytoplasmic membranes of the archaebacteria, microorganisms that live in extreme conditions of temperature, pressure, or pH. To withstand these extreme living conditions, the hydrophobic chains that form the lipid bilayer of their cell membranes are fused, creating macrocyclic molecules which are not easily synthesized in the laboratory. For over 30 years, the scientific community has been trying to synthesize lipid analog capable of forming vesicle membranes. Considering the slower rate of the membrane fusion phenomenon in monolamellar membranes, the bolaform surfactants represent a good alternative, and archaeal membranes have begun to be used in the pharmaceutical preparation of liposomes [18]. Figure 7 shows some examples of symmetrical and unsymmetrical bolaamphiles that are used for nanotube synthesis.

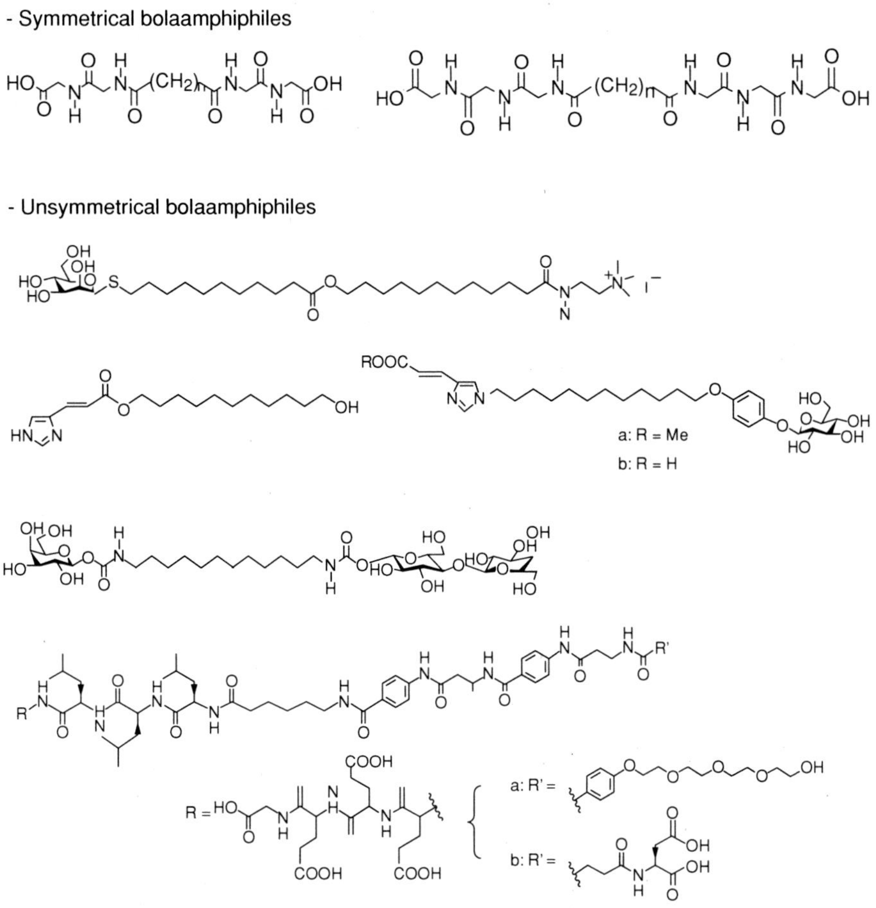

- Symmetrical bolaamphiphiles

- Unsymmetrical bolaamphiphiles

FIGURE 7 Chemical structures of bolaamphiphiles.

3 BIOMEDICAL APPLICATIONS

Peptide molecules in general have been investigated intensively for biomedical applications, including artificial organs, drug delivery, and antibiotics. Furthermore, as stated above, they have been valuable tools for studying transmembrane protein functions and onset mechanisms of diseases such as Alzheimer's and cancer.

In particular, the peptide nanotube occupies a special position in biomedical investigations, because of its unique structure. Applications of PNTs to the biomedical field include, but are not limited to, their use for pharmaceutical purposes, regenerative medicine and scaffold fabrication in tissue engineering, antimicrobial coating, and medical diagnosis. The advantages of the definitive tube structure of PNT in regards to these applications are (1) its encapsulating effect, (2) the dual properties of the inner and outer walls, (3) easy control of mass diffusion through the tube, and (4) its well-ordered alignment. These advantages are discussed here in more detail for each biomedical application of PNTs. In this discussion, the basic chemistry related to peptide nanotubes and peptides in general is not the focus. Rather, the discussion is focused on current and possible future applications.

3.1   Drug Delivery

Almost all the materials currently used for drug delivery were developed originally for use in the textile, adhesives, construction, electronics, or polymer industries. It appears that most attempts to exploit the drug delivery potential of these materials often fail to provide the specialized delivery functions required for drug formulation in the pharmaceutical or biotech industry. Both synthetic and biopolymers have been investigated intensively for drug delivery, especially for controlled drug release and targeted drug delivery. For example, biodegradable temperature sensitive polymers such as polyesters, polyphosphazenes, polypeptides, and chitosan, and pH/temperature-sensitive polymers such as sulfamethazine-, poly(β-aminoester)-, poly(aminourethane)-, and poly(amidoamine)-based polymers, usually form spherical hydrogel vesicles to encapsulate drugs. At a certain temperature and pH, these polymers degrade and release the drug.

Liposomes are a good example of the many materials that are used to develop therapeutic agents capable of specifically targeting cancer cells and tumor-associated microenvironments, including tumor blood vessels. These therapies are regarded as holding the promise of high efficacy and low toxicity. In these therapeutic strategies, anticancer drugs are encapsulated by liposome vesicles that bind to the cell surface receptors expressed on tumor-associated endothelial cells. These anti-angiogenic drug delivery systems could be used to target both tumor blood vessels and the tumor cells themselves.

Encapsulation of drugs with polymeric substances has to overcome several challenges for wider use in drug delivery. One of the challenges is the control the permeability of polymer capsule. Liposome capsules, for example, have very low permeability, due to the lipid bilayer. The bilayer may be modified with OmpF, a membrane protein from the outer cell wall of *E. coli*, to open up a channel for mass transfer (Figure 8) [19,20].

A peptide nanotube has two open ends, which facilitate the diffusion of drugs. The diffusion rate can be controlled readily by adjusting the length and diameter of the PNT. There may be diffusion through PNT walls, but usually the permeability of PNT walls is very low. When peroxidase enzymes (soybean peroxidase) were encapsulated inside dipeptide PNTs, as shown in Figure 9, no diffusion limitation was observed, while enzyme activity lasted longer and was higher than that of free enzymes in the bulk phase.

In addition to easy drug loading and dissolution, another advantage of PNT in drug delivery is found in its effective surface functionalization. To target specific cells or a specific area, various chemicals, liposomes, DNA, aptamers, and ligands can readily be bound onto the functional groups of PNT on its outer surface. Compared to carbon nanotubes and other biopolymer gels, such as alginate, chitosan, and liposome, PNTs

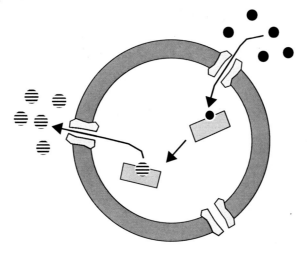

FIGURE 8   Liposome nanovesicles with encapsulated enzymes (squares). Substrates (dots) diffuse into the bilayer wall through porin channels, react with the enzymes, and then the products are released back into the solution through diffusion. (From [21].)

have well-defined functional groups that enable high affinity molecules to bind target cells (e.g., cancer cells) or an endothelial area at precise positions.

To help better understand the concept, an example of surface modification is introduced here. Compared to carbon nanotubes, PNTs have more functional groups available that can be used for surface modification under mild conditions. Peptide molecules, especially bolaamphiphiles, have amine and carboxyl groups that may be readily form covalent bonds with other organic molecules, organometals, or metal ions. Figure 10 shows the structure of a bolaamphiphile peptide monomer, bis (N-$^\alpha$-amido-glycylglycine)-1,7-heptane dicarboylate, and its nanotube formation.

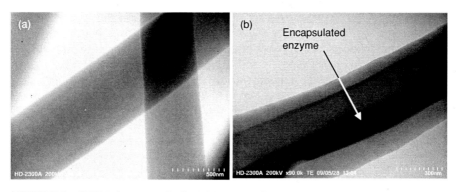

FIGURE 9   STEM images of dipeptide nanotubes. (a) hollow PNT; (b) enzyme-encapsulating PNT. (From unpublished data of the authors.)

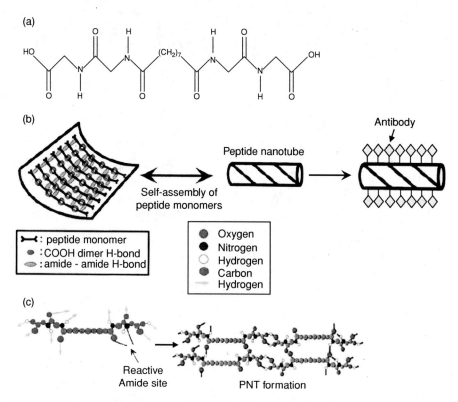

FIGURE 10 (a) Peptide monomer, bis(N-$^{\alpha}$-amido-glycylglycine)-1,7-heptane dicarboxylate; (b) peptide nanotube formation and immobilization of antibodies; (c) reactive sites on the monomer and structural illustration of PNT formation. (Excerpted from [22] and [23].)

Note that there are oxygen and nitrogen reactive sites. Hydrogen atoms also participate in binding with neighboring peptides via hydrogen bonding.

Bolaamphiphile PNTs are formed as peptide monomers bind through three-dimensional intermolecular hydrogen bonding in NaOH/citric acid solution via three-dimensional intermolecular hydrogen bonds [23]. It is relatively easy to immobilize various proteins and peptides on the PNT surface, as this mimics reproducibly the biological reaction occurring at normal temperatures. Antibodies or peptide may be bound to the surface of PNTs and delivered to target cells such as cancer and tumor. Anti-mouse IgG and anti-human IgG were bound successfully to free amide sites on the nanotube sidewall via hydrogen bonding by means of simple incubations [24]. Nanotubes were incubated with goat anti-mouse IgG in a pH 7.2 phosphate buffer. The attachment of anti-mouse IgG on the nanotube was confirmed by fluorescence microscopy.

Monoclonal antibodies are a rapidly growing class of drugs used for the therapy of human cancers and other diseases. They can be used effectively to target tumor-specific molecules and thereby modulate key signaling pathways that play a role in tumor growth, survival, and metastasis. An antibody-modified PNT may be loaded with other drugs that facilitate anticancer effects with controlled release. Clinical success of novel antibodies has stimulated great interest in the promise of antibody therapeutics for cancer, and the development of PNTs will play an important role in antibody and protein therapeutics with significant synergistic effects.

3.2 Gene Therapy

Chemical stability and excellent biocompatibility of self-assembled peptides are the two major advantages in many biomedical applications. In addition, when the peptides form a tubular shape, the result is amazing versatility, due to its shape and functional groups inside and outside the tube walls. When positively charged surfactant-like peptides form PNTs, they can encapsulate negatively charged DNA and RNA for gene delivery. Genes are introduced into tissues or cells via gene transfer, to derive a therapeutic or preventive benefit from the function of these genes. These genes are inserted into the cells and tissues of a patient to treat a disease, such as a hereditary disease, by replacing a deleterious mutant allele with a functional allele. The technology is still in its infancy, but it is one of the most promising and active research fields in medicine and biomedical technology.

In their pioneering work on September 14, 1990 at the U.S. National Institutes of Health, Anderson and his colleagues, Blaese, Bouzaid, and Culver, performed the first approved gene therapy procedure on a 4-year-old girl, Ashanthi DeSilva. She was suffering from a rare genetic disease known as severe combined immune deficiency, caused by a defective gene, and was vulnerable to every passing germ or infection. To cure her, Anderson and his colleagues removed white blood cells from the child's body, cultured the cells in the laboratory, inserted the missing gene into engineered viruses that could not reproduce, and then infused the genetically modified blood cells back into the patient's bloodstream. This operation was successful and Ashi is now a 16-year-old, leading a totally normal life.

Despite substantial progress, a number of key technical issues need to be resolved before gene therapy can be applied safely and effectively in the clinic. One of the critical challenges is the gene delivery vector. When vectors come into a human body, the inflammatory response causes multiple organs to fail. As in Ashi's case, genetically altered virus is used as a vector, but generally, vectors applied in gene therapy can be classified as viral or nonviral, and they may cause severe problems. All the details of these vectors are not discussed here because they are not within the scope of this chapter. But oligonucleotides, lipoplex, and dendrimers are just a few examples being studied as nonviral vectors. Cationic peptide nanotubes might be a good candidate as a nonviral vector for gene therapy because of their great biocompatibility and because they can penetrate the cell membranes. Carbon nanotubes have been investigated for gene therapy, but PNTs offer unsurpassable opportunity because they do not require as much surface modification as carbon nanotubes require, and as mentioned earlier, positively charged PNTs attract the negatively charged phosphate groups in the DNA backbone, which makes it easier to combine PNTs and genes.

3.3   Tissue Engineering

In tissue engineering, basically, tissue cells are grown in a scaffold that serves as a mechanical and biological support for cell growth. Growing within the scaffold matrix, tissue cells could be regenerating or could replace skin, bone, cartilage, or part of an organ. Tissue engineering develops materials and methods to promote these cells' differentiation and proliferation toward the formation of a new tissue. Various materials have been proposed for use in the processing of scaffolds, (i.e., polymers). Biopolymers offer the advantage of being similar to biological macromolecules, which would not incur incompatibility problems in a biological system. Due to their similarity to the extracellular matrix (ECM), biopolymers can probably avoid the possibility of chronic inflammation responses or immunological reactions and toxicity which are often detected when using synthetic polymers. Another important aspect to be addressed is the processing of the materials into porous matrices, a task that usually requires technologies other than those usually employed in the processing of conventional synthetic polymers. There is a need to make biomaterials in various shapes and structures for clinical reasons, including processing of nano- and microparticles (for controlled-release applications), and into two-dimensional structures such as membranes for wound dressing.

The major challenges in tissue engineering may be summarized as (1) mechanical strength for proper function of the body part to be treated, (2) porosity for lighter weight and sufficient blood and oxygen supplies and waste transport, and (3) biocompatibility to prevent immune rejection and inflammation.

PNTs make a great candidate for scaffold synthesis and cell culture with regard to these major challenges. The cylindrical porous structure of PNTs offers a greater possibility for exceptional mechanical strength and porosity control than do any other biomaterials in tissue engineering. Furthermore, reactive surface functional groups of PNTs make it possible for various multifunctional organic, inorganic, and metal ions to be included in the scaffold formation and cell culture. They can also contain cells and bioactive agents inside the tube that help antibiotic, nutrient, and hormone supplies. Many self-assembling materials aimed at mimicking the extracellular matrix (ECM) have been proposed for tissue engineering applications: fibrin, alginate, hyaluronic acid, chitosan, and collagen from a variety of origins, alone or combined, to name a few.

In addition to mimicking ECM, peptide nanotubes offer (1) controlled release of regenerative medicines, nutrients, growth hormones, minerals, and so on, during tissue formation, and (2) can contribute to porosity manipulation. In addition, PNTs can align better than spherical vesicles or fibers in forming a three-dimensional structure, which helps reinforce the mechanical stability of the tissue. Surfactant-like PNTs were used to encapsulate neural progenitor cells that show the ability to induce very rapid differentiation into neurons while discouraging the development of astrocytes. Figure 11 shows the process of tissue regeneration by amphiphilic PNTs.

3.4   Antimicrobial Surface Coating

As shown in Figure 12, PNTs can align in a well-ordered uniform direction, forming a crystalline PNT coating on a solid surface. This particular characteristic can be

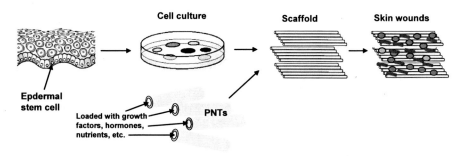

FIGURE 11   Wounded skin regeneration using PNTs.

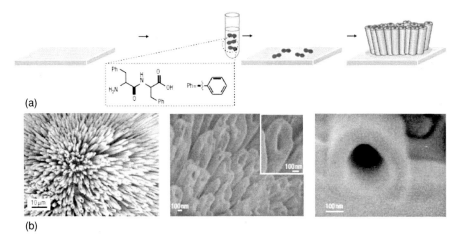

FIGURE 12   Vertical alignment of PNTs. (From [25].)

used to prevent the microbial attachment and surface colonization that results in biofouling. Therefore, PNT's nanoscale surface alignment can be used to protect medical devices, in ultraclean water treatment system and underwater devices, and to protect artificial organs from being contaminated by microbial fouling and protein attachment.

3.5 Concluding Remarks

Over the past years, carbon nanotubes (CNTs) have been studied extensively for many applications, due to their mechanical stability, conductance, and large surface area. In particular, they have a tensile strength about 10 times greater than that of steel at about one-fourth of the weight. Another potential application is for energy storage, such as in hydrogen storage media. However, CNT characteristics are easily affected by exposure to humidity, oxygen, N_2O, and NH_3. Lack of uniformity of CNTs is also believed to pose problems in device fabrication. In addition, hydrophobicity, thus limited solubility, lack of reproducibility of precise structural properties, high cost, and limited opportunities for covalent surface modification are drawbacks of CNTs. With regard to these issues, PNTs offer an attractive alternative for device fabrication in biomedical applications. As indicated earlier, cationic dipeptides, NH_2—Phe—Phe—NH_2, self-assemble into PNTs at neutral pH. The tubes can be absorbed by cells through endocytosis and deliver oligonucleotides in gene and drug delivery. Other possible applications were also described in this section.

In addition to biomedical applications, PNTs have been used in nanoelectronics. Twenty-nanometer silver nanowires were fabricated successfully, with PNTs acting effectively as degradable casting molds [26]. Silver ions are reduced to metallic silver in the lumen of the tube, and the peptide template is removed by enzyme degradation. Such a small nanowire cannot be made by conventional lithography. Silver nanowire will be of great value in biomedical material development due to its excellent antimicrobial effect, allowing for its use in sutures, gauze, scaffolding, and artificial organs.

4   ENVIRONMENTAL APPLICATIONS

PNTs may be of great value in environmental fields. Possible applications of PNTs are to environmental sensors, in bioremediation, and in environmental protection. This great potential is discussed in this section. PNTs have found a sizable niche in environmental biosensor development. In detecting environmental hormones, food toxins, and various environmental pollutants, PNTs can be a strong assisting component in enhancing sensor durability, sensitivity, and selectivity. The versatility of PNT's surface functionalization and its high porosity can offer great potential for bioremediation. Zero-valent nano iron (nZVI), for example, can be combined with PNTs for groundwater remediation. As introduced earlier, PNTs can be used in drinking water distribution systems, ships' hulls, food containers, and various electronic devices to protect them from microbial contamination.

4.1   Biosensors

The PNT's role in biosensor applications is twofold: (1) encapsulation of sensing elements and (2) surface modification of sensor electrodes. Sensing elements are the part that binds with target molecules, and typically, antibodies, DNA, RNA, ligands, and enzymes are used because of their high affinity toward the target molecules. To protect the sensing molecules from the environment, many encapsulating materials have been developed; hydrogel, liposome, polyelectrolyte, and polar lipid are a few examples among many.

Major problems of these materials are related to mass transfer and enzyme activity. Although the encapsulation shields the sensor from harmful environments, at the same time it slows down the diffusion rate of target molecules and products toward and from the sensing elements. Even worse, active sites of the sensing elements can be blocked, altered, or even damaged by encapsulation that results in deactivation of the sensor.

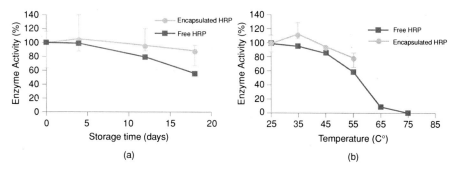

FIGURE 13   (a) Activities of free HRP and encapsulated HRP; (b) comparison of thermal stabilities. (From unpublished data of the authors.)

These two challenging issues can be resolved by encapsulating sensing elements with PNTs. PNT has two open holes at both ends, and marginal diffusion could take place through PNT walls. Therefore, the diffusion rate can be controlled readily by adjusting the diameter of the open holes and the length of the tube. Furthermore, compactness of encapsulated elements inside the tube can be adjusted to control the diffusion rate of the target molecules. Functional groups, surface charges, and/or the hydrophobicity of the PNT wall surface can be used effectively to immobilize sensing elements without compromising their activity.

For example, horseradish peroxidase enzyme (HRP) can be encapsulated effectively by PNTs. HRP has become one of the most important enzymes in immunoassay and industrial uses for its strong redox potential. HRP is used for sensors to detect H_2O_2, environmental hormones, and food toxins. Figure 9(b) shows encapsulated HRPs in a PNT. The activity of the encapsulated HRP was measured and compared with free HRP [Figure 13(a)]. Compared to the free HRP, the encapsulated HRP showed higher activity for a longer time and higher thermal stability [Figure 13(b)]. The thermal stability measurement of the encapsulated HRP stopped at 55 °C because the PNTs began to disassemble.

The procedure for the PNT encapsulation is simple. PNTs are formed and separated from the solution by centrifugation and dried. Encapsulation is performed in a phosphate buffer solution by adding dry PNT and HRP. HRPs are thought to be introduced into the tube by the capillary effect and attached to the wall by either van der Waals force or hydrogen bonding. The solution is kept at room temperature for one week (Figure 14). Encapsulated enzymes are separated by centrifugation and then assayed for activity and deposited on a gold surface to measure the electrochemical characteristics.

The enzyme-encapsulating PNTs may be deposited on the sensor electrode to improve its electrochemical characteristics, as shown in Figure 15. Depositing PNTs between the sensing element and the electrode is like inserting a dielectric material, as shown in Figure 16. Placing a dielectric material between the two parallel electric plates of a capacitor is equivalent to adding a capacitor and results in an increase

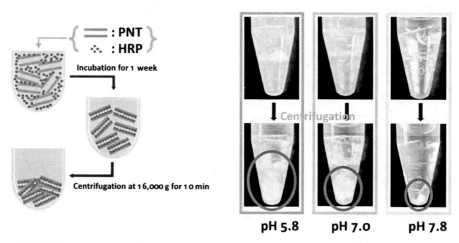

FIGURE 14   Encapsulation of enzymes in PNTs and separation of encapsulated enzymes by centrifuge.

in the overall capacitance value of the sensor system (Figure 16). If the dielectric material completely fills the gap between the sensing element and the electrode, the capacitance is increased by a factor of ε, the dielectric constant, a dimensionless value dependent on the nature of the material. So the higher the dielectric constant, the higher it can store more electric charge for later use.

When a dielectric capacitor is inserted between the electric plates, the electric field applied decreases by a factor ε ($V = V_0/\varepsilon$), where V_o is voltage without a capacitor and V is voltage with a capacitor. When an electric field (i.e., voltage) is applied, dielectric molecules are aligned in the direction of the electric field, and the voltage across the plates decreases by the factor ε (relative electric constant, or dielectric constant):

$$V_{dl} = \frac{V}{\varepsilon} \qquad (1)$$

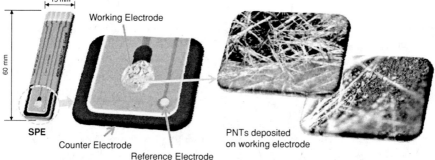

FIGURE 15   Deposition of PNTs on a screen-printed electrode (SPE).

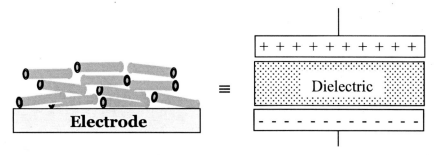

FIGURE 16 How a PNT-modified electrode can be modeled approximately as a simple parallel-plate capacitor with a dielectric. The capacitance of the PNT system is lowered drastically by both the increased distance between the planes of charge that is created by the PNT spacer, and by the low dielectric constant provided by the peptide density between the plates that acts as the capacitor's dielectric material.

where $\varepsilon > 1$. The capacitance of the capacitor is described, before considering the dielectric, as

$$C = \frac{Q_0}{V} \tag{2}$$

where Q_0 is the original amount of charge on the capacitor. Including the effect of dielectric effect, we have

$$C_{\text{adjusted}} = \frac{Q_0 \varepsilon}{V} \tag{3}$$

Because $V = Qd / \varepsilon_0 A$ (d is the distance between the plate and A is the surface area of the plate, ε_0 is electric constant $\approx 8.854 \times 10^{-12}$ F/m),

$$C_{\text{adjusted}} = C_{\text{dl}} = \varepsilon \cdot \varepsilon_0 \frac{A}{d} \tag{4}$$

Therefore, applying the dielectric capacitance principles to the PNT in chemical or biochemical sensors, a PNT-modified electrode produces a profound effect on capacitance values and, as a result, in sensing performance. In other words, the dielectric decrease in C_{dl} with the PNT actually represents a critical advantage for electrochemical performance involving electron transfer in the system. Because the magnitude of the charging current is so markedly lower with the PNT than with the bare electrode, the signal/noise ratio of any faradaic current is improved significantly. The PNT can actually help to suppress the background signal and simultaneously provides a biocompatible surface for biosensing elements. Figure 17 shows improvement in the electrochemical characteristic of SPE when PNTs are deposited on its surface. The electric potential change, ΔE_p, at the maximum current peak for the

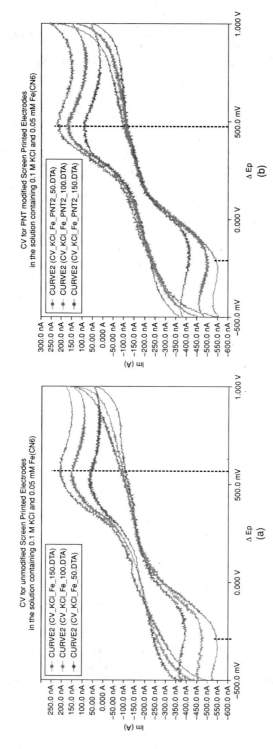

FIGURE 17 Cyclic voltammetry of the bare SPE and PNT-deposited SPE. Electric potential difference, ΔE_P, for the maximum redox current peaks is narrower for the PNT-deposited SPE in (b) than from the bare one in (a).

PNT-covered SPE is narrower [Figure 17(b) than the bare one [Figure 17(a)], which demonstrates that PNT enhances the current response.

Consequently, the advantages of PNT in sensor performance can be summarized as follows: (1) it can immobilize sensing molecules without compromising their activity; (2) it protects the sensing molecules without compromising the mass transfer rate; (3) it enhances the sensitivity of the sensor as it reduces background noises; and (4) it improves response time.

4.2 Environmental Remediation and Protection

PNTs can be used as a valuable tool in environmental remediation. In both abiotic- and bio-remediation, remediating agents (i.e., metals, chemicals, enzymes, and microbes) should be immobilized on a porous substrate during the cleanup process. PNT can provide a stable place for these materials to bind without reducing the diffusion of contaminants. Good stability of PNT against pH and chemicals is another advantage. More than anything, an excellent binding capability through many functional groups located on the PNT walls is the most notable advantage compared to other nanomolecules such as carbon nanotubes. PNTs are readily thiolated, and then the thiolated PNTs can bind to gold. Pt, Fe, and Ni can easily be associated with PNT as well [27,28] and PNT can be used directly to grow nanosilver crystals as shown in Figure 18.

Zero-valent nano iron (Fe^0) is investigated for remediating TCE (1,1,1-trichloroethane)-contaminated groundwater. When contacting with carbonate species [i.e., C(IV)], reduction of TCE is accelerated as the Fe is precipitated as $FeCO_3$, and nontoxic ethane and ethene are produced as by-products [30]. Fe^0 can be bound to PNT (Figure 19), and then the Fe-bound PNT may be embedded into a permeable ion barrier in a treatment zone, and effective remediation can be achieved (Figure 20).

4.3 Antimicrobial Antifouling Agent

Interaction of PNT with a lipid bilayer can cause cell death. Antibiotic effect of cyclic PNT was demonstrated by Ghadiri group in 2001 [32]. Protection of various surfaces

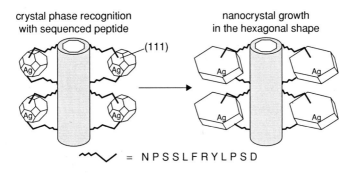

FIGURE 18   Sliver nanocrystal growth in a hexagonal shape on a PNT. (From [29].)

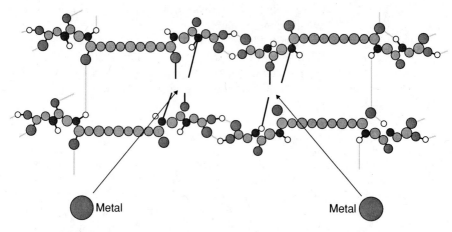

FIGURE 19   Metal binding of heptane bolaamphiphile PNT. (From [31].)

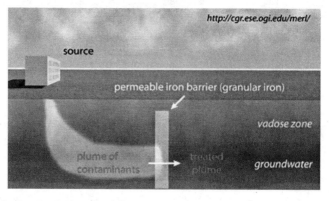

FIGURE 20   Permeable iron barrier for remediation of contaminated groundwater. (From http://cgr.ogi.edu/merl/.)

from microbial colonization is a critical issue in microbial water contamination, food manufacturing businesses, health care products manufacturing, and ship building. PNT can be applied to a bulk fluid phase to kill the problematic microbes, or it can easily be coated on the surface to protect the surface from biofouling.

5 CONCLUDING REMARKS

Reactivity of the functional groups of PNT is a valuable asset for combining various materials for environmental applications. For example, nano silver is being commercialized for the antimicrobial purpose in the apparel and home appliance industries and in biomedical devices such as antifouling urinary catheters. Zero-valent nano iron can be used for remediating contaminated groundwater. These nanomaterials should be immobilized stably in the active area, and PNT can hold them stably and place them in a desired location. With mechanical strength and porosity, PNT can enhance the reaction efficiency by holding them without reducing mass transport.

Environmental sensors can be benefited by PNT, as PNTs effectively encapsulate sensing elements or increase the sensitivity of a sensor by acting like a capacitor when deposited on the sensor electrode. When a specific functional group is needed to immobilized enzyme or ligand, bolaamphile PNT with proper functional groups may be selected. For the electrode surface modification, diphenylalanine PNT of the proper length may be used because length and concentration of PNTs on the electrode surface determine the sensitivity and signaling rate of the sensor.

REFERENCES

1. Ghadiri, M. R., Granja, J. R., Milligan, R. A., McRee, D. E., and Khazanovich, N., Self-assembling organic nanotubes based on a cyclic peptide architecture, *Nature* 366:324–327 (1993).

2. Adler-Abramovich, L., Reches, M., Sdeman, V. L., Allen, S. Tendler, S. J. B., and Gazit, E., Thermal and chemical stability of diphenylalanine peptide nanotubes: implications for nanotechnological applications, *Langmuir* 22:1313–1320 (2006).

3. Santoso, S., Hwang, W., Hartman, H., and Zhang, S., Self-assembly of surfactant-like peptides with variable glycine tails to form nanotubes and nanovesicles, *Nano Lett.* 2(7):687–691 (2002).

4. Zhang, S., Fabrication of novel biomaterials through molecular self-assembly, *Nat. Biotechnol.* 21:1171–1178 (2003).

5. Zhao, X., and Zhang, S., Fabrication of molecular materials using peptide construction motifs, *Trends Biotechnol.* 22(9):470–476 (2004).

6. Cherny, I., and Gazit, E., Amyloids: not only pathological agents but also ordered namomaterials, *Angew. Chem. Int. Ed.* 47:4062–4069 (2008).

7. Shao, H., and Parquette, J. R., Controllable peptide–dendron self-assembly: interconversion of nanotubes and fibrillar nanostructures, *Angew. Chem. Int. Ed.* 48:2525–2528 (2009).

8. Song, Y., Challa, S. R., Medforth, C. J., Qiu, Y., Watt, R. K., Peña, D., Miller, J. E., van Swol, F., and Shelnutt, J. A., Synthesis of peptide–nanotube platinum–nanoparticle composites, *Chem. Commun.* 2004(9):1044–1045 (2004).

9. Görbitz, C. H., The structure of nanotubes formed by diphenylalanine, the core recognition motif of Alzheimer's β-amyloid polypeptide, *Chem. Commun.* 2332–2334 (2006).

10. Gazit, E., Self-assembly of short aromatic peptides: from amyloid disease to nanotechnology, *Nano Biotechnol.* 1:286–288 (2005).

11. Soto, C., Protein misfolding and disease; protein refolding and therapy, *FEBS Lett.* 498:204–207 (2001).

12. Rochet, J. C., and Lansbury, P. T., Jr., Amyloid fibrillogenesis: themes and variations, *Curr. Opin. Struct. Biol.* 10:60–68 (2000).

13. Sunde, M., and Blake, C. C. F., From the globular to the fibrous state: protein structure and structural conversion in amyloid formation, *Q. Rev. Biophys.* 31:1–39 (1998).

14. Reches, M., and Gazit, E., Casting metal nanowires within discrete self-assembled peptide nanotubes, *Science* 300:625–628 (2003).

15. Zhao, X., and Zhang, S., Self-assembling nanopeptides become a new type of biomaterial, *Adv. Polym. Sci.* 203:145–170 (2006).

16. Hartgerink, J. D., Granja, J. R., Milligan, R. A., and Ghadiri, M. R., Self assembling peptide nanotubes, *J. Am. Chem. Soc.* 118:43–50 (1996).

17. Liu, L., Xu, K., Wang, H., Jeremy Tan, P. K., Fan, W., Venkatraman, S. S., Li, L., and Yang, Y.-Y., Self-assembled cationic peptide nanoparticles as an efficient antimicrobial agent, *Nat. Nanotechnol.* 4:457–463 (2009).

18. Conlan, J. W., Krishnan, L., Willick, G. E., Patel, G. B., and Sprott, G. D., Immunization of mice with lipopeptide antigens encapsulated in novel liposomes prepared from the polar lipids of various Archaeobacteria elicits rapid and prolonged specific protective immunity against infection with the facultative intracellular pathogen, *Listeria monocytogenes*, *Vaccine* 19:3509 (2001).

19. Nikaido, H., and Rosenberg, E. Y., Porin channels in *Escherichia coli*: studies with liposomes reconstituted from purified proteins, *J. Bacteriol.* 153(1): 241–252 (1983).

20. Winterhalter, M., Hilty, C., Bezrukov, S. M., Nardin, C., Meier, W., and Fournier, D., Controlling membrane permeability with bacterial porins: application to encapsulated enzymes, *Talanta* 55:965–971 (2001).

21. Park, B. W., Yoon, D. Y., and Kim, D. S., Recent progress in bio-sensing techniques with encapsulated enzymes, *Biosens. Bioelectron.* 26:1–10 (2010).

22. Matsui, H., and Douberly, G. E., Jr., Organization of peptide nanotubes into macroscopic bundles, *Langmuir* 17:7918–7922 (2001).

23. Matsui, H., and Gologan, B., Crystalline glycylglycine bolaamphiphile tubules and their pH-sensitive structural transformation, *J. Phys. Chem. B* 104(15):3383–3386 (2005).

24. Zhao, Z., Banerjee, I. A., and Matsui, H. Simultaneous targeted immobilization of anti-human IgG-coated nanotubes and anti-mouse IgG-coated nanotubes on the complementary antigen-patterned surfaces via biological molecular recognition, *J. Am. Chem. Soc.* 127:8930–8931 (2005).

25. Reches, M., and Gazit, E., Controlled patterning of aligned self-assembled peptide nanotubes, *Nat. Nanotechnol.* 1:195–200 (2006).

26. Cheng, Y.-H., and Cheng, S.-Y., Nanostructures formed by Ag nanowires, *Nanotechnology*, 15;171–175 (2004).

27. Yu, L., Banerjee, I. A., and Matsui, H., Incorporation of sequenced peptides on nanotubes for Pt coating smart control of nucleation and morphology via activation of metal binding sites on amino acids, *J. Mater. Chem.* 14:739–743 (2004).

28. Yu, L., Banerjee, I. A., Shima, M., Rajan, K., and Matsui, H., Size-controlled Ni nanocrystal growth on peptide nanotubes and their magnetic properties, *Adv. Mater.* 16:709–712 (2004).

29. Yu, L., Banerjee, I. A., and Matsui, H., Direct growth of shape-controlled nanocrystals on nanotubes via biological recognition, *J. Am. Chem. Soc.* 125(48):14837–14840 (2003).

30. Agrawal, A., Ferguson, W. J., Gardner, B. O., Christ, J. A., Bandstra, J. Z., and Tratnyek, P. G., Effects of carbonate species on the kinetics of dechlorination of 1,1,1-trichloroethane by zero-valent iron, *Environ. Sci. Technol.* 36(20):4326–4333 (2002).

31. Matsui, H., Pan, S., Gologan, B., and Jonas, S. H., Bolaamphiphile nanotube-templated metallized wires, *J. Phys. Chem. B* 104:9576–9579 (2000).

32. Fernandez-Lopez, S., Kim, H. S., Choi, E. C., Delgado, M., Granja, J. R., Khasanov, A., Kraehenbuehl, K., Long, G., Weinberger, D. A., Wilcoxen, K. M., and Ghadiri, M. R., Antibacterial agents based on the cyclic *d,l*-α-peptide architecture, *Nature* 412:452–455 (2001).

INDEX

Nanoscale Multifunctional Materials: Science and Applications, First Edition.
Edited by Sharmila M. Mukhopadhyay.
© 2012 John Wiley & Sons, Inc. Published 2012 by John Wiley & Sons, Inc.